W9-AGM-374

# North Carolina Yeoman

# North Carolina Yeoman

*The Diary of*
*Basil Armstrong Thomasson,*
*1853–1862*

Edited by Paul D. Escott

*The University of Georgia Press*
Athens and London

*This book has been supported by a grant from the National Endowment for the Humanities, an independent federal agency.*




Athens, Georgia 30602

Designed by Kathi Dailey Morgan
Set in Bakersville by Tseng Information Systems, Inc.
Printed and bound by Thomson-Shore, Inc.
The paper in this book meets the guidelines for permanence and durability of the Committee on Production Guidelines for Book Longevity of the Council on Library Resources.

Printed in the United States of America
00 99 98 97 96 C 5 4 3 2 1

Library of Congress Cataloging in Publication Data

Thomasson, Basil Armstrong, 1829–1862.
North Carolina yeoman : the diary of Basil Armstrong Thomasson, 1853–1862 / Paul D. Escott, ed.
p. cm.
Includes bibliographical references and index.
ISBN 0-8203-1755-1 (alk. paper)
1. Thomasson, Basil Armstrong, 1829–1862—Diaries. 2. Farmers—North Carolina—Diaries. 3. Farm life—North Carolina—History—19th century. 4. North Carolina—History—1775–1865. I. Escott, Paul D., 1947– . II. Title.
F258.T48 1996
975.6'03—dc20 95-8429

British Library Cataloging in Publication Data available

*To Grace,*

*with love*

# *Contents*

*Acknowledgments* / ix

*Introduction* / xi

*A Note on Editorial Methods* / lxvii

*Map of Thomasson's Community, 1860* / lxix

*Book of Remembrance 1853* / 3

*Book of Remembrance 1854* / 20

*Diary 1855* / 60

*Diary 1856* / 114

*Diary 1857* / 160

*Diary 1858* / 187

*Diary 1859* / 225

*Diary 1860* / 266

*Diary 1861* / 297

*Diary 1862* / 321

*Other Documents* / 330

*Index* / 353

# *Acknowledgments*

I am particularly indebted to others who helped with this book, for without their contributions and cooperation the project would not have been possible.

Two of my graduate students made vital contributions. Elizabeth Kelly Kaufman first brought Basil Armstrong Thomasson's diary to my attention, and Darlene Woltman saved me many hours of clerical labor through her expertise with software. Mary Belle Weaver Allred of Hamptonville, North Carolina, supplied me with helpful information about local history in Yadkin and Iredell Counties and told me how to get in touch with the Thomasson family. The owner of the diary, Martha Lou Cole Thomasson, of Forsyth County, North Carolina, has been extremely helpful and supportive throughout the project and made a variety of valuable materials available to me. The fine work of her late husband, Jean Harris Thomasson, who conducted extensive genealogical research and carefully prepared a typescript copy of the diary, proved invaluable. William C. Thomasson of Chattanooga, Tennessee, was very generous of his time, constantly encouraging, and ready to answer my questions on a number of occasions. L. Q. and Jean Thomasson of Hamptonville welcomed me with delightful hospitality and treated me to a tour of the area. Kate Thomasson of Winston-Salem, North Carolina, helped me answer a number of worrisome questions.

Finally, I am grateful to the staffs of the North Carolina Division of Archives and History and Wake Forest University's Z. Smith Reynolds Library and its Rare Book Room.

# *Introduction*

It is a striking fact that historians know relatively little about the yeoman farmers of the antebellum South that comes *directly* from primary sources. The rich and stimulating scholarly literature on this period is replete with outstanding studies of the planter class and of the slaves. The first of these groups left voluminous personal papers, and although the second did not, nineteenth-century autobiographies and twentieth-century interviews have done much to fill the gap. Even less numerous social types, such as slave traders, overseers, and free black people, have received treatment made possible by the existence of primary sources that they themselves created or helped to create.[1]

However, primary sources in which southern yeomen farmers of the antebellum period speak for themselves are either rare or relatively unknown. In twenty years of professional work this editor has come across only six yeoman diaries, although there must be more.[2] This fact confronts scholars with a great irony, one that should be troubling. The yeoman farm families of the antebellum South were one of its largest social groups—more numerous throughout the upper South than planters, slaves, or any other category. Yet our direct knowledge about them, information from their mouths or pens, is exceedingly slim. Only in regard to the landless whites do researchers confront a similar lack of data.[3]

As a result, the honest historian finds it difficult to speak with assurance and in detail about the attitudes or outlook of nonslaveholding small farmers. In fact, even the definition of the yeoman class is in dispute, as some would lump together the small farmers who owned no slaves and those who owned only one or two bondsmen, whereas others argue that slaveholding—even on a small scale—made an important difference.[4]

Without a storehouse of diaries and personal papers from the yeoman class, scholars have turned to a variety of other sources to learn about this group. The records of the United States Census have been mined, along with travelers' accounts, books written by the yeomen's contemporaries, and descriptions of the yeomen or comments on them by planters, politicians, and others. One recent study has shown that folk songs and folklore can shed light on the value system of the yeomanry.[5] Much can be learned indirectly about the yeomen from reading newspapers, speeches, and other elements of the public discussion of their day. By examining what was spoken *to* the yeomen it may be possible to infer what they were thinking on certain questions. Reports of their actions invite similar inferences about their motives.

Despite obstacles, historians have produced some landmark works on the yeoman class. Two books that focus entirely on the yeomen are Frank Lawrence Owsley's *Plain Folk of the Old South* (1949) and Bill Cecil-Fronsman's *Common Whites* (1992). Several other books treat important aspects of their experience or devote substantial sections to the yeoman class, and a variety of articles and other studies have produced valuable information.[6]

Nevertheless, it is a signal fact that only two well-recognized, full-length works have been devoted to the antebellum southern yeomanry. Could one imagine so few scholarly studies of the slaves or of the planter class? The conclusion is unmistakable: our knowledge of nonslaveholding small farmers and their families in the antebellum South remains inadequate, and our understanding of southern society therefore remains incomplete.

## *The Thomasson Diary and Its Setting*

This book, and the diary that it reproduces, make a small contribution toward filling that gap. They document the life of Basil Armstrong Thomasson—a yeoman farmer, one of the unheralded members of North Carolina's largest social group. In many ways Thomasson was typical of yeoman farmers, particularly in his simple lifestyle, humble economic status, and concern for maintaining his independence. Yet in his character, zeal for learning, and interest in personal and community improvement he proved to be a remarkable individual. His diary suggests both the common characteristics of the yeoman class and the potential that resided in its individual members.[7]

Thomasson was born on September 9, 1829, and survived only until September 1862, when disease caused his early death. Throughout his thirty-three years, he lived in the western part of North Carolina's Piedmont. Born in what is now Forsyth County, Thomasson and his parents moved briefly to Stokes County and then to Yadkin County, where he began his adult life as a farmer and schoolteacher. After his marriage, he settled a few miles away in Iredell County, where he secured a teaching position and bought a small farm.[8]

Between 1853 and 1862 he kept a fairly extensive diary. Since the first entry (for August 23, 1853) begins without announcement or comment on his decision to become a diarist, it is possible that Thomasson started keeping records sometime before this date. If so, no entries from earlier years have survived. Perhaps because he had little money and wanted to economize, Thomasson did not buy a printed diary book with blank pages. Rather, he made his own diary, using a needle and heavy thread to sew together sheets of paper. The entries for 1861 were written over the cut-up pages of a map. Inevitably, the words recorded on many individual days lack historical importance, but overall the diary makes possible a rare understanding of his life and beliefs.

Basil Armstrong Thomasson lived in a part of the South where few were prosperous and slavery was not dominant. The seventy-one acres that the census recorded beside his name were typical of the small farms in his area.[9] Although he worked hard as a farmer and schoolteacher to support himself, his wife Mollie, and two children, he accumulated little property. If riches were uncommon in his world, so too was slavery. In 1860 slaves made up only 13.4 percent of the population of Yadkin County and 27.2 percent of the population of Iredell County. In Thomasson's immediate neighborhood, only 15 of 86 households held slaves; 144 of 612 people were unfree. In his parents' neighborhood, across the line in Yadkin County, only 2 households out of 67 owned slaves, and the unfree population accounted for only 8 of 328 people.[10]

For his part, Basil Thomasson (or Strong, as his family called him) never owned a slave and never desired to become a slave owner. In fact, slavery was in conflict with his value system. The diary makes clear that he revered freedom as America's signal virtue and disapproved of slavery. Unconcerned about the wrongs supposedly inflicted upon the South during the sectional crisis, Thomasson deplored the breakup of the Union. To him the Civil War was a tragic example of human folly, and he no doubt gave thanks, after the fighting began, that his occupational status afforded him an exemption from military service.[11]

Although Thomasson's life was simple, it was not devoid of significant human achievement. As he worked and raised his family, he strove to improve his education by wide and constant reading. Just as assiduously he strove to improve his soul and to lead a godly life. In material possessions or worldly power Strong Thomasson was hardly noteworthy. But in character and thought he was an enlightened and impressive person who deserves our respect. His life and his attitudes were far more interesting than any recital of the bare facts of his existence could ever suggest.

## *Family and Faith*

The life of Basil Armstrong Thomasson, only four generations removed from us, remains familiar to many Tar Heel natives because it exemplifies values and traditions that persist among them—especially the importance of family, church, and local attachments within a life of few things. Yet it may seem unusual and strange to those who have known only the individualized and secular values of a mobile, materialistic society in the late twentieth century.

Basil Armstrong Thomasson dreamed, sought opportunity, worked to better himself, and enjoyed good times just as individuals do today. But his world was rural and his surroundings plain. The self-sufficiency that he prized would now seem a life of poverty. Moreover, two powerful influences held central importance for him: family and faith. These shaped, directed, and constrained his life to a remarkable degree, in ways that were common then but relatively uncommon now. They imparted a distinctive quality to his experience.

For Basil Armstrong Thomasson, family had an emotional and economic meaning that was far reaching and enduring through life. Like most Americans of his day, he came from a large family. Andrew, his father, who also had been born in North Carolina, was one of eight children, and through his marriage to Leah Hauser he had eight children of his own. Strong was the second of these, younger than Wiley but older then Martha, William Clark, Mary, Nancy Eliza, Caleb, and Joseph.[12] His parents and siblings, along with his wife, in-laws, uncles, aunts, and other relatives, were the primary personages in his world. He lived his life in sustained connection with them.

Powerful emotions bound the family together. One of the first entries in Basil's diary records a family crisis and reveals some of his feelings and assumptions about family ties. "This is a day long to be remembered," Strong wrote in the summer of 1853. "Sister Martha got married last night without the consent of Pa and Ma, and such a time as we have had! Oh! it will never be forgotten."[13] Strong's choice of words and use of exclamation marks indicated his shock and discomfort about an event that shattered the family's unity and calm. He expected his parents and siblings to be closely tied together through life. Parental approval of a child's marriage partner, prior to the ceremony, was virtually required. Thus Martha's precipitate action at the age of twenty-one brought the Thomassons to an abyss, and initially Strong could not see how they might bridge it.

The family's feelings, however, were too strong to allow this chasm to become permanent. Martha, moreover, had neither left Yadkin County, where the family lived near Hamptonville, nor chosen an unacceptable mate. Her husband, Alexander Johnson, was a young man, two years older than Strong, who lived nearby with his parents. On the second day after the unexpected marriage, "Alexander came over," undoubtedly to pay his respects to Andrew and Leah Thomasson, pledge his love for their daughter, and seek their blessing. With evident relief Strong reported, "'[P]eace is made;' so 'we are at peace with all the world, & the rest of mankind.'"[14] Two days later family routines had resumed as Strong "half soled Martha's shoes." Several weeks later, when relatives from Guilford County paid a visit, Strong "went over [the creek] after Martha, as the kinfolks wanted to see her."[15] Thereafter the Thomassons and the Johnsons remained in close contact, visiting frequently and sharing their lives and work together.

Strong cared deeply for his relatives, and he wrote frequently in the diary about his contacts with them. He recorded his brothers' and sisters' birthdays and commented on letters that he received when one or another was away from home.[16] Visiting with relatives at festive times of the year was a common practice. In December 1854, for example, Strong left Yadkin County with his sisters Martha and Mary and brother-in-law Alexander and traveled first to Salem, in neighboring Forsyth County, and then on to Guilford County. Over the holidays they stayed with one cousin and three uncles and in the process enjoyed "a Christmas-eve romp," a Christmas night buggy ride, more than one "'big dinner,'" some flirting, and much visiting.[17] Gone seven days in all, they brought home abundant news for the rest of the family and memories to last the coming year.

In the routines of daily life, close contact with one's parents or parents-in-law was the rule for the Thomassons, as for many other yeoman families. Relatives settled the same areas in North Carolina and tried to stay in close proximity as the generations advanced.[18] Often a network of parents, children, uncles, aunts, and cousins almost blanketed certain localities. In recognition of this fact, the records of the United States Census identified some districts by the name of the family that was predominant there.[19]

Basil Armstrong Thomasson did not live permanently in Yadkin County, but he chose to stay in close proximity to his father and mother. After he married and began a family of his own, Strong wanted to purchase land. A schoolteaching job, plus the existence of a parcel of land for sale just across the county line, lured him to Iredell County.[20] But Strong and his wife Mollie were never far from their parents' homes, and they kept in close contact. One Sunday Strong and Mollie left home to visit *his* parents, but before they could "get far" they "met Mollie's father & mother coming" to visit them. Such family visits were frequent on weekends, and Strong's diary noted with surprise the rare occasions when weeks went by without seeing one another.[21]

In sickness or trouble, family members were an indispensable resource and support. When Andrew Thomasson cut his leg "badly" one winter, Strong promptly "went over to see Pa" and "cut some wood" for the fireplace. Three days later he and Mollie again "went to Pa's," and

Strong continued to help with chores until his father had recovered.[22] On another occasion Andrew Thomasson felt so ill that he "sent after" his son. Strong spent the night with his parents and then traveled seventeen miles to get a doctor, only to find that the physician declined to come. Pressing on, Strong went "14 miles farther" after another doctor, who "would not come either." Before returning home, however, Strong sent word of their father's illness to his brother Wiley and sister Mary, both of whom were married and living several miles from their parents' farm. The three children, plus Mary's husband, all hurried back to Hamptonville and stayed there until Andrew's health improved.[23]

When the children were ill, their parents assumed the responsibilities of caregiving. More than once Strong's father came to his house and wrapped him in a "wet-sheet pack" to induce the chills and sweating that both believed would drive out a fever.[24] When other children in the family became seriously ill, Andrew and Leah Thomasson took them into their home for the duration of the illness. In 1861, for example, William Clark Thomasson contracted typhoid fever, and for five weeks his parents nursed and aided him.[25] The family unit was the primary source of both physical care and emotional comfort.

The uncertainty of life and health in the nineteenth century accentuated the importance of family bonds. Basil Armstrong Thomasson's life was short, yet before his early death he lost two of his siblings: Clark and Martha. Sister Martha was the first to pass away. In November 1858 Strong noted in his diary that she was "quite ill" with "something [like] the consumption." By early February 1859 she was staying at their father's, and one of his brothers "came down . . . to get money to buy medicine" for her. Four days later Strong wrote, "Sister Martha is very low. Dont think she can get well." Later that month she herself recognized that she could not live, and on March 3, 1859, she died.[26]

It was two years later that William Clark Thomasson and his wife simultaneously contracted typhoid fever. On August 9, 1861, Strong learned from his brother Caleb that both were "at Pa's sick." In the succeeding days Strong visited his father's frequently, but not until August 24 could he record the words, "Nancy is on the mend." Clark, however, continued to be "not much, if any, better" and by early September was definitely "getting worse." On the morning of September 12, despite the best efforts of his family, "brother Clark departed this life."[27] By this time Strong had "lost so much sleep" that he himself fell ill and did not even "get to see Brother buried."[28]

Such tragic losses reminded all members of the family that their hold on one another was fragile. "The Flux" was a recurrent threat, especially to younger children, and doctors could do little for infected wounds, respiratory diseases, and a variety of other illnesses.[29] The Thomassons relied on Ann Benbow, a midwife, for assistance with births, but for other problems they often treated themselves with store-bought medicines and with home remedies such as poultices and syrups.[30] Strong believed many of " '[t]he Claims of Hydropathy' " and perceived that cleanliness promoted good health, but he had little faith in doctors.[31] Care of the sick and burial of the dead both took place, for the most part, within the circle of the family.

The joyous rituals of life also unfolded within the bounds of family life. When young Strong went to church with his family, he found many opportunities to talk with the young ladies. Once he walked one girl to the meeting and accompanied a different one home afterward.[32] Serious courtship took place within the girl's home. After Strong began to woo Mollie Bell ("Miss M. A. B."), who ultimately would become his wife, he spent many Sunday evenings—and early Monday mornings—at her house. After one such Monday in 1853, the weary diarist recorded, "I was up last night till four o'clock. This way of staying up so late is rather a bad business, tho' it's quite pleasant at present time."[33] Strong's diary did not specify how he and Mollie talked and courted all night without bothering the other members of the Bell family. It is possible that her sleepy parents followed the custom of "bundling," which allowed the couple to lie down and whisper together on a bed equipped with a divider down the middle. The wording of one diary entry, however, suggests that Strong went to sleep in a bed by himself.[34]

Even married couples began their life together within the parents' home and under their gaze. Strong and Mollie said their vows at the Bells' home, where family members constituted the audience. They spent their wedding night there, and on the following day Strong brought his wife "home to Pa's." The couple did not come alone, however, for Mollie's parents and assorted brothers and sisters came with them to continue the family celebration of what seems today, in our more individualistic society, a private moment.[35] For months the newly married couple continued to live at one or the other parents' house, only slowly moving into a separate home of their own.

Family also had a vital economic significance, just as encompassing in the world of production as family ties were in the realm of emotions. Economically, the family functioned almost as a unit across two generations. Although Strong considered himself "independent" at twenty-one years of age, his economic affairs remained entangled with those of his parents for years thereafter.[36] This economic connection was by mutual choice and brought mutual benefits.

For more than five years after his twenty-first birthday, Strong continued to live at home (where on cold nights he looked forward to finding "a warm place" in bed beside his brother Caleb).[37] During this period he kept track of his own finances; he earned money, made purchases, and paid his own debts. But he also continued to help with the work of his parents' farm and in turn made use of their resources. While thinking about having his own home someday, Strong used his parents' land to pasture a red heifer and two hogs that he had purchased.[38] Not until five months after his marriage was Strong finally established in that home of his own.

Even then the farms of Basil and Andrew Thomasson continued to be two parts of one connected enterprise. Before Strong purchased land, he sought out his father's opinion and advice.[39] Thereafter, sharing of labor, draft animals, seed, and advice went on continuously as the two households aided each other. Together the Thomasson men slaughtered hogs, hauled rails, mended fences, sheared sheep, washed wool, and sometimes helped one another in planting and cultivating their crops.[40] Strong borrowed wagons, tools, and draft animals from his father, used his parents' land as pasture, and gratefully received grafts for apple trees from them.[41] In return he "bound wheat for Pa," "helped Pa mow," plowed the fields, stacked hay, and rived boards for a new barn.[42]

In short, Strong and his father's family tackled the work of their farms cooperatively. To a lesser extent, Mollie's father also worked with Strong, especially when the skills of a carpenter were required.[43] There was always plenty of work to do, and despite their labors the Thomassons never lived in material comfort. At times even their supplies of food were slim. But they worked together; emotionally and economically, they made the journey through life in common.

The second pillar of Basil Thomasson's life was religion, for along with the members of his family he firmly believed that this life was merely a journey toward a better place. The Thomassons were devout Christians, and Andrew was a Methodist minister who preached most Sundays in addition to carrying on the work of his farm. To them the earth, despite the sometimes absorbing nature of its strivings, pleasures, and pains, was only a temporary abode. Their goal lay beyond death in an eternal life with God. For Strong this was more than an occasional thought that presented itself during Sunday services. It was a conviction that shaped his whole approach to living. It anchored his beliefs and values, influenced the way he lived his life and used his time, and shaped his ultimate aspirations.

The sense that a follower of Christ should despise the world and view it as a potentially dangerous source of corruption and wickedness developed early in the history of the Christian church. Many of the first Christians had belonged to a Jewish culture whose members were a self-conscious minority struggling against its Roman rulers. These Christians in turn came into conflict with the Jewish community around them, and then for more than two and a half centuries the early church lived in opposition to the pagan values of Rome. Its members had good reason to feel that they were a small and threatened band of dissenters.[44] Accordingly, the idea that one should be in, but not of, the world took deep root and remained an influential form of Christian piety.

Basil Thomasson shared in this outlook. Though he had a great zest for learning and ambitions to improve both himself and his community, Strong also stood apart from his surroundings. By conscious effort he kept himself from prizing too highly the pleasures of the world. On his twenty-fourth birthday he reflected on his "appearance into this 'dark and gloomy world' " and quoted some lines of poetry:

> "Few are the comforts" I've enjoyed,
> Fewer the hours of bliss;
> The promise of a better world,
> Is all that's bright in this.[45]

He assumed that a believer might come into conflict with the world and was unlikely to enjoy great worldly success. Reflecting on the life of an American preacher and missionary, Strong once concluded, "The wicked world will per[se]cute the Christian."[46] The text of a sermon by John Wesley, "The Danger of Riches," prompted him to comment in his diary, "It is, doubtless, very dangerous to be rich, or even to desire riches." Because human beings "are never satisfied with gain" and tend to become overly concerned with earthly treasures, it is safer to avoid the temptations of wealth.[47]

On numerous occasions he reminded himself that his goal lay beyond the temporal world. When an elderly neighbor died, leaving some property "for his children to [w]rangle over," Strong questioned the purpose of "treasure in this world" and rejoiced that the deceased man had "also laid up [treasure] in heaven." On more than one occasion he expressed the wish that he would "so live this life" that when it came to a close, he would leave the world "without regret." In 1854, after observing Good Friday, Strong wrote in his diary that Jesus had died in the spring, "when this earth seems most enticing." He then prayed, "Oh that I may be prepared to meet death, at its coming, whether it be in the spring time, summer, autumn or winter, and be fully able to 'go up and [to] the goodly land,['] of which I read in the Bible."[48]

The frequent deaths in the Thomasson family were surely enough to summon thoughts on the fate of one's soul. In 1857 Strong constructed the coffin for his Aunt Mary; a year and a half later it was his sister Martha who passed away, followed the next year by his sister Mary's husband and the succeeding year by his brother Clark. Each death caused Strong to consider the character of his loved one's life and faith. He believed that his aunt had died "easily" because "she lived 'the life of the righteous.' " Martha spoke of Jesus in her last hours, saying, " 'He is close by,' " and Strong was glad that his sister had "lived a Christian and died happy." Mary's husband, Ezekiel Williams, suffered greatly before he died, but he had been a church member "for some years" and expressed confidence "that he was going to live with the Good Man." When Clark died, Strong wrote that he and his family "hope to meet him" in heaven. Partly as a result of these experiences, Strong took his religion seriously and fervently hoped that his "last end" might be happy too.[49] Even as a young man of twenty-five, he had not expected to live to age fifty.[50]

Consistent with these beliefs, Basil Thomasson approached life with an underlying seriousness of manner. He was not a dour person—in fact, a positive, optimistic spirit comes through his diaries. But he believed that life had a higher purpose, and he could be austere. Sundays, he firmly held, were for worship, study, or contemplation, not for merry-making. "Sunday visiting is not in accordance with the Divine Law," he once wrote, disapproving of the way neighbors got together and spent the day that was to be kept holy "in worldly conversation." Avoiding such behavior on the Sabbath, he either attended church services or stayed at home reading the Bible or "good books & papers," which he believed were necessary to a moral education.[51]

Early in the years covered by his diary, Strong opposed the use of alcohol, tobacco, and even coffee. "I believe they were made for some good purpose, but not to eat & drink," he once wrote.[52] In 1855 he took a temperance pledge, and thereafter his disgust with drunkenness never abated. "If the money that is anualy expended in the trafic of rum, tobacco, & coffee, was added to the School fund," he wrote in 1856, "what a great blessing it would bring upon our people, where as it now only adds to the cup of their misery and wretchedness."[53] With time, however, he mellowed somewhat in regard to tobacco and coffee. By 1860 he planned to grow and sell some tobacco, though he still opposed the use of the weed. His love for Mollie caused him to relent on his opposition to the consumption of coffee and bring some into the house. By 1860, when he bought a coffee grinder, he expressed amazement that they had been keeping house "over five years without a coffee mill!"[54]

As these habits suggest, Strong believed that a moral and Christian person faced important choices in life. After reading John Milton's *Paradise Lost,* he wrote, "[T]here are but the two parties, viz, God and His angels, the Devil and his angels, and all men that fight against God, fight in the ranks of the Devil."[55] In a lengthy diary entry in September 1858, he discussed his beliefs about moral responsibility. "Man is a free agent," Strong wrote. "He can choose life or death. If he chooses life God will help him to work out his soul's salvation. If he chooses death the devil will help him to work out his damnation." In this "war-fare" between good and evil, he believed that "actions speak louder than words." All evidence in the diary indicates that he earnestly labored to live well.[56]

To further his moral and religious education, Strong read steadily and rather ambitiously in the Scriptures and religious books. At the beginning of 1858 he started a program of Bible reading, planning "to read it regularly" through that year and "every year" as long as he lived. "Every person," he thought, should read "the book of books . . . through" at least "once a year." Although he had fallen behind in his "task of Bible reading" by the end of February, Strong resolved to do better and kept at his study.[57] In addition to the Bible, he read books of sermons, religious texts such as *Clark's Commentaries on the New Testament,* and other literature such as *Pilgrim's Progress, Paradise Lost, History of the Martyrs,* and the ancient Jewish historian Flavius Josephus.[58]

Rarely did Strong miss a quarterly conference of local Methodist churches, at which there was much preaching, or a camp meeting. To attend the latter he traveled to neighboring counties, enduring "a long and hot ride" one September to get to Mount Tabor (now in Winston-Salem, in Forsyth County) and journeying the next month to Union Grove in Iredell County. Through frequent attendance he became a connoisseur both of the worshipers' tents and shelters and of the ministers' preaching. Reacting favorably to the message, if not the appearance, of one minister, Strong concluded, "Holton can preach if he is ugly." Always he hoped that the camp meeting would bring conversions and a changed way of living to those present. "Among all the campmeetings that ever I attended," he once noted in surprise, "I've no recollection of such a one as this. . . . [T]here was not the first soul converted during the meeting!"[59]

Not surprisingly, the church played a major role in the life of this pious man. Although Basil Thomasson did not attend services every Sunday, his diary indicates that the church was a magnet to him. Steadily over time it drew him into a deeper involvement and commitment. Early in the years covered by his diary, Strong was keenly interested in singing. He copied hymns and songs, evidently taking a leading role in a "singing school" at Aylesbury Church.[60]

Before long he became interested in teaching Sunday school. One of his first ventures along this line clearly was not a success, as he recorded that he had "[l]ectured the good people" of Aylesbury "and highly insulted some of them." Apparently, instead of encouraging the people in their good intentions, he had chosen to emphasize "the truth" about their shortcomings.[61] Two years later he again became leader of a Sunday school class. Attendance was poor, however, and this fact may have discouraged Strong for a while, as he devoted several Sundays to study at home.[62]

As a church member he took an interest in governance and was willing to speak up during congregational meetings. The members of Aylesbury Church, like other evangelical congregations, investigated charges against their fellows and disciplined those found guilty of improper behavior. Strong recorded the expulsion of Richard Green in 1854 for "telling one falsehood and sanctioning t[w]o others that his father . . . told." He also reported the trials of J. F. Reinhardt, who "was had up in the church for swearing, and was expelled," and Reinhardt's mother, who was found not guilty of "an evil report against her." When Richard Green "tried to get back into the church" almost three years after his expulsion, Strong "objected." He believed that Green should have made "some acknowledgements before joining the church again." The members of the church agreed, and because Green was "unwilling" to admit that he had lied, he did not regain fellowship.[63] Living an upright and moral life was a matter they took seriously.

After years of reading, study, and attendance at church and camp meetings, Basil Thomasson confronted the choice of playing a larger role in church affairs. At the beginning of 1860 he found his name "added to the list of Class Leaders," and this appointment brought to the surface a cluster of concerns and doubts that he had entertained about his suitability for leadership. "Shall I accept of the office?" Strong asked himself. "There *is* a work for each member of the church to perform, but is it *my* work to lead?"[64]

Strong felt the normal doubts about his goodness: "[I]f I know not the way, as I should, how shall I lead others therein?" But he had also gained enough self-knowledge to sense that his austere personal habits put him at odds with many of his fellow congregants. Asking himself, "[W]ill the sheep follow?" he admitted, "Some will, others will not, or at least very reluctantly, frequently straying from the fold." Strong knew, in particular, that members of his church could be found "with the drunken rabble," at least "on a publick day" when politicians or militia officers treated the crowd to liquor. His own stern habits compelled him to conclude that such men were not followers of Christ but they had "gone over to the ranks of the enemy." It was proper, he felt, to "dismiss them from the church," but he also was reluctant to sow discord or prove an ineffective leader.[65]

In the months that followed, Basil Thomasson wrestled silently with his dilemma. The popularity of liquor among his neighbors genuinely distressed him. After witnessing one militia election, he likened the stampede of men toward liquor to the scuffle of "50 half famished hogs in a lot" fighting over an ear of corn.[66] Few neighbors turned out to help when he hosted a corn shucking or barn raising because he would not serve liquor.[67] Yet clearly he longed to lead, writing in his diary, "Help me, O Lord, to deny myself, to take up my cross, and to follow Thee."[68]

By October 1861 he had made his decision, for a document preserved along with his diary indicates that "at the Quarterly Meeting Conf[erence] held for the Jonesville C[ircui]t" officials of his church licensed him "to Exhort in the M[ethodist] E[piscopal] Church South."[69] Whether Basil Thomasson would have become a successful minister can never be known, for he died the next year, but without question he had followed the urgings of his faith, which was one of the foundations of his life.

## *Education*

Another distinguishing characteristic of Basil Armstrong Thomasson was his thirst for knowledge. Growing out of this personal trait was the high value that he placed on state-supported public education. Throughout his life he devoted as much time as possible to reading and studying, and teaching in North Carolina's common-school system was more to him than just a means to earn some money. He believed that the public needed to be educated and that teachers should be well trained. Education was both a personal mission and a key to social progress.

It is likely that Thomasson absorbed some of his dedication to learning from his parents, for Andrew Thomasson was a minister, and Protestants historically placed great emphasis on the Scriptures and the need for each believer to read and study the Bible. Another sign of the family's interest in education was the fact that Strong's parents sent a younger son, William Clark, to Jonesville Academy. This school, located in a village near Elkin, had opened its doors in 1818 and subsequently developed a considerable reputation. It attracted boarding students and at times had as many as three hundred scholars in attendance.[70]

Yet it is also clear that Strong Thomasson's interest in learning went beyond that of his parents. In fact, he tended to be so bookish in his habits that he was suspected in the family of shunning work in order to read. "You know that it always has been the case," he admitted to one of his brothers in 1856, "that if the farm and house work did not progress as rapidly as [Pa & Ma] thought it should[,] there wer[e] signs of displeasure in their faces. Time spent in study seems to[o] much like a waste, hence we must work, work, work."[71] Strong denied that he was lazy: "I now, and have all the spring[,] worked early and late, yet you say that Pa told you I didn't work much." Still, his words plainly revealed that he valued education more than farmwork. "It is true, I did but little out door work last winter," he wrote; "I was at work on my mind. Now I think more of that [mind,] if it is a little, weak one, than I do of this, or any other farm, and you know that where one's treasure is ther[e] will his heart be also. . . . I intend to study no matter whose face wrinkles, and next year some one may tend the farm that will work."[72]

Neither the diary nor other family documents indicate how much formal schooling Basil Thomasson received, but he gained enough education to begin teaching school early—a pattern his brothers imitated. At least by 1851, when Strong was twenty-two years of age, he taught school in Forsyth County. During the following year he worked for a tobacco farmer and clerked at a store, but by the winter of 1853–54 he was again in the classroom.[73]

By no means, however, did he consider himself educated. He opined, "'Knowledge is power,' and more precious than gold," and he longed to have more formal schooling. Early in 1854 he was thinking of going to school that summer: "I must have an education, no matter what it costs." These thoughts stayed with Strong Thomasson for some time, but he never resumed formal studies. As a result, he had to pursue knowledge through personal reading. He did so whenever possible, reading Milton's *Paradise Lost* while clerking at a store and regretting that his duties left him "but little time for study."[74]

After he married and set up a household of his own, Strong's efforts at self-education continued and seemed to gain momentum. Perhaps courtship no longer consumed so much of his time;[75] certainly as his own employer and director of his own affairs, he could read and study when he saw fit. In any case, the reports in his diary on works read become more numerous. He devoted inclement weather to reading, noting that he had spent one rainy day as he would have liked to have spent "a great many, in reading and writing." He often spent much of the Sabbath Day in study, happily turning his attention to the Bible and "other good books and papers." [76]

Thomasson was an avid surveyor and consumer of newspapers. As an inexpensive means of obtaining information, newspapers were attractive to those, like him, who were short of cash. The subscription rates of many papers were only a dollar or two per year, and the annual postage bill for one of his favorite journals, the Raleigh *Spirit of the Age,* totaled a mere thirteen cents.[77] Moreover, many newspapers offered "specimen" copies, which a potential subscriber could request and examine for free. Writing away often for such specimens, Thomasson sampled a large number of newspapers from various locations, including Cincinnati, New York, and New England.[78] In a typical year he received issues from a dozen different newspapers,[79] and he usually subscribed to three or four papers at one time. Trips to the post office on Saturday were a high point in the week, for some of his beloved newspapers might be there.

Booksellers were also a golden pathway to information and knowledge for Thomasson. He read their catalogs frequently and bought as many volumes as he possibly could afford. The diary reveals that his intellectual interests were broad. In addition to religious subjects, which were always important to him, Strong enjoyed reading history, biography, accounts of exploration, works of literature, and studies of languages and science, including Latin, algebra, geology, anatomy, and "Phisiology." A sampling of the books he read between 1854 and 1858 includes *Paradise Lost,* Seneca's *Morals,* an unspecified work by Byron, Washington Irving's life of George Washington, an account of the Reformation, the *Vicar of Wakefield* by Oliver Goldsmith, Pope's *Essay on Man, Pilgrim's Progress,* and other works in history, mathematics, and science.[80]

From this energetic program of reading Thomasson derived views that were clearly enlightened and progressive for his time. Though religion was of fundamental importance, he also was hungry for knowledge about science. His reading had made him aware of fossil evidence pointing to the great age of the earth—evidence that Darwin was using to shape his theory of evolution—but he feared no conflict between science and religion. After finishing one book, for example, Strong wrote: "Read Geology, which being rightly understood, and the Bible correctly translated, would not disagree." [81]

The notion of evolution (as opposed to an understanding of the mechanism by which it occurred) was fairly widespread, and Thomasson was familiar with it. In one diary entry in 1858 he discussed the ideas of adaptation and extinction: "This earth is, and ever has been, according to Geology, changing. Hence it is not a fit place for animals whose natures are not adapted to it. Hundreds of years ago there were races of animals which are now extinct. When the productions of the earth, the climate, etc. were adapted to their natures they lived, but when these changed they died, and now can nowhere be found." This process of change and extinction, he added, was continuing. Such ideas were intellectually exciting to Thomasson, who firmly believed that society and human knowledge were progressing. He rejoiced that he lived "in these days of knowledge and improvement."[82]

Given Thomasson's passion for education and desire for progress, it is not surprising that he became involved in North Carolina's fledgling efforts to provide free public education. In 1826 the General Assembly established and began to build up a fund "for the Establishment of Common Schools," generally called the Literary Fund. Then, in 1839, a new law authorized voters to decide whether or not to tax themselves for the support of common schools. Each county that voted favorably would receive two dollars from the Literary Fund for each local tax dollar, and local committeemen and "superintendents of county schools" were to be chosen. Several years later the legislature gave local boards of superintendents the power to appoint examiners who would judge the qualifications of would-be teachers, and in 1852 lawyer and state legislator Calvin Wiley was appointed the first state superintendent.[83]

Basil Thomasson took part in these first steps toward a public school system. In the fall of 1853 he went before an examining committee and received his first certificate to teach. As he returned each year to have his certificate renewed, he showed pride that his scores went up from threes and fours (out of five) to "No[.] 1 all round."[84] However, the examiners, like everything else in this new system, fell short of professional standards. Thomasson reported that they had granted a certificate to a man who insisted there were two *g*'s in the word *wagon* and had referred Thomasson "to Webster."[85] In 1853, when Thomasson's diary begins, Yadkin County had established forty-two school districts but only thirty-one schools. Roughly half of the eligible students attended a school term, usually of three or four months, with breaks for work in the fields and cold weather.[86]

Basil Thomasson was passionate about the importance of education, and the new school system needed teachers with his enthusiasm, for there were many problems to overcome. For example, no one took serious responsibility for the schoolhouses, which frequently were in miserable condition. Repeatedly during his career as a teacher Thomasson found his schoolhouse in disrepair and had to threaten and plead for help in fixing it. In November 1857, for example, when he found the building " 'in a bad fix,' " he appealed for workers to help him repair it on Saturday. That day, however, he "[w]ent over to work on the schoolhouse, but found no one there," so he left.[87] A few years later he alone showed up at the start of a school term. Finding that the school's "chair, desk & black board were gone, the house in miserable condition and the door locked," Thomasson "soon concluded it was no use to stay there" and left for home.[88]

The county boards of superintendents put a great many arrangements in the hands of the teacher. When Thomasson began his school in 1853, he hired a neighbor to make him a blackboard, "six feet long & five wide." It also was his responsibility to obtain books and prod the students and their parents to buy them.[89] Even so, only a few books were available for a large classroom of students. If a teacher had books, a blackboard, and chalk, he could try to hold class, despite distractions such as fleas. On one occasion Thomasson stayed at the schoolhouse till after dinner, spending a good part of the time killing fleas. "I gave several of the little *jumpers* their last lesson, and sent them to their long home."[90]

A teacher in country schools of this period always had to struggle to establish discipline. It was traditional for the scholars (who were often large in stature) to try to "turn the teacher out" at Christmas time and thus win a vacation. Thomasson usually maintained control by offering them a reasonable break and some small present, but there were unscheduled troubles as well.[91]

Once, only one week after rejoicing that he had "a peaceable school," he had to "give two little boys a stripe each for playing 'hard knuckle.' " With fifty-seven students crowded into his schoolroom, such behavior could rapidly get out of hand, and Thomasson feared that he would "have to use the gum freely in order to maintain good order." On another occasion, when boys got into a fight over who stole whose watermelons, Strong had to send for the school committee.[92] Usually, however, he managed to maintain order by himself. By the end of his diary he had learned to "read the 'Law of School' most every day" and tell the scholars what he would do if they transgressed that law. The results were encouraging, for his scholars were as "attentive" and tried "as hard to learn" as any he had ever dealt with.[93]

What was not encouraging, in fact what caused Thomasson the greatest despair, was the pattern of attendance. "Free schools," he learned, "are generaly like a wasp, big at both ends and little in the middle."[94] Both extremes were troubling. Noting once that he had "sixty odd scholars" that day, Thomasson understandably wondered, "What can one man do with 60 unclassable scholars?" Large numbers were even more problematic when one considers the range of skills represented, from older learners to "at least a few" just mastering their ABC's.[95] But overall, Thomasson clearly was most upset about poor attendance and the lack of interest in education that it indicated. Frequently he had only a handful of students, and many who had been attending never returned from vacation times.

Believing that education was vital, Strong had difficulty understanding the many parents who did not share his values. "How strange it is," he wrote in 1853, "that people will close their eyes against their own interest! Ignorance is the cause." As the years went by with no improvement in attendance, his comments became sharper and more judgmental. "A man that won't send his sons & daughters to school ought not to have any," he once observed, and later he doubted that the man who refused to send his children to school but spent "30, 40 & even 50 dollars a year for coffee and tobacco" could "escape Divine vengence."[96] From their parents children learned the attitude that education was not important, and perhaps that was the reason why the girls, in particular, would not study.[97]

The attitudes of county leaders and school officials were not much better. In 1854 local officials did not bother to hold a scheduled election for school committeemen in Thomasson's district. "Very little attention," he noted, was "paid to education." Nor were there any tangible inducements for local citizens to become involved with the common-school system. Although the state superintendent received a salary and the county superintendents a small percentage of the money spent, local committeemen got "nothing but the *honor* of the office." That was not enough to be effective in the vast majority of cases. Thus, when one local official, professing great interest in the work of the schools, came out to see him one day for the first time, Thomasson fully expected it to be "his last visit."[98]

Teachers also received little pay. State superintendent Wiley claimed in 1859 that teachers were paid an average salary of twenty-eight dollars per month, but Basil Thomasson never received anything close to that amount. Fifteen or sixteen dollars per month was not unusual for the salaries he was offered, and in some communities there was so little interest that Strong concluded, "There is no chance for a school here unless one would teach for nothing and board himself."[99] Moreover, to derive the full financial advantage of a teaching post, one needed to board with local families, something that was inconvenient for Thomasson after he married and had a family of his own.

However, teaching was more than a source of income to him. "One can make out to live at it," he observed, "and if he does his duty he is laying up 'treasure' in a better world." So Thomasson brought unusual dedication to his work. Sometime in 1856 he drafted a series of primers in reading and spelling, designed for young students, and sought the aid of Calvin Wiley in getting them published. He also wrote on his own to publishers, hoping that advertisements could make the price of his primers so low that teachers in western North Carolina would be able to buy and distribute them.[100] There is no evidence that Thomasson's primers were published and used, but his initiative, and his confidence that teachers would pay to have books in the hands of their students, reflect his dedication.

Teaching school was not always a pleasure, and Thomasson sometimes felt what is today called "burnout." Even in 1853 he remarked that he was glad to be a "free man" once again at the end of a "long & tiresome" school term. After he married and bought his own farm, he sometimes worried that he might "lose as much by teaching school" as he would gain. Improvements could be made on the farm during the winter school term, and he wrote, "I ought to be doing up my fences, ditching and cleaning up the swamp." The simple need for cash sometimes kept Thomasson at teaching. In March 1859 he admitted, "[I]f I were able to pay my debts I dont think I'de teach any more."[101]

Thomasson also loved teaching, however. The beginning of any term always brought him excitement, and it tended to be the distractions and difficulties that wore him down. "I like to teach school," he wrote, "if I could only have things to my notion. Have a good [school]house well furnished, and books of the right kind, and enough of them." The eagerness of some of his classes to learn also inspired him. When Christmas approached in 1859, he "lectured" his students "on the importance of going to school, and learning all they can while young, and then left it to them to say how much holyday [they] should have; They said—*One day!*" Delighted with their attitude, he resolved to "give them some apples" as a "treat." "If they had books enough they would learn a great deal this session," he believed. "If I were able I would supply them with good school books at my own cost."[102]

## *Social Improvement*

Basil Thomasson's belief in the importance of education was part of a broader interest in community progress and social improvement, an interest that he shared with others in North Carolina and the western Piedmont. The *Spirit of the Age,* one of Strong's favorite newspapers, breathed forth this spirit of progress, publicizing its importance every week through its pages. Published in Raleigh, the *Age* was a temperance journal that deplored the waste of resources and damage to human beings caused by abuse of alcohol. But the *Age* also advocated a broad program of social improvement through forward-looking and democratically oriented public policies. It favored improvement of the "moral condition" of the people, the "spread of intelligence and knowledge," better common schools, better transportation, the improvement of agriculture and mechanical arts, and a homestead tax exemption for the state's poorer residents.[103]

In an effective editorial in 1853, editor Alexander M. Gorman took the legislature to task for doing little in its recent session that actually broadened the educational opportunities or advanced the economic life of the average citizen. Larger appropriations for the common schools and better roads and railroads were high on the paper's list of priorities. A representative article in the same year analyzed census figures showing that farmland was much more valuable in the North than in the South. The striking regional difference, argued the *Age,* was due to the use of better methods in the North, and these resulted "from agricultural science." Not surprisingly, the *Age* also showed a keen interest in scientific discoveries and inventions that put scientific knowledge to practical use.[104] Learning, technology, and democratically oriented public policies were the key to social progress.

Many in the western Piedmont optimistically shared this creed. Though Thomasson's region was poor, it seemed to have a considerable number of residents who believed that vigorous economic development and institutional progress were just around the corner. The present rudimentary state of their society need not be permanent, they believed, for with the intelligent investment of human energy and money, society could undergo the kind of transformation that was sweeping through many parts of the North.

Schools seemed a key to community progress in the eyes of many besides Basil Thomasson. In addition to Yadkin County's successful Jonesville Academy, another "college" had recently opened in Iredell County, just to the south of his neighborhood. New Institute got its start in 1851, and a small village quickly grew up around it. By 1856 the private school had reorganized, and both school and village had changed their names to Olin, in honor of Stephen Olin, a respected Methodist educator. Members of the Methodist denomination played a substantial role in starting and sustaining this "college." When Brantley York, a Methodist minister, opened its doors in 1851, he was reported to have one hundred students.[105]

Others thought that an academy or college would spur intellectual and social progress in Basil Thomasson's backyard, and they hoped that combining the support available in several Protestant denominations might bring success more rapidly. Calling themselves the "Union Academy Association," these people met near the end of 1854 and tried to build momentum for their plans. Thomasson was keenly interested in their goal and decided early on to build himself an "office at the Academy." He attended meetings and "subscribed Five Dollars to Union Academy."[106] When the trustees of the association laid out a site plan and offered lots for sale, Strong and his friend Wiley Messick Jr. bought lot "No. 4 for $32.50, the nicest lot sold."[107]

Tangible progress was slow, however, and in a few months denominational rivalry added to the problems of the proposed academy. Rumors circulated that the Baptists were going to take over the organization, and Strong was relieved in April 1855 to hear good news: "The Academy dont pass into the hands of the Baptists, but is to continue on the union flag." However these controversies may have been adjusted, by 1857 the academy still had not come into existence. Its trustees then pressed Thomasson to pay for the lot he had agreed to purchase. He argued with them and offered merely to return his deed, but they made him pay. Although Strong had hopes of selling his lot to someone else, he eventually, in 1861, "swap[p]ed" his lot at the proposed academy (which still had not come into being and was then called Zion Institute) "for one of Colt's Repeaters."[108]

These ill-fated plans for Union Academy may have had their origin in another scheme for community improvement, one that Basil Thomasson helped originate. In December 1853 a neighbor, John E. Grant, visited him, and they talked about putting up "a blacksmith shop, a school house, etc etc." at Windsor's Crossroads.[109] A few weeks later they called a meeting in the neighborhood "to talk about building a School house" and Thomasson drew up "two Subscription lists."[110]

Impatient for progress, however, he and John E. Grant did not wait for others' support to materialize. They immediately acted on their aspirations and set to work building a blacksmith shop. Cutting their own timber, they raised and equipped the shop with help from some family members and opened for business. In a few months they hired a man to work for them regularly as the smith. Soon Thomasson began to suspect that his shop was "rather too lean a business" to support him, but he continued to be connected with it until late in 1855.[111]

Thomasson's promotion of "a blacksmith shop, a school house, etc etc." and the subsequent Union Academy venture are closely connected in time and place, and Strong's energy in developing subscription lists may have sparked wider enthusiasm for a school on the "union" basis. The evidence in the diary is too sketchy for one to be certain whether there was a connection between the two endeavors, but it seems clear that Thomasson had no leadership role in the Union Academy Association once it was formally organized.

Thomasson's hopes of commercial success for his blacksmith shop, though they proved to be somewhat rash and unsuccessful, were only a pale reflection of the dreams that others were pursuing in the area near the Yadkin-Iredell county line. In 1846 Andrew Baggarly bought land on Hunting Creek in Iredell County, very close to the area where Basil Thomasson would later live. Within a few years he had recruited three partners—W. I. Colvert, Andrew Morrison, and A. B. F. Gaither—to help him pursue aspirations for economic development on a grand scale. Advertising for capital in the Salisbury *Carolina Watchman* in 1849, Baggarly claimed that his property had "the most valuable Water Power in the Southern Country." There was sufficient power "to drive any number of mills or machinery . . . adapted to the manufacture of paper, to calico printing, to bleaching, etc."[112]

Infusions of capital were not forthcoming, so Baggarly and his partners went ahead on a more modest scale and began to build a cotton mill, gristmill, blacksmith shop, store, and houses for employees. In 1854 Baggarly again put his dreams before the public in a long-running series of advertisements in the Charlotte *North Carolina Whig.* He declared, "The proprietors of Eagle Mills and South Troy, . . . nothing daunted by a degree of public neglect, are pushing forward." In only eleven months, he boasted, they had made $4,000 "clean cash." With a "liberal and full handed company" much more could be accomplished, for the "intrinsic and prospective value" of the site had been estimated "at $2,700,000." Again he appealed for investors, but if large sums did not materialize, he promised that he and his partners would go on as they had been doing and make money. Brimming with confidence, Baggarly declared, "Eagle Mills and South Troy are bound to go ahead just as sure as the sun shines."[113]

A year later Baggarly repeated his ad, this time with a subtitle regarding his venture: "A Mine of Wealth—A little more help and its riches shall be developed to the astonishment of the world and to the enrichment of the South." As excitement about building a western extension of the North Carolina Railroad—from Salisbury to Statesville and on to Asheville—began to build, he assured readers that his development would not be in competition with those plans. He foresaw a "North Western Road from Salisbury" to Eagle Mills that would open up the "millions of dollars of produce in that direction, and mines of mineral wealth beyond calculation."[114] In 1855 Baggarly also announced the founding of Eagle City, which he called "The Great Point of Attraction,

Destined to be the great center of manufacturing interests in Western North Carolina and perhaps the United States." With striking showmanship, he deeded to the president and Congress a ten-acre tract of land in Eagle City that would be called Eagle Square.[115]

Thus, exceedingly ambitious hopes of community progress and economic development flourished among some of Thomasson's neighbors, despite meager results and repeated discouragement. Baggarly and his partners never gained the backing that they sought, and the 1857 panic nearly ruined them. Baggarly was forced to sell his interests in Eagle City, and the 1860 census reported him as a forty-nine-year-old mechanic with no real property and only $50 worth of personal property. W. I. Colvert remained at work, however, as the agent of the cotton factory, which now was owned by Morrison, Gaither, and Company. The factory's real property was worth $12,500, and both Colvert and A. B. F. Gaither were fairly wealthy people, compared with their local peers.[116] Their labors had produced a factory where none had been before, and the number of mills and factories in the immediate area showed that entrepreneurial ambitions were very much alive.

Of course, just as the promoters of Eagle City had to give up their dream to be "the great center of manufacturing interests in Western North Carolina," Basil Thomasson had to relinquish his hopes for commercial and educational enterprise. His dreams and his labor for community progress had few tangible results. The blacksmith shop never flourished, and the trustees' plans for Union Academy seem never to have gotten off the ground. Similarly, Thomasson rarely was able to find paying students outside the common-school system. Perhaps the people of Yadkin and Iredell Counties were simply too poor, and their numbers too small, to support the progressive ventures of which Strong and others dreamed.

Certainly Basil Thomasson and all his close friends and relations remained poor in terms of the world's goods. Although his simple life had many rewards, he never was wealthy enough to know a substantial difference between ease and discomfort, sufficiency and hardship. The walls of his home had wide cracks between the logs, and "daubing" the cabin—pressing mud into the cracks—was a yearly chore. Evidently it did not make an airtight home. After one snowy night Strong noted that he took about two bushels of snow out of his loft. During another winter, when the Thomassons were hard pressed to buy enough firewood,

Strong made for his family "a bed of straw and 16 lbs. of cotton" to keep out the cold. Although his fields and garden produced some abundant meals, there were also hungry times. In October 1860 Strong cheerfully remarked that cooking was easy because they had little to cook. "We have been living for several days on little roots, many of which are so small it takes two or three of them to make a mouthfull, which we call 'sweet potatoes.'"[117] In the midst of dreams of progress, he and many others still faced the reality of doing without.

## *Honor and Southern Character Types*

Basil Thomasson's sense of conscience, dedication to self-improvement, and desire for community progress all suggest that he was not a typical antebellum southerner—at least as that typical southerner may sometimes appear in modern historical studies. The seigneurial, precapitalist, hegemonic planters described by Eugene Genovese loom large in historical literature, but Strong Thomasson certainly was not one of them.[118] Nor was he a typical representative of the culture of honor described in the influential work of Bertram Wyatt-Brown.[119] Thomasson's character and behavior diverged sharply from the belligerent, self-regarding style of manhood that Wyatt-Brown described as characteristic of the culture of honor.

Bertram Wyatt-Brown's impressive analysis of the culture of honor dissects a complex of social norms, societal values, and personality traits that may well have characterized many southerners. The diary of Basil Armstrong Thomasson, however, forces us to ask how widespread and determinative that culture was. Wyatt-Brown describes a pattern of values and norms that promoted particular behaviors and beliefs. According to his analysis, the culture of honor encouraged and supported the following: sociability, conviviality, and the use of alcohol (as opposed to privacy, abstinence, and the use of time to read or think); patriarchy, male dominance, and fear of female power (as opposed to companionate marriage); "haphazard" father-child relations; "irregularity" of habits in which "the notion of keeping strict [financial] accounts" was "rare"; belief in a society of "ordered ranks"; and a conviction that slavery was honorable.[120]

We have seen that Basil Armstrong Thomasson was a teetotaler who discouraged socializing and loved to read and think. He also deviated from every other pattern or belief in the above list. To Bertram Wyatt-Brown the norms and values of the South's system of honor stood out in sharp contrast with Yankee values.[121] Basil Thomasson certainly regarded himself a southerner, but in terms of Wyatt-Brown's archetypes he resembled the conscience-driven, self-regulating Yankee far more than the honor-seeking, self-regarding southerner.

## *Views on Women, Marriage, and Gender Roles*

Strong Thomasson's attitudes toward women were responsible rather than self-serving, governed by standards of right and wrong, and surprisingly egalitarian for the mid-nineteenth century. For many planters, influenced by the culture of honor, one's manhood was something to glory in, display, and prove through conquests of the opposite sex, particularly of slave women. No set of attitudes could have been further from the mind of Basil Thomasson, who strove to be moral and upright in his sexual life. He was neither self-absorbed nor self-indulgent, and his relations with his wife Mollie revealed more interest in egalitarianism and companionate marriage than in patriarchy.

As a young man Strong approached courtship with enthusiasm and took a lively interest in the young women of his neighborhood. His diary for 1853 and 1854 contains frequent comments on their beauty (or lack of it) and his contacts with various girls. After completing some business with one neighbor, Thomasson playfully noted in his diary: "We are now even; or will be, after I get one of his girls. I have bought his hogs, bees and cow; the girls come next."[122]

Excitement tinged his comments on a brief correspondence he had with "Miss S. A. S." "What a letter!" he wrote after receiving one of her missives, and another "interesting" letter moved him to remark, "Miss Sarah is the girl for me."[123] This enthusiasm for "Miss Sarah" did not prevent him, however, from looking at other girls. After accompanying "Miss Mary B." (his future wife, Mollie) home from church, he confided, "I expect I shall have to go again. Miss M. B. is the *gal*." Even after Strong began seeing Mollie regularly, he obviously enjoyed the duty of driving "Miss Nancy Wilkason" home when she visited in the neighborhood. And on a visit to Salem he appreciated the charms of "Miss Mary Hughs—a good-looking girl."[124]

Strong, however, never aspired to cut a wide swath as a suitor or break a large number of hearts. As early as January 1854 he was hinting in his diary that Mollie Bell was the girl for him and linking her name with the word *marry*. He even began to collect plates, candlesticks, and other items he would need for his own household, while assuring himself that he was "only buying a few dry goods, thats all."[125] Moreover, he believed that he had a duty to act responsibly toward women, whose gender and position in society placed them in a vulnerable position. "I am not like some boys who go to see 1/2 a dozen girls at once," Strong wrote. "Woman's love should not be trifled with," he declared, "for to them, it is life." Then, quoting from something he had read, he added, "'[I]t is all they have to live for; and when it is taken away, they have not, like man, a profession, business, travel, and pleasure, to divert and occupy their minds. It is a sorrow they can never tell, to seek the healing balm of sympathy—they have only to sit down and endure.'"[126]

Thus Thomasson rejected a double standard in relations between the sexes and bound himself to a code of moral conduct. But he did not accept society's definition of woman's place. Rather, he questioned and challenged the social norms that relegated women to an uneducated and subordinate role in society. Strong belonged to a debating society that once discussed women's role and influence in society, but the greatest impact on his attitudes probably came from his family. In his parents' home, he "had a regular chat" one night "about woman's work etc." The conversation must have explored the subject thoroughly, because Strong recorded a radical and uncompromising conclusion: "There are more slaves in the U. S. than most of us are aware of. Freedom is a great thing, but woman cant be alowed to enjoy it; they are slaves to men & fashion." Moreover, Thomasson made it clear that he opposed this situation. He added, with feeling, "[t]he time is coming when they will be free; may that time come quickly. Amen."[127]

His convictions about the education of women were especially strong. When Thomasson made a list of local children entitled to schooling, he always included girls along with boys, for the importance of learning extended to women as well as to men.[128] Women, he believed, should raise their voices for social improvement, and education was the tool that would enable women to play a larger role in society. In one diary entry he lamented, "Our females generally are raised up in stupid ignorance, hence they are not aware of the extent of their influence, and of the good they might do." Hoping that southern women would become active in temperance reform, he wrote with emotion, "Oh! that the female part of our community would rouse up . . . and enforce a liquor law among us as the ladies (be it said [to] their honor) of the far west have done in their country."[129]

Strong acted on these ideas and made them a part of his relationship with Mollie Bell. While they were courting, Strong was glad to find Mollie reading, and he persistently encouraged her efforts at education. Evidently he wanted an intelligent and educated wife, for he "tried hard to get Miss B. to read" for him. Initially she balked and refused to cooperate. Although Strong seemed frustrated by her refusal, he did not give up his efforts, and later that month he sent Mollie "two No.'s of the *Age*," his favorite newspaper. In a matter of weeks he had convinced her to suscribe to a newspaper of her own, which he picked up from the post office and brought to her.[130]

After his marriage to Mollie, Strong continued to promote her education and intellectual growth.[131] By the middle of 1858 his diary indicates that he had been teaching Mollie to write. "Why did'nt you write more Mollie?" he asked on one occasion. Later he reported, "Mollie is going ahead with her Grammar," and praised her ability to phrase sentences. In May 1859 he asked her to write the heading for a page of his diary and noted that she was responsible for the completed job.[132] By encouraging his wife to develop her skills and read regularly, Thomasson gave proof of his desire to have a wife who was his intellectual companion rather than subordinate.

In other ways as well Strong and Mollie developed a surprisingly egalitarian and nonpatriarchal marriage. Although they generally followed the traditional division between outdoor work for husbands and indoor work for wives, the boundary between their spheres was neither rigid nor impermeable. For his part, Strong frequently lent a hand and helped with housework. He made the brooms that Mollie would use in sweeping the house, and on one occasion he boasted, "Mollie now has sixteen nice, corn brooms,—enough I think to last her one year at least." Strong also helped with the washing of clothes, sometimes combining his work in the fields in the morning with helping "Mollie wash in the p. m." Occasionally, also, he "helped Mollie quilt."[133]

Although Mollie usually did the cooking, Strong took a turn in the kitchen whenever she was sick or tired during pregnancy. On these occasions, he developed a thoroughgoing appreciation of the scope of Mollie's household duties. One Sunday Strong simply made note of the fact that he had cooked their "*vittles,* and performed the work of house-maid in general." But on another occasion—a Saturday when he had gone to the mill and planted corn—he complained that with Mollie sick, he "had to feed the pigs & chickens, milk the old cow, *etc, etc, etc.*"[134]

Usually he tackled household duties with good humor; in one diary entry, for example, he recorded these sentiments: "It is well enough for a man to know how to cook, wash dishes, etc. Had I . . . been ignorant of these things, I would have been likely to have gone to bed to-night minus my supper. 'Knowledge is power.'" But he knew enough of the demands of housework to resolve, early in their marriage, that Mollie should not have to weave cloth, which was difficult for her and extremely time consuming.[135]

For her part, Mollie occasionally helped her husband in the fields. On separate occasions she helped him plant potatoes, dropped the seeds of corn while he plowed and covered them over, and helped him gather fodder. Strong boasted that Mollie "proved to be a good hand in the fodder field." One September she also climbed up onto "the first *hay stack* she was ever on" and packed the hay around the pole as Strong threw up it to her. Clearly he appreciated her help and was proud of her adaptability. In a pinch he could call on Mollie for help, such as when he needed an extra set of hands to help him doctor a sick cow.[136]

Such sharing of work was probably not uncommon among yeoman farmers and their families. Historians have recognized that small farms sometimes needed all the labor they could call upon, and recently one scholar has seen women's work in the fields as a form of patriarchal domination.[137] But it is equally possible that the sharing of work, such as occurred in Strong Thomasson's household, represented a companionate and egalitarian type of marriage, at least among some people and in some locales. Anne Radford Phillips has studied the farm women of Stokes County, North Carolina (which adjoined Yadkin County on the northeast), and she found that in the early twentieth century "patriarchal authority simply did not exist as a rigid, dominant pattern in Stokes." Neither did she find "a system of paternalism" flourishing "among the small independent farm families." Both genders cooked, washed, worked in the fields, and did the many other chores as part of a "concept of shared work." Many of the women she studied relished outdoor labor and even made the key economic decisions for their families.[138] These men and women paid less attention to traditional gender roles in work than to the common needs of the family. It is easy to imagine that such a pattern would have been possible with Basil Thomasson.

Strong's diary gives fewer details in regard to his style of child rearing, but it is evident that he was a doting parent rather than a distant one or a figure of patriarchal authority. He kept close track of the growth and development of his son, Jodie, and worried when he was ill. In an entry from January 1859, Strong sounds very much like a concerned modern parent: "Jodie seems better this morning. I took him up and washed him in tepid water. He kicked and tried to scream, but he is so ho[a]rse he could not do much at it, and I washed. He ate a few mouth fulls of breakfast, and is now, 1/4 past 9, driving nails in a stick of wood. He delights in a hammer and nails. Hope he'll soon be as well as ever."[139] Such anxieties over Jody's health appear several times in the diary, and Strong obviously took delight in the fact that Jodie was "a great papa boy."[140] As Jody became old enough to learn to spell and read, his proud father chronicled his progress and noted the first day his son accompanied him to school.[141]

Perhaps Strong was absorbing from his newspapers and wide reading some of the ideas of affectionate family life that were spreading in America at this time. Indeed, Jane Censer's careful research indicates that such patterns were common among North Carolina's upper class in this period.[142] That his ideas were not patriarchal is also suggested in the notice Strong took of events in Mollie's family. His parents-in-law continued to have children after Mollie had grown up and left their house. In fact, only three months after the birth of Jodie Thomasson, Mollie's mother gave birth to a daughter. Four years later, after Strong and Mollie visited her parents, he noted, "They have there now two babies. One is nearly as old as Jodie and can neither walk nor talk, tho' its a very fat baby. The[ir] other one was just born a few days since." Strong was too polite to criticize explicitly, but he seemed to raise his eyebrows at the pattern of long-continuing births, with heavy maternal responsibilities, that he saw in the Bells' home.[143]

There is also evidence in the diary that Thomasson applied his egalitarian attitudes toward women beyond his own household. Ann Benbow, who served many families in the neighborhood as a midwife, engaged in a number of activities that were not customary for women. From time to time she tried her hand at both preaching and practicing medicine. Strong did not disapprove of her efforts. Once he criticized the quality and spelling of some poetry that Ann Benbow had circulated, but his objection was to the quality of her work, not to the fact that she had undertaken it. "See, what spelling," he wrote. "I should hate for such *stuff* to be afloat in the neighborhood bearing my name." When he went to a Quaker meeting and listened to her presentation, he reacted much more positively: "Heard Ann Benbow preach. She gave us, as I think, very good advice."[144]

Of course, Strong Thomasson was not a liberated man of the late twentieth century. He was a product of his time and place, which assumed that women were weaker, less responsible, and rightfully subordinate. It is not surprising that he echoed some of the attitudes common to his day. His diary contains references to husbands who have to " 'bear the everlasting din of woman's tongue' " and to women's responsibility to keep a neat house, be " 'help meets' for man," and economize rather than spending money "trying to keep 'in fashion.' " He even echoed the supposedly biblical notion that woman was the "weaker vessel."[145]

However, the diary shows that Strong did not make denigrating attitudes his own. He hoped as a prospective husband to "get an Angel" with whom he could "enjoy heaven on earth," and indeed he seemed to be very happy with Mollie. A "help meet" in his mind was not a slave, and he was greatly upset when he found his mother doing the washing for her daughter Mary, who was ill. Strong and the other children had always taken on that task—"Ma has not, since I can remember been used to washing for her own family"—and he wanted neither his mother nor Mollie to be overworked and "run down."[146] He regarded both his mother and his wife as equal human beings and treated them with dignity and respect. His personal life showed that he was not patriarchal but supportive of greater rights for women.

## *Economic Attitudes*

In economic life Basil Armstrong Thomasson also contradicted Bertram Wyatt-Brown's description of the typical southern character type. Thomasson's economic life was not characterized by "irregularity," and "the notion of keeping strict accounts" was certainly not "rare" with him. In fact, his habits were so completely the opposite that his behavior probably puts him in the opposite camp—the ranks of frugal, cost-conscious Yankees whose habits of saving and regulating their impulses shaped their character.

Thomasson was one southerner who revered and sought to emulate Benjamin Franklin, a man whose life and pithy maxims made him virtually an exemplar of the northern character type. To Strong Thomasson, the advice found in *Poor Richard's Almanack* was economic gospel. Early in his diary he lamented the fact that he mailed a letter to his Uncle Flem, only to have that uncle pay an unexpected visit later in the day. "If I had just have waited a few hours I might have saved three cents," Thomasson wrote. "[T]hree cents *aint* much," he admitted, "but 'a penny saved is a penny made.'"

In 1860 he reminded himself of the truth of this maxim in order to argue himself into a better humor about repairing the family's shoes. "I dont like to mend old shoes," he wrote, "but . . . I can save a goodly number of pennies in a year, by making and mending our own shoes at odd times. Then why not do it? We have to do many things which we do not like."[147] Later that year he decided that he had not paid enough attention to "Dr. B. Franklin." After reading "the *Life and Essays*" of Franklin, Thomasson exclaimed that "every young person in the world should read that book." He hoped it would encourage every reader to "be industrious, honest and frugal" every day.[148]

Like Franklin, Thomasson believed in the importance of hard work and the necessity of disciplining one's desires. Among his papers was a page on which Strong, in 1850, had copied the resolutions of the Reverend Joseph Benson of England. Writing in 1748, Benson had determined to rise at five in the morning, spend his time well, avoid useless conversations or visits, retire at nine at night, and pray four times during the day. Franklin's diary, of course, recorded careful use of the time in each day, so that more work would be accomplished. Although

Thomasson does not seem to have been lazy or sedentary, he aspired to accomplish more and wished that he was not "such a dull, sleepy-headed animal."[149]

He believed firmly in the necessity of being economical. Writing in 1853, Thomasson admonished himself with the words, "I must curtail my expense, and live on economy awhile," and he repeated such advice often. "I intend, in the future, to spend less money in proportion to my income," he wrote early in 1855. "This way of spending all one makes, and a little more, is a bad way." In 1859 he even spoke of eating less to save money—although he weighed less than 134 pounds—because Franklin advised, "Better to go to bed supperless, than get up in debt." "I've set in this year to get out of debt," Strong wrote. "I do not intend to buy anything on time this year, unless its a bit of land, if I can help it."[150]

Clearly Thomasson did not escape debt and get ahead in the way that Ben Franklin had, but the reason was not that he was lazy, luxurious, or devoted to sensual pleasures. He made many efforts at economic self-improvement, including the building of a wagon that he hoped to sell for profit. But this venture, like his attempts to launch an academy or open a blacksmith shop, suffered from the poverty of his rural neighborhood and the consequent weakness of its internal market. His largest expenditure was the purchase of his farm, and his only indulgence was the buying of books and newspapers. These, of course, Thomasson intended not for sensual gratification but for self-improvement.[151]

As for "irregularity" in the keeping of accounts, Thomasson would have none of it. Even the casual reader of his diary will note the frequency and meticulousness with which he kept track of such things as the items he purchased, their cost, and the amount of interest he had to pay on notes. It would appear that the diary comes close to giving a complete record of Thomasson's financial affairs. Yet, even though the diary's information is quite detailed, its pages were not intended to record his financial affairs. He maintained a separate book called "Cost of Living," which unfortunately has been lost. It undoubtedly provided a complete accounting of all Strong's income and expenses.[152]

In regard to his economic goals, it is clear that Thomasson was not attracted by slaveholding society's love of wealth and aristocratic display. Since one's goals should be spiritual, it was "very dangerous to be rich." He disliked gaudy displays and aristocratic pretension and was drawn instead to ideals of progress, democracy, and Jeffersonian agrarianism. One telling indicator of his values was his dismissive attitude toward militia musters. These events were prime opportunities for the assertive display of manhood, and for the elite they were a means to declare and confirm one's status. To Thomasson they were useless and a waste of time, for these were not his means of judging an individual's worth. "Mustering appears to me to be a childish practice," he wrote. The drunkenness that resulted when upper-class officers treated their men to liquor made musters still worse.[153]

Rather than deferring to his social betters, Thomasson made up his own mind and was sensitive to any attempt of a squire or aristocrat to throw his weight around. He wanted to be an autonomous and independent-minded citizen, and he was not afraid to speak up for himself. When he clerked at a store in East Bend, he actively participated in a debating society and relished the exploration of issues. Seeing himself as one of Jefferson's sturdy yeomen, he praised agriculture "as an honorable" occupation.[154]

Thomasson valued the economic, social, and political independence that rested on the ownership of land. By owning one's own land, growing one's own food, and supplying one's own needs, the yeoman farmer was able to stand on his own feet and conduct his affairs without dependence on anyone else. Thomasson and his neighbors did not take this venerable American ideal for granted, because there were many landless people among them and many "out land hunting." "Land! Land!" exclaimed Thomasson. "All want land."[155]

The 1860 census provided evidence of the fact that landlessness was common. In the neighborhood where Basil's parents, Andrew and Leah Thomasson, lived, there were thirty-two farmers but also fourteen "farm laborers," fourteen "day laborers," and two "farm tenants" in addition to those in a handful of other occupations. Most of those listed as "farm laborers" appeared to be sons in a landowning family, so their chances of owning land in the future were good. Yet even if one does not consider those younger sons among the landless, 25 percent of the agriculturalists, and nearly a quarter of the households, were landless. Similarly, in Basil Thomasson's neighborhood in Iredell County, there were forty-seven farmers (the census taker included six sons in the category) but also eleven "day laborers," two "laborers," and one "Hired man."[156] The number of landless men (if one does not count the sons in that category) amounted to 23 percent of all agriculturalists. To put the same facts another way, almost a third of those who worked on farms in the neighborhood did not own land.[157]

Thus it was a high priority for Thomasson to acquire and own land. After moving away from his parents' home in 1859, he evidently rented land from others, farming it while he taught school. Unfortunately he did not record the cost of renting land in his diary, but he complained about the high cost of renting a house and buying firewood. Early in 1859 he developed a persistent interest in a parcel of land offered by a neighbor, Milus Dobbins. Strong was skeptical about the quality of this land at first, calling it "a mixture of flint rocks and thin soil" that would not pay. But the price was "cheap, $2 1/2 per acre" for seventy-one acres, and Thomasson thought that prevailing prices for land might buoy this parcel to five dollars per acre "in a few years." After his father looked it over, Strong closed the deal and was excited about moving to his family's new home.[158]

Another aspect of maintaining his independence was settling his debts with stores and neighbors in a responsible way. As scholars have pointed out, yeoman farmers were *not* completely independent; in fact, they were interconnected with one another in many ways. They shared or swapped skills, tools, and labor, maintaining an assurance of their capacity to be independent, rather than actually living without help from anyone.[159] In the early years of his diary, Thomasson occasionally worked for someone else, helping with a variety of agricultural tasks. Sometimes he hired neighbors to work for him, paying them at an agreed-on rate or swapping the work. With gratitude he sometimes recorded an act of neighborliness that brought him some help for free or at a reduced rate.[160]

The diary makes clear that he lived in a practically cashless economy. The settling of small debts with friends and neighbors could become quite complicated because of the shortage of cash, but all the parties kept track of their obligations and were flexible about the times and means of settlement. In 1856 Thomasson described the history of some small debts that were owed him: "I handed 3 notes on the boys Andrew, James & Write Snipes to brother Wiley more than a year ago. He has collected 2 of them, and there is one of $4.00 yet due."[161]

A few years later Thomasson needed money and "[w]ent to J. S. Hane's near Hamptonville, to see if he would not take up the note" Thomasson held against him. Hanes was hard pressed, however, and could only offer to pay some wheat in settlement. "The amount of the note is $10.73," noted Strong, "and I agreed to take wheat for it at $1 per bushel. Well, I got two bushels only, this time."[162]

Strong was also careful to keep up with his accounts at stores. Merchants extended him credit on account and through notes, but the occasions on which he had money were rare, and when he tried to borrow some, he encountered great difficulty. Thus, when Thomasson acquired some cash, he promptly paid it out to settle his most important debt or the ones longest outstanding. In 1856, for example, he sold his wagon on September 9 and the next day paid off notes with two merchants for $30 and $23.08. Similarly, in early 1860, after receiving his pay of $45 for teaching school, he "went over and paid it all over to Mr. Dobbins" to reduce his debt on the farm.[163]

It was necessary to have some means of raising cash, and that was part of the reason Thomasson taught school even when he was discouraged, or tried other ventures, such as making and selling a wagon. Another common strategy for nonslaveholding farmers was to plant a small portion of a market crop, such as cotton or tobacco. Antebellum yeomen were wary of the market. Because they feared its ups and downs, they usually were reluctant to abandon their food crops and plunge wholly into cash crops. But by giving only a small portion of their time and effort to the cash crop, they could protect their self-sufficiency in food and earn a little money too. In 1859 Thomasson grew a little cotton, although he seemed to wish that he could sell vegetables instead. One year later he overcame his scruples about tobacco and "left a patch" for the weed. When it proved difficult to find plants, however, he sowed the ground in corn.[164] In all these economic decisions, Thomasson sought independence, respectability, and progress rather than the values of aristocracy.

## *Political Attitudes*

In politics, as well, his thinking diverged sharply from the values described by Bertram Wyatt-Brown. Thomasson's attitudes were not determined by the "politics of slavery," and his feelings did not mirror the outrage of proslavery politicians. Nor did he believe that slavery was "honorable."[165] He had an independent interpretation of the issues and his own set of priorities. It was clear that he disapproved of slavery and loved the Union. As the South rushed headlong into civil war, Thomasson even proved willing to leave his home and region, if that would be a means to save his family from violence and disruption.

Perhaps the most striking aspect of his diary's political comments is the omissions. Throughout the 1850s the sectional crisis was becoming deeper and its rhetoric more inflammatory. Southern political leaders focused relentlessly and ever more vociferously on the issue of slavery and the South's rights in the territories. Repeatedly they decried the dangers to the South and the wrongs it was suffering. News of the Kansas-Nebraska bill, the Supreme Court's decision in the Dred Scott case, Bleeding Kansas, and the alarming rise of the Black Republican Party (as proslavery southerners called it) filled many newspapers in both sections. Yet Basil Thomasson's diary is silent on all of these issues. He showed no interest in the controversial political events that supposedly were degrading him, as a southerner, and threatening his rights.

The reasons for his disengagement are not hard to find. Beyond the religious and educational priorities that Thomasson held, he was a strong Unionist. He loved the nation and the historical reputation of the United States as a champion of human rights. On July 4, 1856, he joined in spirit with the citizens "throughout the U. S." who were "celebrating the Declaration of Independence." His next detailed and explicit comment on political issues did not appear until December 27, 1859. By this time the clouds of war were becoming visible to Thomasson, and he reacted with fear and an allusion to the Devil that underlined his love of the Union: "The days of 1859 will soon be numbered. Then 1860. O! who can t[e]ll what will take place in these United States during the year 1860. The enemy of all that's good seems to be stretching every nerve for the destruction of the Union."[166]

Surprisingly, he did not vote for Stephen Douglas, the northern Democrat, or for John Bell of the Constitutional Unionist Party in 1860. Rather, his diary records that he voted for "J. C. Breckenridge & Joseph Lane," southern Democrats who were especially popular in the Deep South.[167] Nevertheless, when one considers local political conditions, Thomasson's vote was not one cast in defense of slavery or against the

Union. If many slaveholders in the Deep South saw Breckinridge as the champion of states' rights and slavery interests, that certainly was not how everyone viewed him, particularly in the upper South. Breckinridge benefited greatly from the fact that he ran as the candidate of the South's Democratic Party, an organization that was strongly identified since the days of Andrew Jackson with the interests of the common man. Moreover, Breckinridge explicitly disavowed his connection with secessionists. After being challenged by several newspapers, the Kentucky slaveholder altered his plan not to speak and delivered a single address. In that speech he flatly denied that he was a disunionist or belonged to any organization whose aim was secession. After this statement, Breckinridge's supporters in the upper South "went out of their way to stress their loyalty," in the words of historian Avery O. Craven.[168]

Basil Thomasson soon expressed his Unionism as he watched—with dismay—the course of political events. In February 1861 he voted against the proposition that North Carolina should hold a convention to consider secession.[169] At the end of March 1861 he wrote, "These are gloomy times, and seem to be growing darker and darker every day. Every mail brings us some bad news—news of disunion and war." He faced with anxiety, but also with determination, the question that troubled Unionists everywhere: "[I]s this union to be utterly and forever destroyed? This glorious union—the price of the blood of our fathers—is it to be abandoned as a thing of no worth? No. 'Never give up the ship.' *Never! Never!*" It seemed unbelievable to Thomasson that the nation was "going to war with itself!" and he appealed to "the God of heaven" to save the nation. Deploring the outbreak of war, Thomasson felt that "all men ought to love and do all they could to keep peace." But as a southerner, he did not approve of his region's course: "The South against the North! What folly!" he wrote.[170] Consequently he showed absolutely no inclination to volunteer for military service.[171]

If Thomasson was a Unionist, what were his opinions about slavery? The diary rarely comments on human bondage directly, but the feeling that slavery was undesirable, or at least morally questionable, is a given. When Thomasson commented on women's status—"There are more slaves in the U.S. than most of us are aware of"—he went directly on to assert, "Freedom is a great thing." He prayed in this instance that women would obtain it "quickly." On another occasion he joyously celebrated the Fourth of July as a "great day," but then he expressed a wish and a qualification that suggested his feelings about slavery. "May the people of the U. S.," Thomasson wrote, "long remain free and independant—if, indeed, they are free."[172]

It is true that Strong once "made two *thundering* speeches in favor of slavery," but these were part of his participation in the debating society that he enjoyed in East Bend. Apparently the debaters were focusing their attention on questions of oppression and social justice in America, for they also argued the proposition that the expulsion of the Native Americans by European settlers was not justified. By one vote they condemned the treatment of the Native Americans.[173]

Their vote on slavery was even more interesting. Almost half of the debating society voted that slavery should not "be tolerated in America," despite the fact that this formulation of the question was strongly negative and raised important constitutional issues.[174] Many who disapproved of slavery admitted the fact that the Constitution provided for and protected it; toleration of slavery was actually embedded in the Constitution. As a consequence, the institution's continued existence, though undesirable, might be necessary to ensure a great good—the Union. Considerations such as these caused many opponents of slavery to accept its presence. Even Lincoln and the northern Republicans staunchly defended slavery where it already existed.

Moreover, the very idea of a debate on slavery was anathema to its southern defenders, who wanted no dissent or even discussion of the issue. Basil Thomasson, in his debating society and elsewhere, actively questioned the reigning southern shibboleths about slavery. In a letter to one of his brothers he described a proslavery article that asserted, " '[A]t all hazards domestic slavery must be maintained,' " and also, " 'The Southern States stand committed in the providence of God to maintain it.[']" Thomasson questioned these statements directly, saying, "Now are these things so?" On the contrary, it seemed to him that the South was "gathering thorns" for what it deemed "flowers" and that it was killing off its "darling institution" by pursuing a secessionist course. In fact, he judged that course "the shortest way" to guarantee emancipation. His feelings about this prospect appear in the following words: "Then if this be the case they [the slaves] are bound to be freed and that at no very distant day. Well, 'If it be of God,'—if He sees fit . . . to put an end to the institution of slavery there is no power on earth that can prolong its life one hour beyond the appointed time, much less in the fifteen slave states."[175] Certainly Thomasson was not convinced that slavery was "in the providence of God" or should be maintained "at all hazards."

The diary, unfortunately, reveals little about Thomasson's attitudes toward African Americans. He strongly approved of William Penn for treating "red men as brethren," and he mentioned the few black people with whom he came in contact without derogatory comments. Not surprisingly, on the other hand, he used the racist terminology that was common in his day to refer to objects and people—for example, "the 'nigger seat'" for a perch behind the buggy's seat, and "darkie" for a black woman.[176] Perhaps the strongest statement that can be made is that Thomasson did not seem to be driven by virulent racial antipathies.

It is possible, however, that on race he took a position similar to that of Abraham Lincoln in the 1850s. Throughout that decade Lincoln voiced disapproval of slavery but avowed that he was unwilling to make blacks his equals. As the war began, Thomasson looked to the future and asked, "[W]hat will be the result? Freedom to the slaves? I think so. The South has commenced the war, and now I do not believe the North will ever listen to any terms of peace which do not include 'Freedom to the slaves.'"[177] Strong did not couple this observation with applause for the prospect of emancipation. His silence is difficult to interpret, but he may have worried that the end of slavery would bring racial changes, or at least unsettled conditions, that were not in his interest.

Most of his letter of April 1861 to his brother concerned the likelihood that slavery would be ended. At the beginning of that letter he advanced the idea that all the Thomassons should move from North Carolina. Whereas his brother had suggested that the family "sell off" and go "to Tennessee," Strong asked, "had we not better go to Indiana or Illinois?" Then, moving on to discuss emancipation, Thomasson wrote, "These are troublesome times, and from present appearances they will not be any better soon."[178] The content of his letter, and the order in which he put down his ideas, suggested that Strong did not believe that God's purpose was to preserve slavery. But perhaps, like many other southerners, he did not want to live among free blacks or felt that the destruction of slavery would make residence in the South more difficult. The available documents do not offer enough information to be certain.

Soon it became clear, however, that the Civil War would make residence in the South more difficult for everyone. Strong continued to teach school, farm, and study the Scriptures, but occasionally references to the "dreadful war" or to high prices and scarcity crept into his diary.[179] Yet, before he fell sick and died, Thomasson was luckier than most. Although his family had to cope with difficulties on the home front, state and Confederate laws allowed him to escape military service.

Under North Carolina law men could be drafted from the militia when the number of volunteers from the state was insufficient to meet the needs of the central government. This proved to be the case early in 1862, and in March a draft from Thomasson's militia company occurred. The law provided, however, that "ministers of the gospel, regularly ordained," were exempt, so Thomasson would not have been eligible.[180] Several weeks later the Confederate government enacted the first national conscription act in American history. The accompanying exemption act, however, excluded both "ministers of religion in the regular discharge of ministerial duties" and "all teachers having as many as twenty scholars."[181] Apparently these provisions kept Basil Armstrong Thomasson out of the army until he died the following fall.

Some readers might wonder whether Strong had hoped to avoid militia duty or a state draft by obtaining his "License to Exhort" in October 1861. An even more pertinent question is whether he was aware of all the laws' provisions. Definitive answers to both questions are probably unattainable, but the diary clearly reveals that throughout the 1850s he never took advantage of his exemption, as a common schoolteacher, from regular militia duty. In addition, Thomasson apparently did not, after becoming a minister, claim exemption from enrollment in the militia but continued to go to parade and muster. Thomasson's character and profound religious beliefs also argue strongly against the idea that he became a minister merely or primarily to avoid military service.

On these, as on other issues, the diary does not tell us all we might like to know. Rich though it is, it sometimes falls silent, and at other times its entries are too brief. Basil Armstrong Thomasson's diary provides clear evidence, however, that the attitudes of this southern yeoman were little affected by slaveholders' hegemony or by the culture of honor. Strong Thomasson was a staunch Unionist who disliked slavery and did not regard it as honorable. Independent in his economic and personal life, he also was independent of southern slaveholders in politics. Pious and progressive, he was eager for learning and driven more by conscience than by community sentiment. His reality was a very different one from that which historians usually discern through slaveholders' records, and it is precisely for this reason that we need more documents from, and more information about, the nonslaveholding yeomen of the South.

*Notes*

1. Edmund L. Drago, ed., *Broke by the War* (Columbia: University of South Carolina Press, 1991); William Kauffman Scarborough, *The Overseer* (Baton Rouge: Louisiana State University Press, 1966); Juliet E. K. Walker, *Free Frank* (Lexington: University Press of Kentucky, 1983); Michael P. Johnson and James L. Roark, *Black Masters* (New York: W. W. Norton, 1984); Michael P. Johnson and James L. Roark, *No Chariot Let Down* (Chapel Hill: University of North Carolina Press, 1984); Edwin Adams Davis, ed., *The Barber of Natchez* (Baton Rouge: Louisiana State University Press, 1954); Willard Gatewood, ed., *Free Man of Color* (Knoxville: University of Tennessee Press, 1982).

2. Clement Eaton wrote about the diary of Ferdinand Steel in *The Mind of the Old South,* rev. ed. (Baton Rouge: Louisiana State University Press, 1967), 147–50. Steel's diary is in the Manuscripts Department of the University of North Carolina at Chapel Hill. The North Carolina Division of Archives and History owns the diary of John Flintoff, who became a slaveholder after years of frustrating effort. Lacy Ford and Stephanie McCurry have used the journals of James F. Sloan of South Carolina. McCurry has written about him in an essay entitled "The Politics of Yeoman Households in South Carolina," in *Divided Houses,* ed. Catherine Clinton and Nina Silber (New York: Oxford University Press, 1992), 31–34. Bill Cecil-Fronsman, in *Common Whites* (Lexington: University Press of Kentucky, 1992), mentions two diaries: the John Osborne Diary and the Emsley Burgess Diary and Papers in the Manuscripts Department, University of North Carolina at Chapel Hill. The diary of Basil Armstrong Thomasson makes the sixth diary of a yeoman farmer known to this editor, who would be grateful for information about any others.

3. Recently a carefully researched exploration into the social world of the landless whites has appeared. See Charles C. Bolton, *Poor Whites of the Antebellum South: Tenants and Laborers in Central North Carolina and Northeast Mississippi* (Durham, N.C.: Duke University Press, 1994). A recent article on a "poor white" whose clash with the law produced documents useful to historians is Scott Culclasure, "'I Have Killed a Damned Dog': Murder by a Poor White in the Antebellum South," *North Carolina Historical Review* 70, no. 1 (January 1993): 14–39.

4. Historian Bill Cecil-Fronsman includes small slaveholders in the ranks of yeoman farmers. In my study of North Carolina, *Many Excellent People* (Chapel Hill: University of North Carolina Press, 1985), however, I argued that the differences between those who owned slaves and generally farmed for the market and those who did not were so substantial that the line should be drawn between them.

5. Cecil-Fronsman, *Common Whites.*

6. There were some pioneering early studies, but Frank L. Owsley, *Plain Folk of the Old South* (Baton Rouge: Louisiana State University Press, 1949), is recognized as a landmark. Bill Cecil-Fronsman's *Common Whites* did not appear until 1992, forty-three years later.

Another valuable study containing much significant information is Everett N. Dick, *The Dixie Frontier* (New York: Alfred A. Knopf, 1948). An informative book that more recently drew on postwar interviews in Tennessee is Fred Arthur Bailey, *Class and Tennessee's Confederate Generation* (Chapel Hill: University of North Carolina Press, 1987). If one believes that small slaveholders should be included in the ranks of yeomen, then James Oakes's important study, *The Ruling Race* (New York: Alfred A. Knopf, 1982), should be included, along with the recently published diary of David Golightly Harris: Philip N. Racine, ed., *Piedmont Farmer* (Knoxville: University of Tennessee Press, 1990).

Other highly relevant information about the economic decisions of yeoman farmers is found in Gavin Wright, *The Political Economy of the Cotton South* (New York: W. W. Norton, 1978). See also the important but somewhat speculative article by Eugene D. Genovese and Elizabeth Fox-Genovese, "Yeomen Farmers in a Slaveholders' Democracy," in *The Fruits of Merchant Capital* (New York: Oxford University Press, 1983), 249–66. Other works that concentrate substantially on the yeomen include Robert C. Kenzer, *Kinship and Neighborhood in a Southern Community: Orange County, North Carolina, 1849–1881;* Steve Hahn, *The Roots of Southern Populism* (New York: Oxford University Press, 1983); Allan Kulikoff, *The Agrarian Origins of American Capitalism* (Charlottesville: University Press of Virginia, 1992); and Paul D. Escott, *Many Excellent People* (Chapel Hill: University of North Carolina Press, 1985).

Articles on the important question of market participation by yeoman farmers in the South include Lacy K. Ford, "Rednecks and Merchants: Economic Development and Social Tensions in the South Carolina Upcountry, 1865–1900," *Journal of American History* 71 (September 1984): 294–318; Arthur C. Menius III, "James Bennitt: Portrait of an Antebellum Yeoman," *North Carolina Historical Review* 58, no. 4 (October 1981): 305–26; and Paul D. Escott, "Yeoman Independence and the Market: Social Status and Economic Development in Antebellum North Carolina," *North Carolina Historical Review* 66, no. 3 (July 1989): 275–300.

7. For recent studies of a yeoman farmer who lived a very different kind of

life, see Scott P. Culclasure, "'I Have Killed a Damned Dog': Murder by a Poor White in the Antebellum South," *North Carolina Historical Review* 70, no. 1 (January 1993): 14–39.

8. The date of his death remains unknown. Entries in his diary ceased in early September 1862, and family tradition holds that he fell victim to an unspecified disease shortly thereafter.

On the family's places of residence, see Book of Remembrance, September 17, 1853; Book of Remembrance, January 29, 1854; and diary entries for October 24, 1856 and January 8, 1858. Hereafter, all entries will be cited as "Diary," followed by the date.

9. In both Iredell and Yadkin Counties, the dominant categories of farms by size were fifty to ninety-nine acres and twenty to forty-nine acres. In Yadkin, 78.7 percent of all farms were under one hundred acres; for Iredell the corresponding figure was 74.6 percent. These statistics were calculated from data in *Agriculture of the United States in 1860; Compiled from the Original Returns of the Eighth Census* (Washington, D.C.: Government Printing Office, 1864), 210.

10. Thomasson lived in Yadkin County and then Iredell County during the years covered by the diary. For the state of North Carolina as a whole, African Americans constituted about 36 percent of the population. The statistics cited above were compiled from data in the Population and Slave Schedules of the 1860 census, for the areas served by Eagle Mills Post Office in Iredell County and by Zion Post Office in Yadkin County.

11. For details on his military status, see passages below, near the end of this introduction.

12. An abundance of information on the Thomasson family may be found in J. H. Thomasson, *A Partial Listing of the Decendents of John and Edith Thomasson of Stokes County, N.C.* (privately prepared, 1985). Through the courtesy of Mrs. Jean H. Thomasson, a copy of this study is in the author's possession.

13. Diary, August 30, 1853.

14. Diary, September 1, 1853.

15. Diary, September 3, 1853; November 12 and 13, 1853. The visitors were Basil's brother Wiley, who was teaching school in Forsyth or Guilford County; his uncle Fleming Thomasson with his wife Theny, who lived in Guilford County; and a cousin, Andrew Snipes.

16. For example, see Diary, December 10, 1853, December 26 and 31, 1853, November 14, 1854, December 2, 1854, and passim.

17. Diary, December 20–25, 1854.

18. For a careful delineation of these patterns, see Kenzer, *Kinship and Neighborhood.*

19. The census taker who in 1860 recorded Basil Armstrong Thomasson and

his family on a page that he numbered "3" recorded on pages "57" through "59" the residents of "Morrison's District." Nearly one-third of the residents were named Morrison.

Genealogical research by Jean Harris Thomasson, a grandson of Basil Thomasson's oldest child, strongly suggests that the first Thomasson in America came to Louisa County, Virginia, in 1677. By 1790 Basil's great grandfather, John Thomasson Sr., was living in what was then called Stokes County, North Carolina. See J. H. Thomasson, *Partial Listing*, N-1, N-2.

20. Diary, August 4–7, 1858, and February 12, 17, 18, and 23, 1859.

21. Diary, September 23, 1860, and May 28, 1859.

22. Diary, March 12, 15, and 16, 1857.

23. Diary, May 31 and June 1, 2, 3, 6, 8, 9, and 10, 1859.

24. Diary, June 23, 1855, and May 7, 1857.

25. Diary, September 12, 1861. Basil's sister Martha also returned home when gravely ill, as is discussed below.

26. Diary, November 12, 1858, February 10, 14, 28, 1859, and March 3, 1859.

27. Diary, August 9, 11, 13, 19, 24, and 29, September 7 and 12, 1861, and passim.

28. Diary, in entry for September 12, 1861.

29. Diary, August 4–6, 1856, June 3, 1858, September 28, 1861, January 2, 1859, and letter dated February 8, 1859, in Other Documents.

30. Diary, October 22, 1856, January 10, 1855, June 3, 1858, July 27 and 31, 1855, April 14, 1857, and November 30, 1858.

31. Diary, January 9, 1855, September 30, 1855, and November 30, 1853.

32. Diary, September 11, 1853.

33. Diary, December 12, 1853.

34. Dick, *Dixie Frontier*, 134; Diary, April 29, 1855.

35. Diary, July 18 and 19, 1855.

36. On Clark's twenty-first birthday, Basil wrote, "Guess he feels quite independent now as he is 21 years old." Diary, September 10, 1855.

37. Diary, January 1, 1854.

38. Diary, November 4 and September 8 and 9, 1853.

39. Diary, February 15, 1859. For an earlier instance in which Basil sought his father's opinion on some land, see February 22, 1857.

40. Diary, December 27, 1855, March 12, 1856, May 13, 1856, May 8, 1860, May 14, 1856, and September 15, 1856.

41. Diary, October 31, 1857, March 14, 1859, April 18 and 21, 1859, July 20, 1859, March 26, 1859, March 24, 1859.

42. Diary, July 2, 1857, October 1, 1857, April 8, 1858, July 9, 1858, September 10 and 12, 1859.

43. Diary, February 21, 1857, November 13, 1861. It is important to note as well that Mr. Bell cosigned the note with which Basil purchased his farm. See entry for February 23, 1859.

44. John McManners, ed., *The Oxford Illustrated History of Christianity* (Oxford: Oxford University Press, 1990), 40–41; Wayne A. Meeks, *The Origins of Christian Morality* (New Haven: Yale University Press, 1993), 58–65.

45. Diary, September 9, 1853.

46. Diary, October 7, 1856. Basil Thomasson was referring to Lorenzo Dow (1777–1834), an energetic and somewhat eccentric Methodist evangelist. In the first decade of the nineteenth century, Dow traveled extensively throughout the South and later wrote about his career and experiences.

47. Diary, March 30, 1856.

48. Diary, December 1, 1853, October 3, 1858, April 14, 1854.

49. Diary, October 27, 1857, March 3, 1859, October 3, 1860, September 8, 1861, and March 3, 1859.

50. Diary, September 9, 1854.

51. Diary, January 29, 1854, and October 2, 1859.

52. Diary, July 25, 1859. For earlier expressions of his attitudes, see entry for May 31, 1858, in which he condemns tobacco as "the ruin of many a poor soul."

53. Diary, March 4, 1855, and March 1, 1856. For comments against drunkenness and drinking, see December 29, 1854, October 16, 1857, and January 26, 1859. See also the story that he drafted, "Oh Mother, dear Mother," in Other Documents.

54. Diary, April 26 and 27 and May 16, 1860; August 23, 1860.

55. Diary, March 21, 1855.

56. Diary, September 19 and 24, 1858.

57. Diary, January 10 and 31, 1858, February 28, 1858.

58. Diary, June 6, 1858, January 15, 1858, January 24, 1858, March 21, 1855, and August 17, 1856; November 10, 1861; October 6, 1861.

59. Diary, April 1, 1855, September 17, 1853, October 16, 1853, August 25, 1855, August 20, 1855, August 28, 1855.

60. Diary, September 3, 1853. Local custom at this time named the church Aylesbury, although later it evolved into Asbury United Methodist Church.

61. Diary, May 28, 1854.

62. Diary, May 2, June 22, and August 10, 1856; March 22, 1857.

63. Diary, September 8, 1854, May 30, 1856, June 17, 1857.

64. Diary, March 3, 1860.

65. Diary, March 3, 1860.

66. Diary, March 1, 1856.

67. Diary, October 29, 1856.

68. Diary, March 3, 1860.

69. License in possession of Mrs. Jean Harris Thomasson of Forsyth County, North Carolina, dated October 19, 1861.

70. In the 1850s the academy's headmasters were W. L. Van Eaton and John A. Marler. Sometime after the Civil War, Jonesville Academy closed. See Frances Harding Casstevens, ed., *The Heritage of Yadkin County,* published by Yadkin County Historical Society (Winston-Salem: Hunter Publishing Company, 1981), 246.

71. Basil Armstrong Thomasson to "Dear bro.," June 1, 1856, in Other Documents.

72. Ibid. Perhaps Basil influenced other members of the family before he died, because a considerable number of the Thomasson descendants have been (and still are) schoolteachers.

73. Diary, August 28, 1859, and January 29, 1854.

74. Diary, March 18, 1854, April 20, 1854, March 21, 1855, and February 22, 1855.

75. On January 29, 1854, Thomasson wrote in his diary, "William C. is going to school at Jonesville, and I am at home trying to study a little sometimes but spending rather too much of my time in going to see the girls."

76. In Diary, see, for example, entries for February 26, 1854, May 6, 1856, and April 10, 1859.

77. Diary, entries for February 2, 1857, and February 26, 1858.

78. In Diary, see, for example, the entries for January 20, 1854, February 4, 1854, February 20, 1854, January 1, 1856, and March 13, 1858.

79. For example, see the Diary entries for 1853 and 1854, which mention the Raleigh *Spirit of the Age;* the *American Eagle,* later renamed the *United States Intelligencer; Youth's Cabinet; Arthur's Home Magazine;* the *Press;* the New York *Journal; Day's Great Christmas Pictorial Brother Jonathan; Nichol's Journal;* the Greensborough *Ballot Box;* the Cincinnati *Dollar Times;* the *Lady's Wreath;* and the *Georgia Blister and Critic.*

80. See, for example, the Diary entries for September 28, 1854, March 21 and 22, 1855, March 30, 1855, May 30, 1855, January 28, 1856, February 28, 1856, March 14, 1856, September 24, 1856, November 26, 1856, February 1, 1857, January 6, 1858, February 14, 1858, and October 5, 1858.

81. Diary, entry for January 28, 1856.

82. Diary, entries for September 23, 1858, and August 24, 1853. Thomasson was also confident that life spans would increase as people learned "how to live" and conquered ignorance. See entry for February 15, 1854.

83. M. C. S. Noble, *A History of the Public Schools of North Carolina* (Chapel Hill: University of North Carolina Press, 1930), 59–61, 85, 107. On p. 205 Noble

added that in 1852 county boards were required to appoint committees of examination.

84. Diary, entries for October 5, 1853, September 19, 1857, and July 31, 1858.

85. Diary, July 31, 1858.

86. Schoolteachers were almost all male in these early years, and Yadkin County did not have a single licensed female teacher until 1856 (though by 1860 the U.S. Census shows several). In 1862 the county schooled 1,274 male children and 817 female children out of 3,690 eligible students, with an expenditure of $4,683.18 in county and state funds. Casstevens, *Heritage of Yadkin County,* 243.

87. Diary, entries for December 19, 1856, November 2, 1857, and November 7, 1857.

88. Diary, November 19, 1860.

89. Diary, October 8 and 7, 1853.

90. Diary, September 21, 1854.

91. See Diary, December 20, 1853.

92. Diary, August 21 and 28, 1854, August 26, 1858.

93. Diary, January 6, 1860.

94. Diary, November 26, 1856.

95. Diary, August 29, 1854, July 18, 1856.

96. Diary, October 31, 1853, November 6, 1854, April 30, 1858.

97. Diary, December 15, 1856.

98. Diary, October 12, 1854, October 15, 1856, September 1, 1854.

99. Noble, *History,* 223; Diary, entries for January 13, 1857, January 19, 1858, January 18, 1859, and August 15, 1859.

100. Diary, entry for July 21–25, 1856; letters to C. H. Wiley and Messers. A. S. Baines & Co., October 10, 1856, in Other Documents.

101. Diary, entries for August 23, 1853, December 13, 1859, March 10, 1859.

102. Diary, August 20, 1858, December 24, 1859, January 6, 1860.

103. Raleigh *Spirit of the Age,* January 5, 1853.

104. Ibid., January 5, 1863, January 19, 1853, and February 2, 1853. In regard to inventions, note, for example, the article on "Capt. Ericson's Caloric Ship."

Perhaps it is not accidental that analysis of census figures that showed southern agriculture lagging behind the North figured prominently in the famous attack on slavery by North Carolina's Hinton R. Helper. His book, *The Impending Crisis,* which was used by the Republican Party in 1860 as a campaign document, was originally published in 1857.

105. *The Heritage of Iredell County,* published by the Genealogical Society of Iredell County (Winston-Salem: Hunter Publishing Company, 1980), 129. Olin Institute lasted until 1885.

Brantley York wrote an autobiography that also contains an account of the founding of New Institute. See Brantley York, *The Autobiography of Brantley York,* vol. 1 of the John Lawson Monographs of the Trinity College Historical Society (Durham, N.C.: Seeman Printery, 1910), 58–64.

106. Diary, entries for December 16, 1854, January 3, 1855, January 13, 1855, January 18, 1855, and April 10, 1855.

107. Diary, November 23, 1854.

108. Diary, April 24, 1855, February 7, 1857, March 14, 1857, and April 13, 1861.

109. Diary, December 9, 1853. By the turn of the century, some maps were designating Windsor's Crossroads as Winder's Crossroads.

110. Diary, January 17, 1854.

111. Diary, January 30, 1854, February 2, 13, and 15, 1854, April 3 and 10, 1854, May 16, 1854, November 5, 1855.

112. Casstevens, *Heritage of Yadkin County,* 30, 56.

113. Charlotte *North Carolina Whig,* February 7, 1854, 2. This advertisement was dated January 19, 1854, and it ran each week through March.

114. Ibid., January 10, 1855, October 16, 1855. Each of these ads again ran for some months.

115. Casstevens, *Heritage of Yadkin County,* 56.

116. Ibid.; 1860 Census of Population, Iredell County, 19 (Eagle Mills location) for Baggarly, 24 for Colvert and the Morrison, Gaither, and Company cotton factory, 14 for A. B. F. Gaither. Colvert owned real estate worth $2,000 and personal estate valued at $5,885; Gaither was listed as a farmer with $8,800 in real estate and $14,500 in personal estate. The editor has not been able to find Andrew Morrison in the 1860 census; perhaps Morrison had died by this time. Baggarly's name appears in some sources as Baggerly.

117. Diary, November 7, 1859, January 23, 1861, November 20, 1858, and October 21, 1860. In the last quotation, Basil went on to say, "We also have apples, beets, honey, molasses, now and then a chicken, a few eggs, etc. If we only had plenty of good milk and butter to add to our list we could live 'fat and full.'"

118. See Eugene D. Genovese, *The Political Economy of Slavery: Studies in the Economy and Society of the Slave South* (1965; reprint, New York: Vintage Books, 1967) and *Roll, Jordan, Roll: The World the Slaves Made* (New York: Pantheon Books, 1974).

119. See Bertram Wyatt-Brown, *Southern Honor* (New York: Oxford University Press, 1982) and *Honor and Violence in the Old South* (New York: Oxford University Press, 1986).

120. Wyatt-Brown, *Honor and Violence,* 122, 92, 85, 79, 14, ix. Wyatt-Brown seems to recognize that honor may have held its greatest relevance for upper-

class, slaveholding southerners. At one point he remarks, "So little work has been done to describe the ethical and cultural structure of the yeomanry outside the church-going crowd that historians are caught between the romantic and bourgeois presentation of Frank Owsley on the one hand and the familiar stereotypes of poor-white depravity on the other." Yet he also asserts, "Honor in the Old South applied to all white classes, though with manifestations appropriate to each ranking." He later describes pride in an ancestor as a typical means by which "the poor" referred to "the best ideals of their clan." See *Honor and Violence,* 99, 40, 68.

121. See Wyatt-Brown, *Honor and Violence* and *Yankee Saints and Southern Sinners* (Baton Rouge: Louisiana State University Press, 1985).

122. Diary, September 10, 1853.

123. Diary, November 11, 1853, October 17, 1853.

124. Diary, October 9, 1853, October 21 and 28, 1854, and December 25, 1854. For other comments about young women, see also November 3 and 28, 1853.

125. Diary, January 8 and 11, 1854.

126. Diary, January 29, 1854.

127. Diary, May 5, 1854, February 27, 1854.

128. On the pages that followed his diary entries for 1856, Thomasson listed the names of all children eligible for common-school education; this list appears among the documents that follow the diary.

129. Diary, March 4, 1855.

130. Diary, October 8 and 30, 1854, January 20, 1855.

131. This pattern of behavior is inconsistent with the conclusion of another historian, Allan Kulikoff, that "yeomen denied female individualism and insisted that women bend to their authority." See Kulikoff, *Agrarian Origins,* 34.

132. Diary, June 8, 1858, November 3, 1858, and May 13, 1859.

133. Diary, December 17, 1859; August 22, 1861 and May 9, 1856; January 31, 1859.

134. Diary, November 13–19, 1859, May 19, 1860.

135. Diary, January 4, 1859, March 26, 1856.

136. Diary, March 31, 1859, April 18, 1859, September 13, 1859, September 1, 1859, March 25, 1861.

137. McCurry, "Politics of Yeoman Households," 30–36. See also Stephanie McCurry, *Masters of Small Worlds* (New York: Oxford University Press, 1995).

138. Anne Radford Phillips, "Farm Women of Stokes County, North Carolina and the Production of Flue-Cured Tobacco, 1925–1955: Continuity and Change" (Ph.D. diss., University of Maryland, 1990), 143, 144, 145; see also 139–42 and 152–55. She also documented the pattern, evident with Strong and

Mollie Thomasson, of a newly married couple moving in with their parents at first and continuing to visit frequently after they had set up a separate household.

139. Diary, January 14, 1859.

140. Diary, see October 3, 1858, and September 28, 1861, on concerns for Jodie; quotation from October 29, 1858.

141. Diary, January 31, 1861, March 12, 1861, December 2, 1861.

142. Jane Turner Censer, *North Carolina Planters and Their Children, 1800–1860* (Baton Rouge: Louisiana State University Press, 1984).

143. Diary, June 29, 1856, September 29 and 30, 1856, March 24, 1860.

144. Diary, June 8, 1855, July 8, 1860.

145. Diary, December 18, 1853, February 4, 1854, September 30, 1858, and entry that followed 1856 but was dated July 3, 1859 (in Other Documents). See also the draft of a sermon that is reproduced in Other Documents.

146. Diary, December 18, 1853, September 23, 1860.

147. Diary, November 12, 1853, January 31, 1860.

148. Diary, October 28, 1860.

149. See Other Documents; Diary, September 22, 1858.

150. Diary, January 3, 1854, January 8, 1855, January 31, 1859. Thomasson recorded his weight at 134 on January 29, 1860, and described this figure as 16 pounds more than he weighed the previous winter and "nearly as much" as he ever weighed. So, at the time he considered eating less, Thomasson may have weighed as little as 118 pounds.

Thomasson also wrote, on December 13, 1854, that "paying interest is bad business."

151. One can trace the building and sale of his wagon in the following diary entries: June 22, 1854, November 25, 1854, January 27, 1855, February 12, 1855, May 29, 1855, and September 9, 1856. In all, Thomasson spent $56.85 on the wagon and sold it for $55.

152. See Diary, January 3, 1857, for one of several references to his "Cost of Living."

153. Diary, March 30, 1856, October 19, 1855. For other critical comments on musters, see September 1, 1855, and March 1, 1856.

154. For the debating society, see diary entries for February 23, 1855, March 23, 1855, and April 14, 1855. On his sensitivity to aristocratic claims to power, see his confrontation with, and his sarcastic comments on, the "*honorable* and *very wis* (in their own eyes) would be gentlemen, viz. A. W. Martin, *Esq.* Bennett Windsor, *Esq.*" on February 7, 1857. He criticizes those who spend quite a bit for fine buggies on March 9, 1855. His praise of agriculture appears in the entry for March 17, 1854.

155. Diary, October 31, 1856.

156. The data from the U.S. Census of Population for 1860 are drawn from the district served by Zion Post Office in Yadkin County and Eagle Mills Post Office in Iredell County. For Eagle Mills, the figures on laborers do not include the operatives at the cotton factory.

For Andrew Thomasson's neighborhood, the 25 percent represents sixteen of the sixty-two agriculturalists.

157. The 23 percent represents fourteen of sixty-one; one-third represents twenty of sixty-one.

158. Diary, February 7, 1859, January 20, 1859, February 12, 15, 17, 18, and 23, 1859, and March 1, 1859. Earlier, when Thomasson was renting on the Burton farm, his father had been very discouraging about the poor quality of the land. See entry for February 22, 1857.

Thomasson's hopes about the future price of his land were overly optimistic. In 1861 he agreed to sell his land for $180, only a hair more than $2.50 per acre, but the deal fell through. See the entry for October 10, 1861. Had this deal been completed, Thomasson probably would have sought to buy another, and better, farm.

159. For important studies of the American yeomanry's goals and its mode of independence, see James A. Henretta, "Families and Farms: *Mentalité* in Pre-Industrial America," *William and Mary Quarterly* 35 (January 1978): 3–32, and Bettye Hobbs Pruitt, "Self-Sufficiency and the Agricultural Economy of Eighteenth-Century Massachusetts," *William and Mary Quarterly* 41 (July 1984): 335–64. For a good description of the interconnected nature of the yeomen's economic affairs in the South, see Hahn, *Roots of Southern Populism.*

160. For example, see entries for December 28, 1853, July 1, 1858, August 20, 1856, March 11, 1859, February 23, 1854, and November 10, 1858.

161. Diary, August 20, 1856.

162. Diary, April 13–14, 1859. Thomasson had bought the latter note, rather than accepting it from someone he knew in a direct transaction. When he was still collecting part of the debt on January 24, 1860, he resolved that he would not be buying notes again.

163. Diary, July 22, 1861, September 9 and 10, 1856, March 2, 1860.

164. Diary, August 27, 1859, May 16, 1860. Most nonslaveholders in piedmont North Carolina followed the same strategy as better transportation made the market more available to them. See Escott, "Yeoman Independence."

Census data for Yadkin and Iredell between 1850 and 1860 show that farmers in these counties were continuing to grow their usual amounts of corn but were increasing the production of crops that they could sell, such as wheat and tobacco. (The small amounts of cotton grown in both counties actually decreased substantially during the decade.) In Iredell County the number of improved acres increased only 4.4 percent between 1850 and 1860, but production of wheat rose 154.6 percent and production of tobacco soared by 274.8 percent. Because Yadkin County was carved out of Surry after the 1860 census, it is not possible to make a direct comparison for Yadkin. But by combining the totals in 1860 for Surry and Yadkin, one finds similar results. The number of improved acres rose by 14.6 percent, whereas wheat climbed 351.1 percent and tobacco 1,282.5 percent. In 1860 Iredell County produced 504,517 bushels of Indian corn and 130,712 pounds of tobacco. Surry and Yadkin, considered together, produced 561,879 bushels of Indian corn and 607,640 pounds of tobacco. Surry and a line of counties along the Virginia border were by 1860 the center of tobacco culture in North Carolina.

165. On the politics of slavery, see William Cooper, *The South and the Politics of Slavery* (Baton Rouge: Louisiana State University Press, 1978). Wyatt-Brown, *Honor and Violence*, ix.

166. Diary, July 4, 1856, and December 27, 1859.

167. Diary, November 6, 1860. In 1856 Thomasson did not vote; see entry for November 4, 1856.

168. Quoted in Paul D. Escott, *After Secession: Jefferson Davis and the Failure of Confederate Nationalism* (Baton Rouge: Louisiana State University Press, 1978), 22.

169. Diary, February 28, 1861. Thomasson's position was in the majority in North Carolina at that time.

170. Diary, March 31, 1861, April 28 and 29, 1861, and the May 4 postscript to his "April 1861" letter to "Dear Brother," found in Other Documents.

171. Diary, May 6, 11, and 24, 1861, and March 1, 1862. The laws of North Carolina and the Confederacy probably allowed Thomasson an exemption from conscription. See the footnotes to diary entries for 1862 for a detailed discussion of his status.

172. Diary, entries for February 27, 1854, and July 4, 1856.

173. Diary, March 23, 1855, and February 23, 1855.

174. Diary, March 23, 1855, and February 23, 1855.

175. Letter to "Dear Brother" dated April 1861, which may be found in Other Documents.

176. Diary, November 22, 1858, June 7 and 29, 1859, January 21, 1861, December 25, 1854, and October 10, 1859. According to local tradition, Zac Jefferson was a black man rumored to be descended from Thomas Jefferson.

177. Diary, April 20, 1861.

178. Letter to "Dear Brother" dated "April 1861" in Other Documents.

179. For example, see entries for August 1862 and June and July, 1862.

180. See *Revised Code of North Carolina, Enacted by the General Assembly at the Session of 1854* (Boston: Little, Brown and Company, 1855), chap. 70, sec. 75, 422.

The same provision was present in *Revised Statutes of the State of North Carolina, Passed by the General Assembly at the Session of 1836–1837*, 2 vols. (Raleigh: Turner and Hughes, 1837), 1:417, and the practice probably goes back to revolutionary and colonial days.

The fact that a draft took place under state authority has not been generally recognized by scholars. For more detail, see the diary entry for March 5, 1862, and its footnotes.

181. The text of the exemption law appears in an enclosure to a letter from Jefferson Davis to Governor Joseph E. Brown of Georgia, April 28, 1862, in Allen D. Candler, ed., *The Confederate Records of the State of Georgia*, 6 vols. (Atlanta: Chas. P. Byrd, 1909–11), 3:201–2. For more detail, see the diary entry for April 6, 1862, and its footnote.

# *A Note on Editorial Methods*

The text of Basil Armstrong Thomasson's diary, or Book of Remembrance, is presented here with as few editorial changes as possible. For his day Thomasson spelled rather well, so corrections appear only where their absence might create confusion for the reader. Similarly, few additions or changes in punctuation have been made, and these changes are always indicated by brackets. Whereas Thomasson often placed the titles of newspapers in quotation marks, these have been italicized, along with book titles, to accord with today's standard practice.

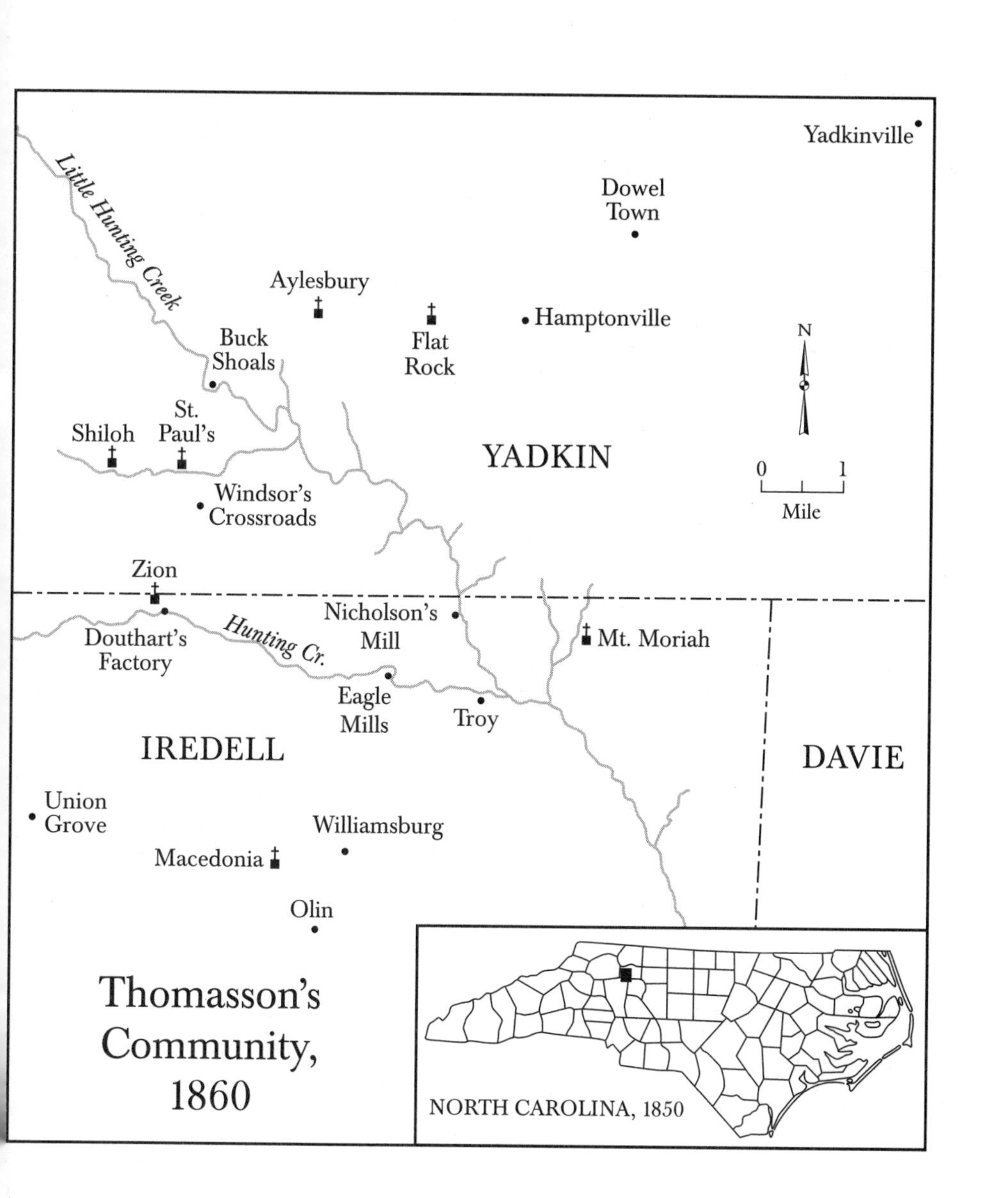


Thomasson's Community, 1860

# North Carolina Yeoman

# Book of Remembrance 1853

*August 23* This day my school closed, and it, being the first *five months* session I ever taught, seemed long & tiresome; it is now out though, & I am, again, a free man. What I shall do next, is uncertain. I have had *two* offers, lately, to teach school in Iredell Co., and one to teach in Wilk[e]s, and one, or two in this county.

*August 24* This is a delightful morning. Last evening, between sun set and dark, a comet made its appearance in the western horizon. Comets, in the days of ignorance, were thought to be harbangers of war, famine, and pestilence; but in these days of knowledge and improvement, their visits create little excitement, since they are not considered the forerunners of some direful event, as in days of yore.

*August 25* This is quite a warm day, though its a little cloudy and rains moderately ever[y] few hours. This is the third evening I have seen the comet, or whatever it is.

*August 26* This is an exceedingly warm day, with but little appearance of rain. Corn is drying up very fast: the fodder will soon do to gather. To day, I went to see about teaching school at Nicholson's.

*August 27* This is another warm day, with some appearance of rain; clouds pass round in the north, and in the south. The comet was visable again yesterday evening. To day, I received the *Spirit of the Age* for this week. The *Age* is one of the best papers in North Carolina; this weeks number contains a sketch of the Rev. John Taylor, who was, probably, as great an orator as the immortal P. Henry.[1]

1. Many of the newspapers mentioned in Thomasson's diary have not survived, but his beloved *Spirit of the Age,* which was published in Raleigh, is extant. It actively supported religion,

*August 28* This is quite a cool morning. Yesterday evening we had a hasty shower of rain, and it was so cloudy, the comet was hid from view. This is Sunday:

"This is the day the Lord hath made,
He calls the hours his own;
Let heaven rejoice, let earth be glad,
And praise surround the throne."

*August 29* This is quite a clear, pleasant day. The comet was seen again yesterday evening, though it was not as bright as usual. It has been quite cool for the last two mornings.

*August 30* This is a day long to be remembered. Sister Martha got married last night without the consent of Pa and Ma, and such a time as we have had! Oh! it will never be forgotten. What a queer world this is!!

The comet was seen last night, and to night. It appears to move in the sun's track. To day I sold Alexander Benbow a box of tobacco at 16 1/2 cents per pound. Bought 9 yds. of cotton cloth, 12 1/2 cents per yard of L. Roughton, engaged to teach the free school in district No. 35, at $18 per month, to commence the first or second Monday in October.

*August 31* This has been quite a warm, dry day. Dock commenced pulling fodder today.

*September 1* This is a very warm day. Dock and Mode Tucker are mowing John E. Grant's meadow. Alexander came over this evening, and "peace is made;" so "we are at peace with all the world & the rest of mankind."

*Sept. 2* This is a very rainy day.

*Sept. 3* This has been a rainy day too. I half soled Martha's shoes to day, sold John E. Grant five plugs of tobacco, etc. I also wrote articles for a singing school at Aylesbury, (a church)[2] and expect to sing there the 24[th] of this month.

---

temperance, public education, and progressive legislation to expand opportunities for the ordinary citizens of North Carolina. This particular sketch appeared in the issue for August 24, 1853, which was no. 5 of vol. 4.

2. Asbury United Methodist Church continues today. It was originally known as Elsberry, after the surname of a local tax assessor, but Bishop Asbury preached at the church in 1785, and variant spellings came into use, including Elsbury, Aylsberry, and others. Mary Belle Weaver Allred, *A History of St. Paul United Methodist Church* (privately printed), in editor's possession.

*Sept. 4* This is a rainy Sunday, though I went to the campmeeting at M't Maria [Mount Moriah];[3] there were but few people out on Sunday, and not a great many on Monday. George W. Farabee preached the first sermon at 11 o'clock on Sunday.

*Sept. 5* A very warm, clear day, and Pa's birth day. I go to the campmeeting, the Rev. James Patterson preaches from this text, "My son give me thy heart." Prov. XXIII, and part of the 26[th] verse.

*Sept. 6* I am sick to day, got the Dispecic [dispeptic] or something else.[4]

*Sept. 7* No better to day, though able to get about.

*Sept. 8* This is a cloudy day, I am but little better. Bought two hogs of Daniel Huchins $2. a piece, a red hiefer at $7. Brought my bees home in the night. Received the first number of the *American Eagle,* being number 9 of volume the X.

*Sept. 9* This is a very rainy day. Received of Huchins Johnson two dollars in full. Brought home my hogs, sold Daniel Huchins 24 lbs of tobacco at .15 per lb. $3.60 paid towards the heifer. This is my birth day. This day 24 years ago I made my appearance into this "dark and gloomy world."

"Few are the comforts" I've enjoyed,
Fewer the hours of bliss;
The promise of a better world,
Is all that's bright in this.

*Sept. 10* This is Dock's birth day. I went this morning and got a little of the "fire water" for medicine. Set a little of the "red eye" on fire and burnt pine and sugar over it, it makes a good medicine for certain ailments. Paid Daniel Hutchins $3.30 which was yet due him for the hiefer. We are now even; or will be, after I get one of his girls. I have bought his hogs, bees and cow; the girls come next. The hogs, $4., bees $4., cow $7.; making $15.

3. Mount Moriah was one of the early Methodist churches established in Iredell County. Located near the line between Yadkin and Iredell Counties, it flourished until about Reconstruction but then became inactive. See *The Heritage of Iredell County,* published by the Genealogical Society of Iredell County (Winston-Salem: Hunter Publishing Company, 1980), 141.

4. Digestive complaints and waterborne diseases were a recurring problem for Thomasson and his neighbors.

*Sept. 11* Sunday, this is quite a nice day. I went to Aylesbury to class meeting; came home, and went back to night meeting, walked with Miss M. B.[5] to meeting, and back with Miss N. W.

*Sept. 12* Went this morning and brought my hiefer home; pulled fodder since dinner.

*Sept. 13* This has been quite a clear, pleasant day. I took the leaves from several corn stalks to day, and tied up 65 bundles, to night, of that I pulled yesterday.

*Sept. 14* Hauled two loads of wood, and three loads of hay this morning. James Martin set in to work for Pa this morning, and cut tops till dinner, and quit as its raining to[o] hard to work out. It rains and rains.

*Sept. 15* This is a beautiful day. I have been pulling fodder & cutting tops.

*Sept. 16* This has been a glorious day. Went this morning to see about my school, agreed to commence on the 26 day of this month. Re[ceived] $3 tuition of Mrs. Williamson, $2.30 yet due. Went to meeting last night at Aylesbury; and after meeting, I went home with Miss M. B.; a charming girl, Miss M. B.

*Sept. 17* A beautiful morning; start to campmeeting, at Mt. Tabor, in Forsyth, have a long and hot ride,[6] get there between sun set & dark, see lots of my old friends, & brother Wiley, hear the Rev. T. Craft preach, etc.

*Sept. 18* This is a glorious day, prayer meeting this morning, the people gather in, there is now a large congregation, and "still they come." Dr. Carter preaches one of his long & plain sermons at 11 o'clock, and Michael Doub at night.

*Sept. 19* This is a pleasant morning, with some appearance of rain. Michael Doub baptizes Philip Mock by immersion, in Silas Creek.[7] Joseph Doub preaches at 11 o'clock & Dr. Carter again at 3. Rather a cool time for a camp meeting.

5. Mary Beall (or Bell) would become Thomasson's wife in 1855. He often referred to her as Miss M. A. B. and as Mollie.

6. Mount Tabor today is part of Winston-Salem. The distance between Winston-Salem and Thomasson's home in Yadkin County is about forty miles.

7. Silas Creek runs past today's entrance to Wake Forest University.

*Sept. 20* Quite cloudy. The C[amp] M[eeting] brakes this morning. I sell a box of tobacco, No. 2, half at 8 cents per plug and half at 12 1/2 cents per pound; amount $13.50. Bought the following goods of Gustin Transou, in Pfaff Town.[8] A pair of fine boots 4.00, waistcoat .75, etc. for which I let the said Transou have the above named half box of tobacco at 12 1/2 cents per pound.

*Sept. 21* Quite a nice day. I go to Salem and get my watch mended. Cost .50 cents. A pocket handerchief .75 cents of Pfohl. Get 4 numbers of the *Youth's Cabinet.* Go up to Winston, see Dr. Hunter, an old friend. Go to M. J. Crews' and stay all night.

*Sept. 22d* A rainy morning. I pay Crews $1.84 cents, which is all I owe him. Stay at his house till after dinner, it quits raining, and I start for home. Stop a few minutes at Sedge Garden, get a tooth brush, shirt, buttons, slate pencils etc. leave S.G. at 3 o'clock for Wm Coltrane's, get there a little after dark, about 8 miles.

*Sept. 23* I make a start for home early. This air is a little cool, cross the Yadkin about 9 o'clock & shove for home, a distance of 26 miles, get here between 4 & 5 o'clock in the after noon, find Uncle Joseph Houser[9] here quite sick.

*Sept. 24* This is a nice morning, though a little cool. Uncle Jo leaves for home. I get Jenny Lind[10] shod, etc, etc, etc, etc.

*Sept. 25* This is a beautiful Sunday. The Rev. T. Nicholson & Pa preached Mrs. Tulburt's funeral to day at Aylesbury, from this t[e]xt: "I have fought a good fight." etc. II Timothy, 7 & 8 verses of the 4 chapter.

*Sept. 26* This has been quite a nice day. I commenced school to day at Felts school house, had 22 scholars the first day! My watch give it up again to day; paid fifty cents for the mending of it, and it run three days! Old John Voglar is done mending watches for me. Bought half quire of paper, for ten cents at Wiley Felts' store.

8. Pfafftown is today a part of Winston-Salem.

9. Joseph Houser was a brother of Thomasson's mother, whose maiden name was Leah Hauser (or Houser).

10. The family's mare, an old but reliable workhorse. The Thomassons obviously named this animal after the famous Swedish soprano, who toured the United States from 1850 to 1852 under the management of P. T. Barnum.

*Sept. 27* This is quite a pleasant day. had twenty five pupils to day.

*Sept. 28* Quite a nice day, had twenty pupils. Swap[p]ed my watch to J. N. Baron for his, giving him seven dollars for the differance. I have a nice start in school, nice indeed! [H]ope I shall get on without difficulty, and have a grand school. Amen.

*Sept. 29* Cloudy in the forenoon, and quite cool. Had 21 scholars to day.

*Sept. 30* Last day of September, quite a cool day. Had 21 scholars.

*Oct. 1* Cloudy, and raining this evening. Bought 17 1/4 lbs. of tallow at 10 cents per lb. of B. A. Johnson Esq. Wrote two letters, one to T. M. Hunter, & one to J. H. White.

*Oct. 2* A little cloudy, and cool. I went to Felts' school house to day & heard Pa preach. He preached near two hours. Received the *Age,* No. 4 of Vol. 5. It contains an interesting piece, "Paul Denton, or the Texas Camp Meeting[.]"[11] The *Age* is one of the best papers in the state.

*Oct. 3* This is quite a nice day, though, a little cool. Had thirty scholars to day. Write to night to Robert Sears concerning books.[12]

*Oct. 4* Quite a nice day. Had 41 scholars.

*Oct. 5* A pleasant day. Went to the Examination[13] at Doweltown, obtained a certifficate good for one year. Bought 6 Arithmetics, 37 1/2 a

11. The date of the issue in which this article appeared was September 28, 1853. The article described the first successful camp meeting in eastern Texas in 1836.

12. Robert Sears is a bookseller. See entry for October 22.

13. North Carolina established the Literary Fund, dedicated to the support of common schools, in 1826 and passed its first school law in 1839. It allowed voters to levy, for the support of schools, a local tax that would be matched on a two-to-one basis from the Literary Fund. Each county court selected a board of superintendents and arranged for the selection of committeemen in districts that were to be no more than six miles square. Beginning in 1851, the legislature required that county superintendents appoint an examining board, and thereafter teachers had to pass an examination and earn a certificate each year to be qualified to teach.

Calvin Wiley, a former legislator, was the driving force behind the system and its first state superintendent. In 1860 North Carolina had 3,484 school districts but only 2,834 functioning schools. Of 200,855 eligible children, 105,048 were attending common schools, whose average length of term was 3 2/3 months. The average salary for a teacher was twenty-six dollars per month. Wiley claimed that the difference in salaries between male and female teachers was not as great as elsewhere. See M. C. S. Noble, *A History of the Public Schools of North Carolina* (Chapel Hill: University of North Carolina Press, 1930), 45, 59–61, 85, 107, 205, 223, 225.

piece. 37 1/2 × 6 = $2.25. 5 readers at .18 3/4 a piece. 18 3/4 × 5 = .95 3/4.[14] 2 readers at .20 a piece. .20 × 2 = .40. 1 reader at dont know what. There were 9 of us examined from about 11 o'clock till half after 4. All obtained certifficates.

*Oct. 6* A beautiful day, had thirty one pupils.

*Oct. 7* Another nice day, had twenty six chaps.[15] Sent out six new books for them to buy.

*Oct. 8* Got John E. Grant to make me a black board, six feet long & five wide; went to Hamptonville, bought a bottle of ink for 6 cents; mended my boot etc.

*Oct. 9* Went to Aylesbury to class meeting, and then home with Miss Mary B. I expect I shall have to go again. Miss M. B. is the *gal.*

*Oct. 10* A pleasant day, had 26 scholars. sold a few books etc.

*Oct. 11* A cool day. Had thirty three scholars. Dock came home to day, from Forsyth.

*Oct. 12* There has been a little frost for several mornings, and a pretty general one this morning; its cloudy & rainy a little this evening. Pa commenced sowing wheat to day; had 28 scholars to day. Sold Columbus Roughton an Arithmetic, 2d part, for 40 cents.

*Oct. 13* No school to day. Went to Joseph Sparks' this morning & got 26 lbs of bacon & 5 lbs of lard at .10 per lb. I and Sparks are now even. Went this evening to the election in District No. 30; it being the second committeeman election[16] I was ever at. There were several out, and the following men were elected: Windsor, Jacks & Myers. It is said that Jacks can not write his name! Such committeemen. Helped Johnny Johnson husk corn till near 9 o'clock; went home & went to bed.

*Oct. 14* A nice day, had 23 scholars. No election in District No. 35. One *good sinner* objects to my having a bell to rattle in school, and to my

14. Despite this error in multiplication, Thomasson was probably among the more able and better qualified teachers in the new common-school system.

15. It appears that Thomasson's pupils were all male. Many families assumed that girls had less need of an education than boys, but Thomasson held decidedly different ideas.

16. This was an election for members of the school committee.

turning out half after four; these are all the objections I hear of, and no gentleman would ever have thought of these. Received of John G. Johnson four dollars & forty cents in full. B. A. Thomasson.[17]

*Oct. 15* A beautiful day, cut some wood this morning, and ploughed in some wheat since dinner at the south end of the School house.[18] Pennsylvania Blue Straw. Jenny Lind, to her practice, is one of the best plough nags I ever saw.

*Oct. 16* A nice day, went to Union Grove to C[amp] M[eeting], heard the Rev. Holton preach a good sermon; the first psalm was his text. He preached at 11 o'clock. He preached doctrines new to me, and contradicted the preaching of many learned divines. He says "We are not sinners by nature"! Many say that all men are sinners by nature, the Rev. H. says it is not so; and I am [of] his opinion. The Rev. Richard Mishew preached at 3. His text was a part of the 10 verse of the VI chapter of Mathew. "Thy kingdom come". He preached quite a good sermon.

*Oct. 17* Another nice day, had only 20 scholars. Received a letter in answer to one I wrote Miss S. A. S. It is quite an interesting letter too. Miss Sarah is the girl for me.

*Oct. 18* A pretty day. Dr. Hunter staid with us to night, on his way to the New Institute[19] in Iredell.

*Oct. 19* Had 21 scholars. I have a nice School.

*Oct. 20* Cloudy, with a little rain.

*Oct. 21* Still cloudy, & raining all day—nearly. Staid last night at Wiley Messick's.

17. Yeoman farmers like Thomasson frequently wrote out IOUs and receipts for goods or money paid on a debt, and apparently the habit of signing one's name carried over into this diary entry.

18. As the teacher, Thomasson was allowed to use the school property to raise a crop for himself.

19. A school called New Institute opened in 1851 with nearly one hundred students, and a village grew up around it. By 1856 the name of both had changed to Olin, after Stephen Olin, a Methodist educator. Olin was southwest of the town of Williamsburg in Iredell County. The school or academy lasted until about 1885. See *Heritage of Iredell County,* 129. Another description of the launching of New Institute may be found in Brantley York, *The Autobiography of Brantley York,* vol. 1 of the John Lawson Monographs of the Trinity College Historical Society (Durham, N.C.: Seeman Printery, 1910), 58–64.

*Oct. 22* A clear pleasant day, sent a letter to Miss S. A. S.; received two numbers of the *Age*, two circulars from Robert Sears concerning his publications. Great books! *Arthur's Home Magazine*, etc.

*Oct. 23* A pretty day, went to Zion[20] to meeting, herd Pa & the Rev. Wm Garner preach.

*Oct. 24* Rain! rain! rain! Snow! snow! snow! It snowed about two hours very fast, and as large flakes as I ever saw, but the ground being so wet it melted as fast as it fell. Since the snow. it is quite cool, or cold.

*Oct. 25 & 26* Nothing of importance.

*Oct. 27* A very rainy day. School small.

*Oct. 28* Clear & warm. Staid last night at old Mr. S. Gentry's.[21] School not large.

*Oct. 29* Tolerbly cool. Sold Moses Tucker 14 plugs of tobacco, or sold him 13 & give him one. $1.30. Took my Black Board to the School house.

*Oct. 30* Cloudy and cool. Wrote a letter to Wiley last night, & think I must take it to the office this evening. Just returned from the office. Received the *Age, Press*[22] and the *American Eagle*. No letters? No, no, no. White, Hunter and Miss S. Why don't they write?

*Oct. 31* A very frosty morning, clear & warm in the evening. School, quite small. How strange it is that people will close their eyes against their own interest! Ignorance is the cause.

*Nov. 1* Another large frost, pretty day. School small. Staid last night at James Brown's. Browns are nice people, every thing looks neatly about

20. Zion Baptist Church in Iredell County started in 1826, with most of its members coming from Union and Flat Rock Baptist Churches. *Heritage of Iredell County*, 143.

21. As was common, Thomasson was staying in the homes of families who lived in the school district.

22. This is probably a reference to the *People's Press*, which was being published at this time in Salem by John Christian Blum and Son. (The town of Winston, incidentally, was founded in 1849 and named in 1851. It was adjacent to Salem and was to be the seat of government for the newly created Forsyth County. Not until much later, however, did people begin to speak of Winston-Salem. See Adelaide Fries, Stuart Thurman Wright, and J. Edwin Hendricks, *Forsyth: The History of a County on the March*, rev. ed. [Chapel Hill: University of North Carolina Press, 1976], 123–30.)

the house, the floor white, yard clean and the spring in tolerbly good order. These are very good criterions by which to judge a nice family.

*Nov. 2* A little cloudy. School small. I was, this evening, at the burying of John Messick's wife; she was just in the prime of life, 24 years old. Received a letter from Dock to night. He dont give a very good account of the Jonesville Academy, says the students do about as they please.[23]

*Nov. 3* Quite a nice day. School increased a little. I took a little hunt with Elbert Messick, John D. Johnson, and Elzy, Pa & Caleb. We had but slim luck; caught only one little oposom. Hunting, in this country, is poor business, without its for a particular kind of game, such as the smoke dried inmates of the cotages in this country, known as the *girls.* Such girls! Some of them are hard favored enough, if it is possible, to turn sweet milk to bonny-clabber.

*Nov. 4* A beautiful, clear and pleasant day. Had sixteen scholars. Nothing of importance took place to day that I am aware of. I have, to day, been reading *Arthur's Home Magazine,* it contains a great deal of interesting matter, and is only $2 a year. I must subscribe for it, if not now, as soon as I get in possession of, a "pretty little wife, and a big plantation". Since "There's no place like *home*" I intend to have a *home* if I live, and then I'll send on for the "Home Magazine for Mrs. ______ to read, and so I will, and I will, & I will.

*Nov. 5* Taught School to day, though its Saturday. This was quite a cold morning, frosty, frosty, icy; the latter part of the day has been quite pleasant.

*Nov. 6* A beautiful day, went to Mount Maria [Moriah] to meeting, heard the Rev. Clegg preach the funeral of Mrs. Gill. His text was the 12 verse of the 90 Psalm.—"Teach us to so number our days," etc. He preached a good sermon to a large congregation.

23. William C., or Dock, Thomasson was attending school in Jonesville, a town in Yadkin County that is north of the Thomasson home, near Elkin. Jonesville Academy opened in 1818 and lasted until after the Civil War. It was well enough known to attract boarding students and at times was said to have as many as three hundred students. See Frances Harding Casstevens, ed., *The Heritage of Yadkin County,* published by the Yadkin County Historical Society (Winston-Salem: Hunter Publishing Company, 1981).

*Nov. 7* School, not very large yet. Quite a pleasant day.

*Nov. 8* Cold and cloudy, but no rain or snow here. I think its nearly cold enough to snow. School small. Too much work to do.

*Nov. 9* Rain, rain. It rained very hard till about 11 o'clock, when it cleared off, cold and windy. Had only twelve scholars to day. Johnny Johnson moved off to his own residence, near Zion, yesterday. Wrote Uncle Flem a letter, etc.

*Nov. 10* Quite a cold morning. School not large. James Martin moved over yesterday to the house Johnny Johnson left.

*Nov. 11* Cloudy, but not very cold. Received, to night, two numbers of the *Age*, the October number of the *Journal*, and a letter from Miss S. A. S. What a letter! I must answer it soon.

*Nov. 12* Cloudy and a little rainy in the evening but not very cool. Went to Hamptonville, bought a new fashioned hat for thirty five cents, a twenty five cent bottle of ink, box of wafers for five cents; mailed a letter to Uncle Flem etc. Wiley, Andrew Snipes, Uncle Flem and Aunt Theny[24] came up this evening. If I had just waited a few hours I might have saved three cents by handing over my letter to Uncle F. myself; but its gone now, and let it go; three cents *aint* much, but "a penny saved is a penny made".

*Nov. 13* Clear, and very windy, and not so warm as you might suppose. Went over after Martha, as the kinfolks wanted to see her, the creek was so full the water almost ran into the buggy. Went to Esquire Windsor's[25] to meeting to night.

*Nov. 14* Clear and warm. Had about twenty pupils to day. I have a tolerably nice School, *considering*. Bought nine chickens of Mrs. Johnson, and caught seven of them to night, and brought them over in Pa's "big basket."

24. Fleming Thomasson Jr. was a brother of Basil Armstrong Thomasson's father, Andrew. Flem and his wife, whose full name was Parthenia, farmed in Forsyth County, near present-day Winston-Salem.

25. Local government in North Carolina was dominated by justices of the peace, often called "squires" for the honorary title of esquire that they could write after their names. As other entries in the diary indicate, the justices of the peace had considerable social status.

*Nov. 15* Clear and warm. Small school. Commenced a letter to Miss S. A. L. S.

*Nov. 16* Warm and a little cloudy. School increased a little. Shucked corn till near 9 o'clock, and did not get done at that. The 17 & 18—tomorrow & next day, are the great feast days at Old Town.[26] It is, I supposed, one hundred years since the United Brethren landed on the waters of Mudy Creek. One hundred years! That is a long, long, time.

*Nov. 17* Got up this morning 1/4 after four o'clock. Early rising is good, if one can stand it. Nothing of more importance than my heifer had a calf; she was only a year old last April. Helped Hutch Johnson shuck corn till nine o'clock, and wore my hands slick, slick.

*Nov. 18* Quite a pleasant day. School small. Bought a field of wheat, that is just sowed, for $5.00.

*Nov. 19* Another nice day. Ploughed an hour or so. Pa finished sowing wheat yesterday evening, and I and James Martin ploughed in the last of it this morning. Went to Benbow's this morning and got 200 oil cakes.[27] Went to Hamptonville, bought *Webster's Unabbridged American Dictionary;* price $6.

*Nov. 20* This is a pleasant day, or rather warm. Went to Flatrock, heard the Rev. Green Brown preach quite a good sermon. Went home with Miss M. B., stayed till awhile in the night & cut for home.

*Nov. 21* Warm, & a little cloudy. Had 21 scholars to day.

*Nov. 22* Warm, cloudy & a little rain. Had 20 scholars.

*Nov. 23* Still warm, though its clear to day. Closed two months of my School to day. Moses Sprinkle and Miss Luzeny Hutchins got married, last night, in the old field, or woods. Its fashionable now-a-days to marry in the woods, and in the dark. Tomorrow is Thanksgiving day, being the 24th day of Nov. 1853.

26. Old Town, a district in today's Winston-Salem, was originally settled by members of the Unitas Fratrum, or Moravians, who came from Germany to Pennsylvania and thence to North Carolina.

27. These are probably cottonseed oil cakes, consisting of crushed cottonseeds and useful as a feed for livestock.

*Nov. 24* Clear & a little cool this evening. Had 21 scholars. Commenced the third month of my school to day. These Common Schools are a vast deal of trouble.

*Nov. 25* Clear & cold. School small. People will not send to School if they can help it; and most of them do help it, for out of ninety five in the District, I seldom have more than twenty five.

*Nov. 26* Clear and coold. One Mr. Kenyon addressed us on the subject of Temperance at Buck Shoal. Kenyon is no "bugeater", that's certain. Though there was but a small congregation, he gave us quite a good talk.

Went to Hamptonville, received the *Age, Journal,* and *Press;* and also a letter, for Mary, from Indiana, and one for James Martin. Bought J. Martin three awls for which I paid five cents, of his own money; mailed a letter for said Martin, and paid two cents of the postage out of my *own* pocket.

*Nov. 27* Not day yet. I guess when day comes, he'll be a little frosty; though, when the "Red Orb of day" makes his appearance, he'll dispel the cold and biting frost, by his bright rays & cheering countinence. Staid last night with Uncle Robin and Aunt Poly,[28] I and Mary. Left this morning about 3 o'clock, came home and took a cold bath; so I feel some better? Yes sir. Went to one Mr. Pardew's to meeting, heard Pa preach, "Teach us to so number our days," etc. Went to the gate with Miss M. B., then on to Alexander Johnson's and dined, then on home and had like to have been thrown off my pony. I let her jump the fence, she got the advantage, & ran off with me & came near doing damage. I escaped narrowly.

*Nov. 28* Cool and cloudy. School small. Went and helpt James Martin shell corn, and saw Miss S. B., who is a tolerbly good looking red-headed girl.

*Nov. 29* Still cloudy, and raining this evening, though not very cool. Nothing of importance.

*Nov. 30* And last day of this month. Clear and warm. School small. Old Mr. Shedrick Jentry died this morning about six o'clock. Poor old man!

28. These were relatives on Basil Armstrong Thomasson's mother's side of the family.

drug-medicines, no doubt, shortened his life many days. How strange it is that doctors try to remedy one evil by administer[ing] a greater! They fall in with the disease and war against nature.

*Dec. 1* A little cloudy, and cool. Dismissed School and went to Mr. Jentry's bur[y]ing. There was but one Messick there, as *much* as they pretended to think of the old man. Mr. Jentry has no doubt gone to a better world; he left some property for his children to rangle over. Such a time as they will have! What is the use for a man to lay up treasure in this world? Mr. Jentry, though, acted wisely, if hed laid up in this world, he also laid up in heaven, and he is now gone to his reward.

*Dec. 2* Cloudy & a little rain. Nothing of importance.

*Dec. 3* Clear and cold. It rained last night. Went to Hamptonville. Took old Coles fifty-one pounds of Tobacco; paid for my Dictionary, and squared up with the old man. Sold C. Rinhart fifty pounds of Tobacco at 12 1/2 cents per pound. $6.25. Took his due-bill. Received the *American Eagle* in a new dress, *The United States Intelligencer,* and a letter from Thomas Hunter.

*Dec. 4 Sunday.* Clear, and not very cold. Went over, but no one at home. Such going, and going, and finding no one, is not very agreeable. Guess I'd better quit going, so I had.

*Dec. 5* A little cool in the morning, though warm in the evening. Had a few new hands at school to day. Nothing more? No sir.

*Dec. 6* Clear, and quite warm. School greatly increased.

*Dec. 7* A little cool this evening.

*Dec. 8* Got up this morning 1/4 before 4 o'clock. Snowing! snowing!! It must have commenced about midnight, as it's two or three inches deep now. It quit snowing about day; the snow, I guess, was about four inches deep. School small.

*Dec. 9* Staid last night at Wiley Felt's. School not large to day. John E. Grant came over to night, and we talked about putting up a blacksmith shop, a school house etc etc.

*Dec. 10* Just returned from Hamptonville. Received a letter from brother Wiley, the *Spirit of the Age* etc. Wiley writes a little good news, and some that's not so good, or at least it dont sound so well in my ears, but "let it went" for what it's worth.

*Dec. 11 Sunday.* This a beautiful day.

"This is the day the Lord hath made,
He calls the hours his own;"

We are therefore not to seek our own pleasure on the Sabbath, but keep it holy.

*Dec. 12 Monday.* A pleasant day. Had 24 scholars. Feel a little sleepy to day, as I was up last night till four o'clock.[29] This way of staying up so late is rather a bad business, tho' it's quite pleasant at present time. How often we purchase present happiness at the expense of future misery. Human nature; how strange it is!

*Dec. 13* Clear, and warm after the morning. Had 24 or 25 Scholars. Nothing of very great importance.

*Dec. 14* A pleasant day. School as usual. This week, so far, has been, I think, as nice weather as I ever saw in December. We have quite a nice frost every morning, and the ground freezes a little every night. Wheat looks tolerbly well. Corn is worth about thirty-five cents per bushel. Pork is worth four, and four dollars & a half per hundred. Wheat, seventy-five.

*Dec. 15* Quite a cold morning, frost! frost!! This is the "big day" of their examination at Jonesville. Look for William C. home to day.

*Dec. 16* Another nice day. Large frost. Dock landed from Jonesville this evening. There were not many people, he says, at the examination. Dry time! Dry time! "Wo unto the world," for want of an interest, and a proper attention to, things of importance.

*Dec. 17* A very rainy day. Went to Hamptonville. Received the *Age,* and a letter and book from Uncle William in Georgia, for Mary. The title of the book is *Heat Blossoms.* I guess its a good Book, as its from Uncle W. H.[30]

29. As later entries make clear, Thomasson was courting and had stayed late at the home of a young woman he was seeing, who was almost surely (see the entry for December 18) Mollie Bell.

30. William Hauser was a brother of Thomasson's mother.

*Dec. 18* Cold and windy. Meeting at Flat Rock & at Shilo[31] but I did not go. I went to see Miss M. A. B. this evening, and staid till awhile after "moon up," then I toddled homeward in a kind of hurry. What's to become of me, I can't tell, but I wont be a bachelor, & if I should be so unfortunate as to have to "bear the everlasting din of woman's tongue, which flies from morn 'till night," I'll try and be content with my lot, and get through this world as best I can; but I trust I shall have better luck, and get an Angel with whom I can enjoy heaven on earth. So may it be. Heaven grant it. Amen.

*Dec. 19* Cold! Cold!! School not large. Pa killed his hogs. Nothing of interest.

*Dec. 20* Very cold. Had an election at the School house. J. F. Hendrax moved that we have holyday from Thursday, twelve o'clock, till Tuesday morning. The motion was put to the house, and received a unanimous vote. Also agreed that we have one bushel of apples, if we can get them conveniently. Some of the boys made bold threats about turning me out a day or so before Christmas, but they failed. They all acted quite gentlemanly.

*Dec. 21* Cold & cloudy. School as usual.

*Dec. 22* Still cold & cloudy, and since twelve its raining considerbly. Dismissed School 12 o'clock till Tuesday morning. Staid last night at James Brown's. I have taught, in District No. 35, just three months, and have an order on James Sheek,[32] the Chairman for fifty-six dollars and fifty cents. They give me $2 for my Black Board and fifty cents for window glass.

*Dec. 23* Cold, and rainy till 12 o'clock, then it blew off very cold. John E. Grant shod my pony this afternoon.

31. Shiloh Church later divided into two churches with two different denominational affiliations: Shiloh Baptist and St. Paul Methodist. Strong Thomasson considered himself a Methodist rather than a Baptist, but he often attended local Baptist churches. For Shiloh Church, see *Heritage of Yadkin County*, 234.

32. Two men, James Sheek and Moses Baldwin, often served as chairman of the board of superintendents for Yadkin County in the 1850s. In 1853 the county had forty-two school districts, but only thirty-one schools, and an enrollment of 1,593 students. Teachers were almost all male; no female teachers were licensed until 1856. Approved textbooks included *Webster's Spelling Book*, Davie's *Arithmetic*, Mitchell's *Geography*, Bullion's *Grammar*, Worcester's *Comprehensive Dictionary*, and the *Common School Catechism*. See *Heritage of Yadkin County*, 243.

*Dec. 24* Not exactly clear and quite cold. Rev. L. Holton preached at J. E. Grant's last night, notwithstanding its being so cold. Nothing of great importance? No sir.

*Dec. 25 Sunday & Christmas day.* Quite cold. Went to old James Godfrey's to meeting. The Rev. Swaim preached young J. G.'s funeral there to day from this text, "And if the righteous" First Peter IV and 18 verse. Went home with Miss M. A. B. again; three Sundays in succession is "doing the thing up brown."

*Dec. 26 Monday.* A cold day. Come home this morning about day; slept last night none at all. Went this evening and dismissed my school for good or for bad; for all time anyway. Went round by Hamptonville; received a letter from brother W. W., two numbers of the *Age* a number of *Brother Jonathan* and the first number (the first number of the volume second, dated Jan. 7, 1854.) of *Nichol's Journal;* it comes out now in a new form; it is henceforth to be a weekly; price $1.50.[33]

*Dec. 27* Clear and not very cold. Received $5 tuition from Charles and Thomas Benbow.

*Dec. 28* Cold snowy day; snow three or four inches deep. Killed a wild pigeon this morning. Hauled three loads of wood for John E. Grant.

*Dec. 29* Clear & cold. Snow melted but little. Went a hunting and caught—just nothing at all.

*Dec. 30* Snowed, hailed, and rained! Such a time!!

*Dec. 31* Last day of Dec., and of the year 1853. Fair-thee-well. Went to Hamptonville; received two letters, one from W. W. T., and one from T. C. Pfohl. It commenced snowing just before night and snowed tremendous. What a snowy winter!

Thus endeth my Book of Remembrance, for 1853.

33. No copies of this paper exist in the extensive holdings of the North Carolina Collection at the University of North Carolina at Chapel Hill. By 1857, however, John Nichols was publishing the Beaufort *Journal,* and it is possible that this may have been an earlier journalistic effort by the same individual.

# Book of Remembrance 1854

*January 1* 1853 is no more; its cares and toils, many dark, and few sunny hours have all passed away. 1853 is gone, forever gone. "Time once passed never returns." 1854 commences with a very cold day. The ground is covered with snow that fell yesterday evening. It also snowed a little this evening. I read aloud to night, an article in *Nichol's Jo.* headed "The Mechanic's Home." Its a good piece. The wind howls so cold without, it makes chills run over me. I must see if I cant find a warm place beside my brother Caleb E., who is now sleeping soundly, and he is evidently warm. I gave him a cold bath just before going to bed, and I tell you, it made him bounce. Cold bath is the idea for a cold, no mistake.

*January 2* A pleasant day over head. Snow melting off very fast. Went out this morning and shot a rabbit. I was walking down the branch that runs through John E. Grant's meadow, and there being no good place to cross, I stopped at the mouth of the branch to see where I should cross, and just then the "old har" hop[p]ed from under the bank as quickly as if he had been bidden; he jumped a step or two and drew up under some trash, thinking, I suppose, that I did not see him; but there he was mistaken, for soon he had a hole through his body; made by a lead bullet; then it was too late to correct his slight mistake, so he "knocked under" without ceremony. Cut wood in the swamp most all day.

*January 3* A clear day and not very cold. Went to Yadkinville, saw James Sheek, who paid me $54 for teaching school 3 months in District No. 35, and $2 for a Black Board. Bought of the firm of Hauser and Wilson in Doweltown, a white handled knife for .80 which was priced $1. Blum's Almanac for '54 for five cents. Bought in Hamptonville, a pair of kid gloves for 75 cts. half quire of paper for .10 cents. Spent to day $1.70. I must curtail my expense, and live on economy awhile.

*January 4* A clear, pleasant day. Cut saw logs in the swamp. William C. came home this afternoon, after taking a long Christmas tour in Davie, Forsyth and Guilford. He brings important news, such as the following:—Isaac Grubbs is married to Miss Selina Tuttle, and J. W. Coltrane to Miss Nance Alspaugh, etc. etc. etc. Received a letter from Mr. J. W. White; he gives lots of news in general, but little in particular.

*January 5* Cloudy and rains a little, cut saw logs again to day.

*January 6* Old Christmas day. Cloudy in the forenoon, cleared off in the afternoon quite cold. The Rev. John Gunn preached at Shilo, and at night here at Pa's; the house was full. Gunn's text was "Whosoever will come after me, let him deny himself, and take up his cross, and follow me." Mark VIII. 34.

*January 7* Held a meeting at Windsor's Cross roads relative the building of an Institute; done but little, as old Leonard Messick was Chairman, more than locate. The house is to be at, or near the X roads. Adjourned to meet Friday the 13 inst. Commenced snowing about 12 o'clock, and is still at it; it is now about 9 o'clock at night. We have *lots* of snow this winter.

*January 8* Very cold. Snow about eight inches deep. Went with Pa to old uncle Ashley Johnson's,[1] and from there to see Miss M. A. B. again. Miss ______ is the girl for me. I do not expect to marry for riches. No sir.

*January 9* Still cold. It is colder to day than it was yesterday, if there is any difference. Snow melted off but slowly. Went a rabbit hunting, and caught nothing. Came home this morning about five o'clock, set up all night, last night. Such work! Such work!! This running after the *galls*.. Pshaw!!! Tomorrow commences old Mr. S. Gentry's sale.

*January 10* Cold and cloudy. Snow melted but little. Looks like it would snow again before morning. Went to the above named sale, but bought

1. This man was not actually Thomasson's uncle but a member of the family of Alexander Johnson, who married his sister Martha.

nothing. Two horses were sold, one for $100, and one for $80; sold well. Pork went at about $5; the old plunder, generaly, sold well.

*January 11* Sale continued to day, though it has been a *very* rainy day. Its a sloppy time, and no mistake, caused by the melting snow and rain; and its tolerbly cool. Property sold near its value. I bought the following articles: One cupboard for $2.00; two candlesticks, brass, one for 30 cents, and the other for 35 cents; one set of plates for 25 cents; half of one beestand for 50 cents; two sheep for 75 cents each; one hogshead for 25 cents. It looks a little like I am fixing to set up on my own hook, but pshaw! I am only buying a few dry goods, thats all. For the above articles I paid the cash, $5.15 cents to Joseph Johnson, the sale clerk. Mr. Gray Hampton is the cryer, and he is a good hand too.

*January 12* Some what cloudy; there fell a tremendous rain last night. The creek is not foardable to day.

*January 13* Clear and frosty, though not near as cold as it was a few days ago. Went to Crouche's store, bought 1/2 quire of cap paper, pantaloons cloth, 2 1/2 yards calico, 1/2 yard linen, 3 yards domestic, 10 pounds of nails, amounting to $3.62 1/2. Came by the "school meeting.["] Nothing done, or but *very* little towards a School house. Went to Hamptonville, received the *Age*, mailed three letters, bought Mary (sister) a pair of gloves for .20 cts. Mary is to drink no more coffee! The wind blows hard.

*January 14 Saturday.* Clear and not very cold. Worked the—or made—or helped to make a foot way across the creek. Nothing more of *very* great interest.

*January 15 Sunday.* A clear pleasant day. Went to Flat Rock and heard the Rev. Brown preach from this text;—"For ye know the grace of our Lord Jesus Christ," etc. II Cor. VIII.9. The Rev. cant do any "big things" in the preaching line. I staid last night at Ashly Johnson's. I read a letter this morning from a man in Minnasotia. It contains "great news", dont know if I shant go out and see those lakes that abound with "fish, and wild fowls."

*January 16* Clear and quite a pleasant day; rove 750 boards, to cover the School house.

*January 17* Cloudy, and rainy; cut shop timber this afternoon. The neighbors met here last night to talk about building a School house, I wrote two Subscription lists, etc, adjourned to meet at the X roads on Saturday 28th inst. Found my knife, which I lost some time ago in the snow. Read five chapters of a *very* interesting story in the *Age*—"A Tale of land and sea."[2] Mr. St. Clair was indeed a hero. H[e] was a printer by trade, and labored for his support. Work is honorable. All must work that wish to be happy.

*January 18* Cloudy in the forenoon and rained a little; cleared off in the afternoon quite pleasant. Hauled shop timber and saw logs. Nothing more of interest.

*January 19* A very rainy day, but not so cold as it has been. Bought a seven dollar note I gave Nute Baron the 28 of last September. Baron sold the note to James West of whom I bought it. The interest on the note was a fraction over 12 1/2 cent. I gave Mr. West a due bill on Christian Rinehart, calling for $6.25, and 87 1/2 cents, in *silver,* for the note. I swaped watches with Mr. Baron, and gave him my note for seven dollars, that being the supposed difference in the value of the watches. My first watch cost me $6.50, and I gave it and $7.12 1/2 for the one I now have, making $13.62 1/2. Dear watch!

*January 20* Cloudy & very damp, though not much rain till night. Went to Hamptonville, received the *Age, Journal,* and *Day's Great Christmas Pictorial Brother Johnathan;* its a tolerbly handsome sheet. I also received a note from I. N. Fowler F. S. for which I had to pay five cents. Such messages are not welcome, well now, they aint. B. A. Thomasson.

*January 21* Rained a great deal last night. Cleared off this morning, and is now quite cold. Nothing more of interest. Wrote for the *Young Brides Book,* B. A. Day, New York.

*January 22 Sunday.* A very cold, clear day. Went to Ailsbury to class meeting, but few turned out as it was so cold. Went home with Miss M. A. B., who is a beautiful girl; so I was "in town," and in agreeable company.

2. This story appeared in the issue for January 11, 1854, which was vol. 5, no. 19.

*January 23* Clear and very cold, indeed. Came home this morning. Slept last night not quite 2 hours. Went to choping and grubing this afternoon, in the swamp.

*January 24* Not quite so cold as it was yesterday. Hewed shop timber this evening. Read the first number of the *Ballot Box,* a very good temperance paper, edited by Rev C. F. Deems, of Greensborough, and published by I. F. Disosway at fifty cents a year, it being a monthly periodical. The *Ballot Box* should be read by every voter in North Carolina *especially.*

*January 25* Cloudy, and hailed some about 12 o'clock, and rained a little. Cut fire wood in the swamp most of the day.

*January 26* Cloudy, rainy day. Cut saw logs. Nothing more.

*January 27* Rained about all night, the creek is up again. Blowed off, quite cold. BAThomasson.

*January 28* A clear, cold day. Went to Hamptonville; received the *Age,* only. Mailed three letters; one to B. A. Day of New York; one to I. F. Disosway of Greensboro; and one to W. C. Thomasson, student at Jonesville Academy. No mail from Salem, as "the river's up." B. A. T.

*January 29* Cloudy and quite cold. Staid at home all day, except a few minutes at John Grant's. Had more company to day than is agreeable on the Sabbath. Sunday visiting is not in accordance with the Divine Law, unless its to do good; this way of neighbors getting together, and spending the day that should be kept holy, in worldly conversation is nothing short of robery. We have six days given us, in which, to do all our labor, and attend to worldly affairs, but the seventh we are to rest both mind and body, and spend that day in the service of Him, who has been so generous as to give us 6/7 of the week, in which to attend to the things of this world, and has only required us to rest from all our labors one day in seven, and to devote that day exclusively to his service. How benevelent is the giver of "every good and perfect gift," and how very wrong it is to rob Him of those few moments he has commanded us to remember, and to keep holy. May the good Lord pardon the writer of these lines for the great sin of Sabbath breaking, and impress upon his mind the importance of keeping the Sabbath day holy, and enable him to do it in the true sense of the word.

This day twelve months ago we left Stokes County; it was a very cold morning too. I left my school yesterday one year ago, and went home. The sun was about an hour high, when I left Conrad's where I was boarding, and rode home, a distance of 15 miles in about 2 hours! I came with Pa, the next day, to the Yadkin River; we reached it about 2 or 3 o'clock; there I left them and returned to the neighborbood of Spanish Grove where I was teaching school, and where my brother, W. W. is teaching this winter. William C. is going to school at Jonesville, and I am at home trying to study a little sometimes but spending rather too much of my time in going to see the girls, or rather the girl, for I am not like some boys who go to see 1/2 a dozen girls at once. Woman's love should not be trifled with, for to them, it is life—"it is all they have to live for; and when it is taken away, they have not, like man, a profession, business, travel, and pleasure, to divert and occupy their minds. It is a sorrow they can never tell, to seek the healing balm of sympathy—they have only to sit down and endure." Woman! Take care. Love those only who are worthy of being loved by angels. "Cast not your pearl before swine."

*January 30* Clear, and very cold in the morning. Helped John E. Grant kill two hogs, and then came home and assisted Pa, as he killed the pig he got of Robin Jones this morning. Pa paid $1/4 for the pig about the last of March, I think, in the year 53 and killed it to day; it would have weighed near 2 hundred; it was up, almost, all the time. Its cheaper to keep a few good hogs, than so many that they cant be half fed. Give one hog as much as he will eat, and he will make more pork than two half fed.

I and J. E. Grant worked on the shop this afternoon.

*January 31* Last day of Jan. 1854, and a nice day too. I and J. E. Grant worked on the shop, framing of it. Received the *Age* No. 21. Put up a mail box at J. E. G 's. Meeting here to night again. The Rev. John Gunn preached, his text was the 6 and 7 verses of the 55 chapter of Isah.—"Seek the Lord while he may be found["] etc. Had quite a good meeting; Miss Setty Baggarly got down to be prayed for, but did not profess, or at any rate not publicly.

*February 1* A very pleasant day; worked on the shop, went to meeting etc. Heard the Rev. J. Gunn preach at Aylesbury from this text:—

"A sower went out to sow his seed." etc. Luke VIII 5, 6, and 7 verses. Bought four books of Rev. Samuel Caloway for 75 cents. *Elizabeth Davidson, Emiley Maria, Golden Treasury,* and *Commandments Explained.* I think, I have bought fo[u]r good and interesting books, all for 75 cents!

*February 2* Another nice, clear day. Raised the shop, etc.

*February 3* A very cold day, wind blows so hard and cold that we cant work on the shop to day, so I set by the fire most of the time.

*February 4* Cold, though the wind does not blow so hard as it did yesterday. Put the rafters on the shop this evening. Went to Hamptonville; received No's 21 and 28 of *Nichol's Journal,* and a No. of the *Dollar Times* from Cincinnatti. The *Times* is a good paper. In the "Lady's Department" is a good lecture by W. G. Eliot. Rev. Mr. Eliot says, "A slovenly house, or a badly ordered table, or ill-clothed children, make an uncomfortable home, and a man must be a saint to resist its unhappy influence on his character." That's "True as preaching."

*February 5 Sunday.* Quite a nice day, and not very cold. Went to see Miss M. A. B. this evening, and found her at home reading a new book. I think it was the *Memmor of Mrs. Grayham,* a truly pious lady.

*February 6* Tolerbly cold, and a little cloudy. Notwithstanding I felt a little sleepy to day, (as I set up last night, all night,) I covered most of one side of the shop; yes I covered most of one side. Pa covered a little.

*February 7* Cold and cloudy. It hailed some this evening. Finished covering the shop, etc.

*February 8* Cloudy and rained some. Hauled a few rocks to build the shop harth. Received No. 22nd of the *Age* and a letter from each of my brothers, W. W. and W. C. The *Age* is rich as usual.

*February 9* Clear and quite a pleasant day. Went to the sale of Roughton, Gentry, and company, whose store goods etc. were sold, or a part of them. I bought a screw-plate for $5.00, and a rasp for 41 cents. Goods sold at high prices.

*February 10* Sale continued, went to day and gave my note for the property I bought yesterday. Jo. F. Johnson signed the not[e] with me.

The note is for the amount of $5.42, to be paid 12 months after date. A clear, warm and somewhat windy day.

*February 11 Saturday.* Clear and pleasant, hauled a few rock[s] this morning to build a harth in the shop, hauled one load of plank, worked on the shop an hour or so, then went to Hamptonville; bought a large file, 10 P.O. stamps; received the *Ballot Box* for which I wrote some time ago.

*February 12* Cloudy and a little cool. Went to Aylesbury to class-meeting, and then home with Miss M. A. B. Miss B is

"The fairest of earth's daughters,
A gem to deck the sky."

*February 13* Cloudy, and rains a little occasionaly. I and Pa worked on the shop furnace, but we did not get quite done. I feel a little sleepy, and so I do.

*February 14* Still cloudy, but not much rain. This is the fourteenth day of February, and is said to be general mate choosing day among the fowls, birds etc, Worked on the shop. Received the *Age,* and *Journal.*

*February 15* Cloudy, and a good deal of rain in the morning. I and Johny Grant got in our anvil block. John done the first work in the shop to night; he made tow nails. Old Mrs. Willard died this morning, at the advanced age of 104, or 105. It is not often one lives to be 100 years old, but I believe the time is coming when numbers will live to be 100 years old. When people learn how to live, they will live a long time. Ignorance has killed thousands, thousands.

*February 16* Cloudy yet. It snowed very fast for an hour or two this morning, though the snow did not stick long as the rain had fallen first, and the ground was quite wet. "Two things cannot occupy the same space at the same time", so says the philosopher; therefore water and snow cannot both cover the earth at the same time. Worked some in the shop, but not a great deal. etc. etc.

*February 17* Clear and cold. Worked on the shop. Received the *U.S. Inteligencer* for February, the *Young Bride's Book,* (which is to be presented to Miss M. A. the day after the wedding,) and a letter from my old friend

J. H. White. I have not examined my new book much, though, I think it's good; it's small, the price being only .12 1/2.

*February 18* Clear, and tolerbly cool. Nothing of importance.

*February 19 Sunday.* Cloudy in the afternoon, and commenced raining in the night, and rained, and rained. Started to Flatrock to meeting but did not get quite there as I heard that the preacher, the Rev. G. W. Brown, took the mumps and went home. I came back and spent the evening with Miss M. A. B.

*February 20* Cloudy, and showering down a mixture of rain and hail. Left Mr. B's this morning about 10 o'clock, came home and found brother Wiley W. at home. He came home this morning. Received a number of the excellent paper, the *Journal* published in N. Y. by T. L. N.

*February 21* A tolerbly pleasant day. Went to Hamptonville; bought 27 cents worth of steel, a bowl for 20 cents, a set of table spoons for 80 cents, making $1.27. Worked on my shop doors, etc.

*February 22* Washington's birthday. A clear, pleasant day. Worked on the shop, making door shutters etc.

*February 23* Clear, and tolerbly cool. Mr. John Messick came and worked in the shop for me to day. We done about $1.62 1/2 worth of work for which he charged me 50 cents.

*February 24* Clear and not quite as cold as it was yesterday. I put up my shop doors etc. Went to J. E. Grant's and heard the Rev. Quinton Holton preach from this text;—"This is the king of the Jews." The sermon was short.

*February 25* Cloudy, and since 12 it has rained *lots.* I went to Couche's store this morning, and bought 5 lbs nails at 7 cts. per pound; 2 lbs & 14 oz. of cast steel at 25 cts per pound; 3 1/4 pounds German steel, at 12 cts per pound. Went to Ham'ville; bought 2 lbs. nails 8 1/3 cts per lb. one pad lock for 20. Mailed 3 letters, one to J. H. White of Forsyth; one to B. H. Day, of New York, one to Burdick, N. Y. publisher of the *Ladies Heart.* Received the *Age;* saw the richest gold ore, found on A. Martin's land, I guess, that has ever been seen in this country. As I came home I saw Miss M. A. B., the prettiest girl in all this country.

"May the Ruler of heaven look down,
And *my* Mary from evil defend."
Amen.

*February 26* Cleared off this morning, after a tremendeous rain last night. The creek is very full, I never saw it so full before. Staid at home to day, as I could not go over the creek, and read my papers, books etc.

*February 27* Clear, and quite pleasant. Cut coal wood in the forenoon & hauled in the afternoon. Had a regular chat to night, about woman's work etc. There are more slaves in the U. S. than most of us are aware of. Freedom is a great thing, but woman cant be alowed to enjoy it; they are slaves to men & to fashion, but the time is coming when they will be free; may that time come quickly. Amen.

*February 28* Last day of February; clear & warm; set up, and fired our coal pit.

*March 1 Wednesday.* March comes in with quite a nice day. Worked on the road, or building a footway across the creek.

*March 2* Still warm, but a little cloudy. Worked on the footway again. John Messick came again and worked in the shop awhile this afternoon for which he charged me 25 cents; he done 80 cents worth of work.

*March 3* Cloudy and rained a little. Went to the School meeting at the X roads, and then to Hamptonville. Received the *Age,* & *Journal* two great papers, perhaps, the best.

*March 4* Still cloudy, and rained some in the forenoon. Went to muster, and such a muster! I never was at just such a one before the 4 day of March 1854.

*March 5 Sunday.* A clear, quite pleasant day. Went to see Miss M. A. B. etc.

*March 6* A nice day; plowed in oats all day; staid last night with Miss M. A. B., so I did not sleep much.

*March 7* Cloudy, & rained in the forenoon; the evening quite nice. My heel strings very sore, ocasioned, I guess, by a hard days plowing yesterday; so I "lay in" till after dinner.

*March 8* A little cloudy, and some rain and quite warm. Ploughed hard again to day, so I am tired.

*March 9* Still a little cloudy, and little rain, and quite warm. Finished sowing oats at the Johnson farm to day. Commenced over at home. Tired again to night, so I must take a cold bath and be off to bed. "Night is the time for rest," so says the poet. I agree with him too, so adieu to night, workers.

*March 10* Cloudy & warm; ploughed till time to start to church, then I left off. The Rev. J. Gunn preached at Aylesbury, his text was the 7, 8, & 9 verses of the VI chapter of Paul's epistle to the Galations. Went to Hamptonville; received the *Age,* two No. of the *Journal,* and *Esoteric Anthropology.* It rained lots this afternoon, and the air is a little cooler.

*March 11* Clear, and cool; went to Couche's store; bought 7 yds cotton cloth, half quire of paper, 7 bridle buckles etc. Dr. J. W. Hunter pluged one of my teeth, which was not very agreeable, though the operation was not painful. I have now 4 plugs of gold in my teeth, which is more than I have in my pocket. I must "turn over a new leaf."

*March 12 Sunday.* A clear, pleasant day. Went to Aylesbury to class-meeting; had quite a good meeting. After services, I went most home with my fair one, Miss M. A. B., then I came home to be with Dr. Hunter.

*March 13* Another nice day. Ploughed hard—too hard. Dr. Hunter left this morning; He staid with us from Friday evening till Monday morning. I paid Hunter $3. for pluging 3 of my teeth, I & Hunter are even.

*March 14* A little cloudy this morning, though it soon cleared off without rain. Ploughed again to day. Ploughing is hard work, but the good book says, "He that tilleth his land shall have plenty of bread."

*March 15* Clear, and quite warm; ploughed in oats, that's all.

*March 16* Got up this morning before day. The road is to be worked to day. Came home just before sun set, quite tired. We worked all day on about one mile of the road. Rained a little this evening—a thunder shower. This has, so far, been the warmest March I ever saw. The peach trees are in full bloom, and the apple trees are putting on their leaves, and the woods begins to look a little green.

*March 17* A little hazy and quite warm, or would be, were it not for the cool south west breeze that fans the cheek. We finished ploughing in oats to day about 12 o'clock. Pa sowed near 20 bushels this year. This is more oats than he has usually planted in one spring. Farming is a delightful occupation, as well as an honorable one. If I had the "needful" to carry on a farm as I would like, it should be my avocation for life.

*March 18* Clear and cool with a brisk wind. Went to Couche's store but got nothing. Went to Hamptonville after the *Journal* but did not get it. I have an idea of going to school this summer. I must have an education, no matter what it costs. I cant get on in this world without an education, so I must, and will have it, if I am blessed with life & health. There is a great call for men of learning; the demand is greater than the supply, and as the number increases there will be a proportionate demand, so a man of learning can alway find imployment, and get good wages. "Knowledge is power," and more precious than gold, but the Bible says, "he that increaseth knowledge increaseth sorrow."

*March 19 Sunday.* Clear and cool. Staid at home and read till late in the afternoon, then I drew on my *go-to-meeting* coat, boots and beaver, and walked over to see Miss M. A. B. the sweet and beautiful, found her at home, waiting for me, and wearing a red dress and black silk apron etc.

*March 20* Clear & quite cool. Came home this morning about day; staid with my fair one last night. Ploughed a little in the bottom, and slept some. Went to the factory to hear a Pis'ian[3] preach, his text was, "Evil communications corrupt good manners," he read out a tolerbly well worded sermon.

*March 21* Cloudy and cold, hailed and rained, went to H.ville; received the *Age*, and the *Dicrip of 100 Citties*. The tale of land and sea is finished in the No. of the *Age* vol V. No 2, Mar 15, '54.

*March 22* Cloudy and cool. James Whitlock worked in my shop to day. We done about $1.70 worth of work, for which he charged me 1/2 dollar. Bought one dollars worth of iron of J. S. Grant, esq.

*March 23* Worked the road again to day. Clear & cool.

3. Perhaps this means Presbyterian.

*March 24* A little cloudy, ploughed, made a yoke for my cow, to keep her from jumping. William C. came home, from Jonesville, this evening; The Rev. Q. Holton was to have preached here to night but did not come, the cause, to me, is unknown.

*March 25* Clear, cold, and very windy. Went to Gentry's sale at Roughton's, bought nothing but 11 sceins of silk at 12 1/2 cents. I think I got a good bargain in thread. Clarke came from Jonesville, and Wiley from Forsyth, or Clarke came yesterday, & W. to day.

*March 26 Sunday.* Still cold & windy. I and W. had an appointment to sing at Aylesbury, but it was so cold we had but few hearers, and sung but little. Clarke went back to Jonesville.

*March 27* Clear and cool. Nothing of importance.

*March 28* A little cloudy, and not so cool as it has been. Bought a buggy of Alexander Benbow for $50. for which I gave him my not[e] six months after date. I and Wiley hitched Jenny Lind to the shaff wagon, and hauled 3 loads of wood with her. She done well, though she never worked in shaves [shafts] before; we also brought the buggy home with her, as she "ploughed so well in the wagon." BAT

*March 29* Cloudy & cold. Went to Aylesbury to the Quarterly meeting; heard the Rev. Peter Doub preach from this text,—"For we walk by faith, not by sight." II Cor. V. 7. It is a cold time for meeting.

*March 30* Cloudy & raining; went to meeting; Peter Doub gave us a talk from this text.—"Not as though I had already attained," etc. Phil III. 12, 13, 14. There were but few out as it was a damp, cold day. Went to Couche's store, in the afternoon, and bought 3 1/2 lbs coffee for .50; one pair of shoes for myself, for 1.25, two lbs sugar for .20; one gallon molasses for .40; sister Mary a calico dress for 1.35. $3.79 spent to day.

*March 31* Last day of March, and a very rainy day. Cut cord wood.

*April 1* Cleared off this afternoon quite cool and windy.

*April 2 Sunday.* Clear & cold. Went to meeting at Aylesbury, and then home with Miss M. A. B. again. Miss M. has received the wreaths I wrote for; they are nice indeed.

*April 3* We have, this morning, a killing frost. Clear & cool. Went to Couche's after my iron, bought 306 lbs at 5 and 6 cents per lb. amounting to $15.56. I paid $10. and gave my note for the ballance.

*April 4* Clear & cool, a tolerbly keen frost. Went to Yadkinville to court as a witness for J. T. Roughton, against C. A. Bell,[4] S. Hicks & others. Got home to night about 8 o'clock, and have to go to court again tomorrow; so I must go to bed. Good night, sir.

*April 5* Went to court. I & Wiley drove Jenny Lind to Yadkinville in the buggy, she worked very well. The will suit was not tried to day, so have to go to court again tomorrow. Got home to night about dark or about the time it would have been dark, had not the "Queen of the night" prevented. Re'cd the *Age.* Clear and pleasant.

*April 6 Thursday.* Clear & pleasant. Went to court, the will suit was tried, and the will established!

*April 7* Clear warm. Cut one saw log, and ploughed some etc.

*April 8* Still clear & warm. Alexander Johnson hauled coal-wood for me till 12 o'clock. I & brother Wiley went to Hamptonville but we did not get any papers or letters. Dry times, no news.

*April 9* Sunday, and quite a nice day. Went to class-meeting at Ayls. [Aylesbury] and then home with Miss M. A. B. again!

*April 10* A little cloudy, and very windy. James Whitlock came to work in my shop; he has set in for one month, and if we both like, at the close of the month, he is to smith it for me one year at $12 per month, or for $144 for the year.

*April 11* Clear & not so windy as yesterday. Went yesterday evening and bought the wood work of a two horse wagon, of Charles Benbow for $20.

*April 12* Cloudy & rains a little, built my coal-house, etc. I have set up with my coal-pit 2 night and have to stay with it again to night. Setting up with coal-pits is not altogether as agreeable as hugging the girls.

4. This is Mollie's father, whose name appears in the 1860 census (under Hamptonville Post Office, p. 666) as Cranberry Bell. See entry for October 29.

*April 13* Cloudy and warm. I and James worked in the shop. Daniel Hutchens died this morning about 1/2 after 9 o'clock. Mr. H was taken sick yesterday about 3 o'clock. It is said he had the colic. Drs. Hough and Hampton were with him, but is seems that his time to die had come. Received of C. A. Bell two dollars for my attendence at court last week as a witness for J. T. Roughton. Bell & Hicks have the costs to pay, poor fellows! Lawing is dear business, and no mistake.

*April 14* Good Friday, and a very rainy day. Daniel H. was laid in "the cold and silent tomb" to day. His soul took its everlasting flight in the spring time when the sweet savor of bright flowers was riding upon the desert air, and when the songs of birds was echoing from hill to hill, and when this earth seems most enticing, and life is, indeed, sweet; but we all have to die, and if we are prepared, it matters not about what time of the year we take our last look upon "our mother earth," and soar away to live with angels in heaven, where spring n[e]ver ends, and where "chilling winds and poisonous breaths" are unknown, and sorrows never come. Oh that I may be prepared to meet death, at its coming, whether it be in the spring time, summer, autumn or winter, and be fully able to "go up and [to] the goodly land,["] of which I read in the Bible Amen.

*April 15* Another flood last night, creek past foarding—Washed away the foot-logs, fences etc. Cloudy to day, but not much rain. Hauled boards and covered my coal house etc, etc.

*April 16* Easter Sunday, and very rainy and quite a cool day. It hailed *lots* since 12—a real north-east hail storm, such storms do not often come in warm weather, or in any other season than winter. "Easter" says Webster, "is a festival of the Christian church, observed in commemoration of our Savior's resurrection, and occuring on Sunday, the third day after Good Friday." and Good Friday, "A feast, in memory of our Savior's suffering, kept on the Friday of passion week." and "passion Week," "The week immediately preceding the festival of Easter; so called because in that week our Savior's passion and death took place." So much for Webster—and he is hard to beat on deffinitions. Easter is toping off with a snow storm. The died eggs are consumed, the pies and cakes have passed around, and now the snow is coming in large flakes. How cold! for the 16th of April. Apple trees look *unique* with their blossoms and green leaves be-doubed with snow. If snow will kill the apples its a case with them now.

*April 17* Monday after Easter. Snowed till about 12 o'clock quite fast. In the evening, the wind blew from the North West, very cold. I think this is the coldest weather I ever felt in Aprill—it feels like winter time—its so cold!

*April 18* Cool and windy. Sold Frederick Reinhart 20 plugs of tobacco at 10 cents per plug. 20 × 10 = 2.00 dollars or 200 cents. Sold A. C. Johnson 10 plugs of tobacco at the same price. I, James Whitlock, Elizabeth Johnson & sister Mary went to old Jacky Johnson's & there spent the forepart of the night in useless chat. I shant do so again, soon, that's certain.

*April 19* Calm and pleasant, though we had a stout frost this morning. I dont think, though, that it has killed the apples. Received the *Age* and Nos. 12 & 13 of the *Journal.* The *Age* and *Journal* are great papers, which I would not be without for double the prices, which is only $1 each.

*April 20* Clear and warm; it begins to look like "spring time." I and Pa helped Moses Tucker roll logs this afternoon. Wiley came from Forsyth and brought news of murder and forged notes; he also brought me $2.25 of tobacco money or money that he got for tobacco of mine that he sold. I have an idea of selling off and going to school.

*April 21* Quite a clear warm day. Nothing of importance took place to day that I am aware of.

*April 22* Clear, with the exception of a few storm clouds that passed around, and quite warm. Split coal-wood in the forenoon, and rolled logs in the afternoon. The old bell ewe has twin lambs—lambs are so innocent; who could stain their hands in the blood of such harmless animals and then eat their flesh? "Thou shalt not kill."

*April 23 Sunday.* A warm day, and a few thunder clouds passing round south. Went to Aylesbury to class meeting, went home with Miss M. A. B. —had quite a pleasant time.

*April 24* Came home soon this morning—went to work in the shop,—

feel like I had lost sleep,—this lying out o'nights is rather a down hill business, and so it is.

*April 25* Clear & warm,—put the tire on my wagon wheels,—mended up my buggy harness etc., preparing to start to Guilford[5] in the morning.

*April 26* Started to Guilford early this morning. Clear and quite warm.

*April 27* Staid last night at Tim Conrad's. Went through Winston—rode on the plank road about 3 miles, bought a buggy whip at Gray's store for 80 cents, went on out to old Johny White's, and while we, and my brother Wiley, were there a tremendous storm blew up from the west, which prostrated the mighty oaks, uncovered houses, blew down fences etc. etc.

*April 28* Staid last night at old Mr. J. White's,—Quite cool, and cloudy.—Started on for Guilford this morning. Got to Mr. Daniel Pegrams about 1 or 2 o'clock P. M.—Cold and raining. What a time! for April.

*April 29* Still cold. Came up to uncle F. Thomasson's—staid there till Sunday morning.—Quite unwell to day.

*April 30* Last day of April,—cold & cloudy—came on up to the widow Snipe's to day, though it was so cold.

*May 1 Monday.* Started for home this morning.—Came by Pfaff Town—sold 31 1/2 lbs of tobacco at 18 cts per lb. $5.67; bought a trunk for $3, and then struck for home. Crossed the Yadkin River about 12 o'clock, and drove home, a distance of 30 miles, by about 8 o'clock in the night. Jenny is a great buggy nag.

*May 2* Clear & cool. Nothing of importance.

*May 3* Clear, cool & windy. Helped Pa plant corn.

*May 4* Clear and warm. Nothing of interest.

*May 5* A little hazy, but not much appearance of rain. Went to Hamptonville; bought a new hat for $1.25. Staid till after the debate. The question was "Has man more influence over society than woman?" Myself &

5. Guilford is the next county east of Forsyth.

Daniel Bell were appointed to decide the question. Roby (the Hamptonville teacher) [and] Miles M. Cowles argued against A. C. Cowles & W. Santford. The females in this, as in all other cases, came out conquerors.

*May 6* Clear and warm,—Went to the school meeting at the X roads etc.

*May 7 Sunday.* A nice, warm day,—went to Aylesbury to Sunday School, preaching, class-meeting etc. We just commenced our S. school to day. Pa preached. His text was. "Wherefore seeing we also are compassed about with so great a cloud of witnesses," etc. Hebrews XII. 7. The sermon was quite a good one too. After class-meeting I went home with Miss M. A. B., that charming little girl with whom I have spent so many pleasant hours.

*May 8* Clear and warm. Feel a little like a nap would help me. Wonder how Miss Mary feels after spending a night of waking?

*May 9* Hauled sawlogs. B. H. Johnson Esq. helped me, so we worked my filly; she works well—hard to beat. Went to James Whitlock's wedding, which came off this afternoon at five o'clock. Went down with B. A. Johnson Esq. It rained on us very hard, or very soft, at any rate the rain fell so fast our umbrella done us but little good, so we came near getting wet. The worst of it was we got lost and did not get to the wedding at the appointed time. They waited for the Esq. though, with as much patience as could be expected. Soon after we came in sight we saw heads in the door. How the hearts of James and Martha fluttered when it was announced "the Esq. is coming." Soon we drove up to the old rough bars, which were down at one end, as though a horseman had passed through. Here the old man & one son met us. After the usual "how do you do" etc the old man to[ok] charge of our nag while the son led us in to the room where the bride & groom were seated side by side. At our appearance the male half of the two that were to be sewn together turned pale as though he had been standing waist deep in the cold stream of death; while the female blushed & redened up, as only timid females can do. Things all ready, the Esq called them out and tied the untiable knot. "In the knot there's no untying." That done, we marched out to the wedding supper, which was quite common, though good enough. There is no use in spending a fortune in preparing a wedding feast. After taking refreshments etc, I and the Esq. took our leave, wishing the new married couple "much joy" long, and interesting lives, rolled off toward home, and left them (the new one made of a fair and tender girl about 16; and a rough, slovenly consumer of tobacco, coffee etc., who was about 21 years old) "alone in their glory." May they do well.

*May 10* Hauled logs till 12 o'clock. A little cloudy.

*May 11* Cloudy. It rained last night "lots." Hauled rails, coal-wood etc. with old Med and Jenny Lind. Went to Couche's store,—bought 1 1/3 gallons molasses .60; 1 yard calico, 12 1/2; 1 bottle copal varnish .20; .05 worth of candy for sister Eliza, who went with me, and 1 box of matches .03; making in all $1.00.

*May 12* Warm & cloudy. Went to Aylesbury to meeting. Heard the Rev. John Gunn reach from this text; "There hath no temptation taken you but such as is common to man:" etc. I Cor. X 13.
Went to Hamptonville; received No. 17 of the *Journal,* 2 No's. of the *Lady's Wreath,* just alike, etc. Stoped and took a peep at my fair one. The old saying "beauty is only skin deep" is an egregrious falsehood.

*May 13* Cloudy & rained some. Alexander helped Pa finish his new fence. I feel quite unwell.

*May 14 Sunday.* Still cloudy and rains a little; still unwell, not able to go to Sabbath school.

*May 15* Cut coal wood, and set it up ready for burning.

*May 16* Fired my coal pit, etc. I do not expect to follow burning coal,—its rather too lean a business. Teaching seems to suit me the best, so I must try it again.

*May 17* Cloudy and windy. Tom Ham, or Sig Ham as he calls himself, was to have shown his "slight of hand" etc. to night at the Shilo School house, but I suppost he is locked out and will have to show in the woods, if anywhere. Ham, I think, had much better go to building mills again or engage in some honorable business. A show also at the Factory, by Mr. Susseroff, or Shoe-suff, of Salem, N. C. He has a little magic lantern with which to humbug the people, and get their money. I did not go into the show, but saw a few of the girls and others who tried the "table moving." That's another humbug, and no mistake.

*May 18* Clear and cool. Great weather for wheat. Made a wash box, ploughed etc. Went again to the "table moving", there were six females and six males, and I tell you, they made the old table move.

*May 19* Clear and cool. Samuel Caliway staid with us last night. He is selling books for the American Tract Society. Went to Hamptonville—received the *Age*, No. 36, & a letter from brother Wiley. The *Age* is rich and brother's letter is good; its one thing, though, to preach and another to practice.

*May 20* Clear and pleasant. I and sister Eliza went to Flatrock Church and heard the Rev. Zachariah Adams preach quite a good sermon from this text: "though I speak with the tongues of men and of angels," etc. I Cor. XIII. 1st verse.

*May 21* Clear & warm. Went to Flat rock to meeting—walked there and back with my little Mary A. B. Took dinner at her pa's and staid till night. I and Miss Mary set out in the shade of a little oak that stands in the yard, and had quite a pleasant conversation. We talked of news papers, flowers, honey bees, and—I shant say what, but will save that for the reader of these lines to guess.

*May 22* John Messick came & worked in the shop, and I worked with him. We done about $1.75 worth of work. Had a light shower this afternoon. The wheat look well, while the oats are quite backward for this season.

*May 23* Rain this afternoon in abundance.

*May 24* Rained last night, I think all night, & quite fast, and is still raining.

*May 25* No important event, that I am aware of.

*May 26* Clear and very warm. I and Alexander Johnson cut fifteen saw-logs, working about half our time, for it was too hot to chop steady. Saw the eclipse of the sun, which commenced a little before 4 o'clock, and went off about 6, in the afternoon, lasting about 2 hours, or over. There were 9/12 of the north side of the sun eclipsed, therfore, he shown but little from 1/2 after 4 till 1/2 after 5 o'clock.

*May 27* Clear, and warm—planted my mellon seed. I have just read the II letter of the Brittish Spy; its a letter of queer ideas. It speaks of stumps of trees, bearing the impression of the axe, and of bones, shells etc, being found deep in the earth.

*May 28* Sunday, and quite a nice day. Went to Aylesbury to Sunday school,—Lectured the good people, and highly insulted some of them, I suppose! Human nature likes to be flattered, but at the truth it takes offence. How strange! Went home with Miss Mary.—had quite a pleasant time, etc, etc.

*May 29* Cloudy. Went to J. Green's store & bought .75 cents worth of sundries. Green's is *some* store, and no mistake.

*May 30* I & Pa hauled four saw logs to the mill this morning; since dinner we have had quite a nice shower, and it looks like raining more. Quit raining, and we hauled five or six logs this afternoon.

*May 31* Last day of May. Wednesday. Hauled logs, nine. It clouded up about 12 o'clock, and commenced raining just before night, and rained very hard. I and Pa got quite wet, as we were in the rain for near an hour.

*June 1 Thursday.* Cloudy and quite cool. I and sister Mary were going to Jonesville to the examination to day if the weather had have been favourable, but as it is we cant go to day. I had rather be at home in bad weather. Home is a great place, even if its a poor one.

*June 2* Clear and cool,—ploughed till dinner,—went to Jonesville in the afternoon. Jonesville is quite a small village, situated in a mountainous country. I got tired of the place in a short time, though I saw lots of girls there.

*June 3* Quite cool. Left Jonesville quite early, and glad enough to get away. Such a bustle as students have in *getting off* at the close of a session soon tires me out. Came by Hamptonville; received 2 *Ages* and 2 *Journals.*

*June 4* Clear & pleasant. Went to Sunday school—then home with Miss M. A. B., a beautiful girl.

*June 5* A beautiful morning. Clear and pleasant.

*June 6* A cloud rose in the southwest, and give a nice shower. Clark came home from Jonesville this afternoon.

*June 7* A pleasant day. Sold John Johnson 6 plugs of tobacco for .60 cts.

*June 8* Clear & windy,—ploughed like "all the world, and the rest of mankind."

*June 9* Clear, and so cool this morning I wore my coat to ploughing. Went to Aylesbury to hear the Rev. John N. Gunn preach; he had taken his text before I got there. Bought me a bible of Samuel Caliway for 25 cents! a nice Bible too.

*June 10* A little cloudy and not so cool as yesterday. I done a hard half day's ploughing. Received a rich No. of the *Age*.

*June 11 Sunday.* "This is the day the Lord hath made," etc. Cloudy and rains some in the forenoon; Went to Aylesbury to class-meeting; there were but few out. My little girl was not there so I came home after meeting.

*June 12* Cut wheat to day. May wheat ripe the 12 of June! Wheat is very good, though it has some smut in it.—the May wheat, I see none in the Blue Straw. I dont think I ever saw better wheat than we have this year, or at least better looking wheat. This is a great wheat country or would be if farmers would take the necessary pains, sow it early etc.

*June 13* A warm day. I & Clark hauled plank from the saw mill to build a barn floor. Cut wheat in the afternoon. The May wheat is royal.

*June 14* Got John E. Grant to shoe Jenny L. this morning. Cut & bound wheat this afternoon,—broke a finger out of Wesly Pardue's wheat cutter,—then tried Alexander Johnson's and broke that. I am not much to cut wheat, but a pretty good hand to break cradles.
A nice shower of rain fell this afternoon.

*June 15* I & Clark hauled from the mill, and packed away in the shop my timber-plank etc. We cut & bound wheat this afternoon. This is a cool summer, quite cool. The wheat is good; oats look promising; corn is small.

*June 16* This is a cool morning. The sky is clear, except around in the northwest, where there are a few thin white clouds. No rain to day. I & sister Mary went to Couche's store. Mary bought her a dress. I bought one dollar's worth, and we left for home. Ploughed some this morning, & bound wheat this evening.

*June 17* A little cloudy, and warmer than usual. A trip to the mountains. Elzy Messick, Pa, and I started for Howel Barker's to day about 9 o'clock A. M., and got there about 2 p.m. We traveled over some rough road, and saw several rusty cabbins whose inmates looked as though they knew little of the cleansing effects of pure cold water. (I can't write now.) How much better it would be for people to live in villages, and enjoy the advantages of society, than to live here and there among the mountains, in little smokey huts, where they never [see] any one, and are very seldom seen. They know nothing and care less.

*June 18 Sunday.* Rose this morning tolerbly early, as I reposed last night in a "sky parlor" with windows thrown open to admit the free passage of "the mountain breeze," I slept soundly. I walked out this bright morning feeling very much refreshed. H. B. lives on the waters of Rocky Creek. His house stands on a considerable hill and is hedged in by a range of mountains. Mr. B. has, no doubt, a healthy place, but a very ill convenient, broken one. Let me live in a level country.

Pa preached to a small, ill[-]behaved congregation, which assembled at the School house near Mr. B. After dinner we took our leave of the kind Mr. and Mrs. B. & family, and made off, in full speed for our own poor, though quite comfortable, homes. "There's no place like home." At home I found the *Journal,* the *Age,* and a long & interesting letter from brother W. & J. H. White; also "The Hive and Honey Bee," a pamphlet of 72 pages, by H. D. Richardson. I have read all of Mrs. Nichol's life contained in the *Journal.* Its interesting indeed. Mrs. N. is one of the greatest writers of the age. The *Age* is rich as usual; "the Hive and Honey B.["] I guess, is well worth .25 cents.

*June 19* A pleasant day. No great event.

*June 20* The warmest day, I think, we have had this season. I and Clark took a ride. We went to Hamptonville and around by the School house near Wiley Felts'.

*June 21* I dreamed last night of finding beautiful sound apples in the sand, under a little tree which grew upon the bank of a very swift-running, muddy stream, and in picking up the apples I let one fall in the water, which carried it swiftly away. My little brother Caleb was with me. We ran down the creek some distance in order to save our apple, but it was gone. Caleb came near falling down the steep steep bank of the creek into the water, in looking after the lost apple. I dreamed t[w]o other dreams, one that Alexander Johnson came over to cut wheat with a mowing sythe; and the other, that I went to see my fair Mary A. B. and found her in not a very good humor. What do such dreams mean?

Bound wheat a little of the hardest.

*June 22* Quite a warm day. Jim Whitlock worked on my wagon half the day. He is to come again to work when I let him know. I guess that won't be soon.

Feel quite unwell to day. Took a cold bath this afternoon, which, I think, helped me some.

*June 23 & 24* Nothing of importance.

*June 25 Sunday.* Went to Aylesbury at 9 o'clock to class meeting,—then home with Miss M. A. B and enjoyed myself quite well.

*June 26* I & Clark started west this afternoon.

*June 27* Staid in Jonesville last night, at Van Eaton's. We rolled out tolerbly early and reached Frank Bryant's, in Ash, a little before night. We crossed the Blue Ridge at Elk Spir[e] Gap, near Trap Hill. We saw some rough country and a few people that did'nt look very smooth. We stopped at James Robert's, near the top of the mountain, & watered.

*June 28* I and Clark staid last night at F. Bryant's. We had been there but a short time before we had to walk with the old man to see his great spring. It is a spring indeed! Its bold & cold. Old Frank told us many strange things—scenes that took place 40, 50, & 60 years ago. He said he had lived at that spring 60 years and would like to live there 100 years yet! He said he was 84 years old the 25 inst. The man who reaches the age of 84 is considered *very* old now-a-days, but in the days of yore, when men lived 400, 500, 600 & even 900 years, the man of 84, would have been looked upon as a youth, & probably as not being far enough advanced in years to do business for himself. What a change![6] We went to Rev. Morgan Bryant's, 2 1/2 miles from old Frank's and staid there till after dinner. Morgan lives on a high & dry ridge, Morgan is a fine man, in one sense of the word at least. I & Clark left about one o'clock for Thompson Robert's,—distance about 18 miles which place we reached just as the red orb of day was hiding Himself from the view of the green twigs of the oak, which seemed to be leaning forward to bid him 'Good night' as He went down behind the western mountains.

*June 29* I & Clark went out to church, and to see about school. We heard the Rev. Edwards preach at a new meeting-house near G. Anderson's Esq., I forget its name. Edward preached a short & dry sermon,— then tried a few of the members for not attending church, and left. I & C went to Anderson's and ate dinner, and then on to T. Roberts' again. Anderson is a friendly man. We soon found there was no chance for a school at that place.

*June 30* Clear & hot. I & Clark left T. Roberts' and went 11 miles above the old C[ourt] H[ouse?] on T. Creck. Roberts went with us. We stoped at a cabin, or near it, went to the spring & took dinner. The good woman of the house gave us milk, and we made a hearty meal on biscuit & milk.

*July 1 Saturday.* Staid last night at Col Cawett's in the Summerfield neighborhood. I & Clark tried for a school. We spent the day in visiting the neighbors, and soon found that they cared but little about education. We took dinner at a farmer's house who milk 19 cows, and makes 3 cheese a week. Mr. Eli Perkins, living a short distance from the road near McClain's store which is immediately on the road.

*July 2* Staid last night at Col C's again. On my return home I came by one Mr. Trenda's, near Grayson C. H., and made an effort for a school but did not succeed. I came on home by way of Fisher's Gap. I drank from the head spring of Fish river, or rather from the branch as it poured over the ground rail of a fence about 8 or 10 steps from the spring, and about 150 or 200 yards from the top of the Ridge. Fisher's Gap is rough, but well watered. I found the road very rough till I reached the state road near Gunn's factory.

6. See, however, the entry for October 9, in which Strong advances another theory to explain the exceedingly long life spans of biblical figures.

I am 3 Or 4 days ahead. On Sunday the 2d of July I and Clark traveled down New River. The road runs close along the water on the bank of the river for some 12 or 15 miles, and is very rough. At several places there is just room for a wagon. On the right a precipice reaching quite to the water's edge, which is many feet below the road, and on the left a mountain reaching almost to the blue sky. There are but few settlers on the road. A few small huts stuck on the side of the mountain, and very much resembling a swallow's nest, are all the dwellings to be seen for many miles down N[ew] river. The Grayson springs are a few feet from the river, near the Wythe [County, Virginia,][7] line. I and Clark passed them about 3 'clock P.M. We stoped for a few minutes and drank of the water which smells & tastes like a hard boiled egg. I drank 7 glasses & felt no weight at all from it. The water is good, tho' not very cold, and possesses great medical properties. We learned that board could be had at the springs, at the *very moderate price* of $20 per month.

[There are no more entries until August 7.]

*August 7 Monday.* This day I commenced school in Dis. No. ___. I board at home and walk about 3 miles to the school-house. I had to day 18 scholars.

*August 8* To day I had 25 scholars.

*August 9* And to day about 30—I forget exactly the No. I have a nice start, and if I have no bad luck, I'll have a first rate school.

*August 10* Very warm. Had 34 scholars to day.

*August 11* Went to Aylesbury (I and 8 or 10 of my scholars) to meeting, and heard the Rev. John M. Gunn preach from this text;—"Render therefore unto Caesar," etc. XXII. 21 Matt. The Gunn shot us all. Not a member of the congregation but what got his "portion in due season."

*August 12* Warm! hot!! Looks a little like rain—and-so-forth.

*August 13 Sunday.* Cloudy and rained some. Went to Terry Burton's to meeting. Pa and Elzy Messick preached old Mr. Willard's funeral at the spring under the branches of a large white oak, to a small congregation

7. Strong and his brother have traveled northwest into Alleghany County, North Carolina.

seated around said oak on very low and wet seats, from this text.—"Set thine house in order, for thou shalt die and not live."

*August 14, 15, 16, 17* had a full school, there being from 30 to 56 scholars each day.

*August 18* Cloudy, and rained slowly most of the day. Only had 45 shavers to day. This the last day of school this week.

*August 19* Rained, I think, all night, and is still at [it],

"The rain comes down with all its might."

Wrote brother Wiley a letter, but fear I shall not get it to the office to day. Alexander hauled over a load of his plunder this afternoon.

*August 20* Went to Flatrock to preaching. Heard the Rev. Green Brown preach quite a good sermon from this text,—"Acquaint now thyself with him, and be at peace; thereby good shall come unto thee." Job XXII 21. Brown is a small or medium sized man, has light hair, a keen eye, high forehead, a swarthy complection, and is *minus* front teeth in the upper jaw; but notwithstanding all these disadvantages Brown can, no doubt, like others, preach better than he can practice. Practice is a hard *rule* to understand.

*August 21* Staid last night at Miss' ______ house or her Pa's. Had 40 odd scholars to day. I have a peaceable school.

*August 22* Clear and quite warm,—had about 54 or 55 scholars to day.

*August 23, 24, 25* Very warm. Had a full school.

*August 26* Very warm, the Thermometer standing at 94.
Went to Hamptonville—received a letter from brother Wiley with a catalogue of Oak Ridge Institute, also the *Age* N. 51, and 2 Nos. of the *Georgia Blister & Critic* published in Atlanta, Ga.

*August 27* I and sister Mary went to the Temple Hill Campmeeting. Heard Rev. W. L. Van Eaton A. M. preach at 11 o 'clock from this text:— "And if the righteous scarcely be saved, where shall the ungodly and the sinner appear?" I Peter IV 18. The Rev. didnt preach much. Rev. Thomas Nicholson preached at 3 o'clock, and his text was "Come for all things are now ready." Luke XIV and the latter part of the 17 verse. I left before the sermon was ended, and came on to Esquire T. Roberts' and stoped there during a heavy rain. It lightened, and thundered loudly. The glass in the windows rattled, and the house tremble.

*August 28* Clear, and not so hot as it was last week. Had to day 57 scholars. Had to give two little boys a stripe each for playing "hard knuckle." I fear I shall have to use the gum freely in order to maintain good order.

*August 29* Cloudy, but no rain. Had sixty odd scholars to day. What can one man do with 60 unclassable scholars?

*August 30* A pleasant day. Had sixty scholars to day.

*August 31* School quite full. The day pleasant.

*September 1* Quite a warm day. School not quite so large—only 40 some odd out to day. One of the Com[mittee] men—G. D. Holcomb—came out to see me to day for the first time during the school, and I fear it will be his last visit. *He's so much interested.*

*Sept. 2* A very warm day. Went to Doweltown, Yadkinville, James Sheek's, Hamptonville, etc. Received No. 52 of Vol V of the *Age;* bought, at Willson's & Hauser's store in Doweltown, a linen coat for $1.37 1/2. Received of James Sheek, Chairman of Superintendents, 18 dollars.

*Sept. 3* Sunday, a very warm day. Staid at home till evening and then went to see Miss M. A. B. This going to see the girls don't pay very well, so I intend to quit it.

*Sept. 4* Still warm. Had only 23 scholars to day. I gave Pa 10 dollars in paper, for 10 in silver, all in half dollar pieces, and rolled them up for future use.

*Sept. 5* Pa's birthday. 50 years have passed away since Pa commenced his earthly pilgrimage. 50 years, to look into the future, seems like a long time, but to look back, I guess, it seems

. . . "as yesterday,
Nor scarce so long ago."

Pa is more than twice as old as I am, but if we should live ten years yet Pa will then lack ten years of being twice as old as I, for he will be 60, & I shall be 35.

I only had 20 scholars this hot day. Came by Aylesbury night meeting. Old Uncle Elzy Messick done the preaching. The house was crowded. I did not go home with Miss M.!!!

*Sept. 6* Hot! School small. Went to Billy Pardue's & got a real *mess* of watermellons. Stoped school for 2 weeks on account of fodder pulling. No more news of very great importance? No sir.

*Sept. 7* Worked on the meeting ground at Aylesbury. We built a stand & bush arbor in the yard, and set things "in order" for the approaching 3 day's meeting. Rain in the p.m.

*Sept. 8* Circuit preaching day, and also the day on which the 3 day's meeting commenced. Went out and heard Uncle Gunn preach quite a plane sermon. His text was the 23, 24, 25, 26 verses of the II Chapter of II "Timothy."—"But foolish and unlearned questions avoid," etc. Richard Green was tried & expelled from the church. Green was charged with telling one falsehood and sanctioning two others that his father, Jim Green, told. William Wetherman, C. A. Bell and John Nicholson, (the old man), sat on the case.
A fine shower of rain fell about 2 o'clock p.m.

*Sept. 9* A very rainy day. Its now about 9 o'clock at night & very dark.

"And the rain pours down with all its might,
Darkening the face of the sullen night."

I pity the way-worn pilgrim, to night, especially, if he's out in this terrible "North Easter."

To day I am 25 years old. 25 years have I sojourned on earth, but where shall I be 25 years hence? I shall, in all probability, "go the way of all the earth," long before 25 years passes away, and "be rewarded according to the deeds done in the body." Direct my steps, O God, in the path of holiness, for Thou hast said, "without holiness no man shall see the Lord," and help me to take heed to my ways, and sin not.

"Help me to watch and pray,
And on thyself rely."

*Sept. 10* Clear & tolerbly pleasant day. Heard Uncle Gunn preach a funeral sermon at A[y]lesbury from this text,—"But now he is dead, wherefore should I fast?" etc. I Samuel XII, 23. There were a great many people out. After an intermission of about 1/2 an hour T. Nicholson preached; his text was.—"To him gave the prophets witness." etc. X, 43 of the Acts of the Apostles. At night Uncle Gunn again preached; his text was.—"What meanest thou, O sleeper?" etc Jonah I, part of the 6 verse.

*Sept. 11* Meeting still continues. Uncle Gunn preached at 11 o'clock. The first Psalm as his text. T. Nicholson preached at night from this text,—"For the great day of his wrath is come, and who shall be able to stand?" Rev. VI, 17.

Jerry Johnson professed to get religion and joined the church; several others joined also.

*Sept. 12* Warm, rain at night. Wrote most of the day.

*Sept. 13* Still warm. I pulled a few blades of fodder, etc. etc.

*Sept. 14* It rained, I believe.

*Sept. 15* Nothing of very great importance. Warm.

*Sept. 16* Cloudy & quite cool. I went to Yadkinville, was examined—and obtained a new certifficate which will show for it's-self. Got home about 8 o'clock at night.

*Sept. 17 Sunday.* Went to Flatrock to meeting. Pa preached. His text was VI, 17, of Rev. "For the great day of his wrath" etc. There was a large congregation in attendence. Cool.

*Sept. 18* Went again to Flatrock, to meeting. Rev. Z Adams preached from the 4 verse of the 27 Psalm. Brother Wiley came to see us, from Guilford Co. were he is teaching school.

*Sept. 19* Went again to the Flatrock meeting. Rev. Adams' text, to day was,—"How shall we escape if we neglect so great salvation?" Adams spoils his preaching with the old Baptist tune, or ham rather. Brother W. left for his school this morning, and Clark went off with him.

*Sept. 20* Cloudy but no rain. I read to day, some beautiful pieces in the *Age* from the pen of W. M. Johnson, Esq.

*Sept. 21* Cloudy. My school was to have commenced to day after a vacation of 2 weeks, but not a single scholar came out! I staid at the school house till after dinner, spending a good part of the time in killing fleas. I gave several of the little *jumpers* their last lesson, and sent them to their long home.

*Sept. 22* No scholars to day either. Great School! Parents very much interested in the education of their *sprouts.*

*Sept. 23* Quite cool this morning. I and sister Mary went to Doweltown, and bought her necessary dry goods for a term at school. I also bought a looking-glass, the N. C. edition of Mitchell's Geography, etc. with 1 1/2 oz. of spunge *on purpose* to use in bathing. I have great faith in cold water.

*Sept. 24 Sunday.* A pleasant day. Went to Aylesbury to "meeting." Pa preached; his text was, "And the Spirit and the bride say, Come." etc. Rev. XXII. 17. After preaching we had quite a nice class-meeting. After the "benediction" I went home with Miss M. A. B., and spent the evening quite pleasantly. We went to night meeting, after which I saw her safe home.

*Sept. 25* School small—only five scolars.

*Sept. 26* Weather still pleasant, with some appearance of rain. Pa commenced sowing wheat last week. Seven scholars to day.

*Sept. 27* School small. Rained a beautiful shower.

*Sept. 28* Clear, and quite warm. School small—only 7 scholars. I am reading Phisiology, and find it to be a very interesting study. Sponge, the connecting link between the animal and vegatable kingdom, is queer stuff.

*Sept. 29* School small. Weather pleasant.

*Sept. 30* Saturday—last day of Sept. Went to Hamptonville. Received the *Journal, Age,* and a letter from the West and one from the East from brother W., with the *Message.*[8]

*October 1 Sunday.* Quite pleasant. Read and wrote.

*Oct. 2* Cloudy, warm with some rain. No. of scholars increased.

8. Thomasson probably referred to the *Greensboro Weekly Message,* a paper that emphasized religious news and was opposed to alcohol.

*Oct. 3* Pleasant, but quite windy. School small. Meeting at Aylesbury. Went home with Miss M. A. B., and came near getting caught in a heavy rain, stoped at Jo. Johnson's till the shower was over,—got home a little before midnight.

*Oct. 4* Quite cool, and windy. Had 19 scholars to day. Paid James Whitlock $1.00, and still owe him 75 cents.

The pegeons [pigeons] went over this morning in large groups.

*Oct. 5* Died to day my old certifficate. Clear and cool. Saw some frost this morning. School small.

[No entry for October 6]

*Oct. 7* Clear and pleasant. Esquire Roby was to have addressed the good people of this vicinity on the subject of Temperance, to day, at Aylesbury, but the crowd was so small, there being only 10 or 12, that the Esq. concluded to "hold on". Probably he thought it would not be proffitable speculation to "fire off" his heavy loaded musket at so small flock of the feathered tribe. I went on to Hamptonville and rec'd the *Age, Message* and a letter from brother Wiley, and a letter from my old friend, Thomas M. Hunter.

*Oct. 8* Clear and warm. Took Miss M. A. B. to Senter to meeting. We had quite a pleasant ride, it being about eight miles to Senter. Hear[d] the Rev. T. Nicholson preach; his text was;—"For ye know the grace of our Lord Jesus Christ, etc." II Cor. VIII, 9. The Rev. spoke first of Christianity, secondly, of the riches of the Saviour, thirdly of his humble appearance into the world; and fourthly, the object of his mission. The sermon was short and tolerbly good.

I tried hard to get Miss B. to read for me, but not a sentence could induce her to read. Such a *gall!*!!!!

*Oct. 9* Ma's birthday. 47 years of Ma's "earth life" have passed by. Counting 3 months a year, as some writers suppose they were counted previous to the days of Abraham, Ma is 188 years old! Had only 8 scholars to day.

*Oct. 10* Not exactly clear, but quite warm. The pigeons went south this morning in large goings. An army of pigeons marching on in double-file looks beautiful. The pigeons, I suppose, have come over to make war on, and devour the acorn mast. School small.

*Oct. 11* Clear and warm. School small. Came by Hamptonville,—received another No. of the *Age*. I dont so well understand that.[9]

*Oct. 12* The day for electing committee-men in each District for the ensuing year, but we had no election in District No. 31. Very little attention paid to education.

*Oct. 13* Quite pleasant; only had four scholars! Dock came home from Guilford, but he has no important news.

*Oct. 14* Rained some in the forenoon, carried a part of my wagon to Jim Whitlock's shop. Went to Hamptonville,—received the *Age* only. No. 6 is quite good. The *Age* is the paper.

*Oct. 15* Clear and cool. Went to Shiloh to meeting! Went in the afternoon to see Miss M. A. B. who appeared quite glad to see me.

*Oct. 16* We had quite a frost this morning, followed by a cool day. School increased some.

*Oct. 17* Clear and cool. School small.

*Oct. 18* We had, this morning, a killing frost. The day was clear and pleasant.

*Oct. 19 & 20* No important news. School small.

*Oct. 21 Saturday.* Clear & pleasant. Took a buggy drive with Miss Nancy Wilkason to her ma's and back. Its about 12 miles to Mrs. Wilkason's, so we drove about 25 miles. Miss Nancy is one of the *galls.*

Bought, of A. C. Cowles, Hamptonville, a set of waggon boxes for which I paid $1.25.

*Oct. 22 Sunday.* Quite a pleaant day. Went to Mt. Maria to meeting. Heard the Rev. W. L. Van Eaton preach to a large congregation from this text; "For a day in thy courts is better than a thousand . . ." etc. 84 Psalm, 10 & 11 verses. After an intermission of a few minutes old Uncle

9. Perfect regularity in deliveries of mail still does not exist, 140 years later.

James Patterson, an aged minister, preached. His text was, "Turn you to the strong hold, ye prisoners of hope." etc. Zech. 9 ch. 10 verse.

*Oct. 23, 24, 25* Clear and pleasant. School small.

*Oct. 26* A pleasant day.

*Oct. 27* Cloudy, with some rain. This was General Muster day.

*Oct. 28* Still cloudy but not much rain. Took Miss Nancy Wilkason home and had a pleasant trip, too. Miss Nannie lives about 3 miles south of Yadkinville.

*Oct. 29* Sunday—still cloudy and damp. I found myself this morning at the widow Wilkason's about 14 miles from home. I chatted some with Miss Nan and then left for home about 9 o'clock. As I travelled on homeward I read "The Empress of Rusa" and "The Umbrella Girl", two quite interesting pieces in the *Age*.[10] I stoped at Flat rock and heard thc Rev. Green Brown preach quite a good sermon from this text—"The Lord hath done great things for us whereof we are glad." After preaching I came on to Mr. C. A. Bell's where I stoped and spent the evening quite pleasantly with Miss Mary A. I landed home, a little after the bright orb of day had hid himself behind the western hills, glad enough to rest.

*Oct. 30* Still cloudy, had ten scholars to day, and rain in the afternoon plenty. Sent Miss M. A. B. two No's. of the *Age*.

*Oct. 31* School increased some. Cleared off warm.

*November 1* Staid last night at William Pardue's, Jr. at which place I expect to board while my school lasts. The Pardues are a kind but somewhat singular people. The old man is short in stature, grey headed, and quite a fine old man in his way.

*Nov. 2* A pleasant day. School not large.

*Nov. 3* A little frosty early in the day, but the sun with his warm rays soon drives away the frost and warms the air till its almost hot. We have had, this week, beautiful weather.

10. "The Empress of Russia" and "The Umbrella Girl" both appeared in the issue for October 25, 1854, which was no. 8 of vol. 6.

*Nov. 4* Quite a rainy day,—finished sowing wheat—went to Town—received the *Age*, a letter for Dock, the *Blister* & the *Dime.*

*Nov. 5 Sunday.* Went to Flat Rock to hear Pa preach the funeral of old Mr. Nicks, a revolutionary soldier, but he did not get out. Came home with Miss M. A. B. and spent the evening pleasantly.

*Nov. 6* A great frost yesterday morning, and another this morning. School distressingly small. Why will not people send their children to school? A man that won't send his sons & daughters to school ought not to have any.

*Nov. 7* My school was small yesterday, but to day its smaller. Received a letter from brother Wiley with a pamphlet entittled "Love, Marriage, and Divorce." Guess its interesting.

*Nov. 8, 9, 10* Pleasant weather. School quite small.

*Nov. 11* Rained lots yesternight. Went to Hamptonville—received the *Age*. Its good as usual.

*Nov. 12* Cloudy, and quite rainy, but not cold. Read the closing chapters of "Friendship's Triumphs," in the *Age*, which is quite an interesting story.

*Nov. 13* The coldest day we have had this season. School increased some, but small yet.

*Nov. 14* A cold morning. Rec'd a letter from sister Mary in which she writes she is not well.

*Nov. 15, 16, 17* Cold, Nothing worth taking a not[e] of.

*Nov. 18* Cloudy. Took the cotton to Douthart's factory & bartered 52 lbs. of the raw material for six bunches and a piece of yarn, giving 9 lbs. for a bunch. Came thro' Hamptonville and received the *Age*, a letter from brother Wiley, also one from Uncle William Hauser, M. D.—both interesting letters.

*Nov. 19* Sunday; a little cloudy and quite cool. Took a ride over to see Miss Mary A. B. Cool time I had too!

*Nov. 20, 21, 22* Nothing of importance. School small.

*Nov. 23 Thursday.* Sale of the lots to day at Union Academy. Nine lots sold for $303.50. I and Wiley Messick, Jr. bought No. 4 for $32.50, the nicest lot sold.

*Nov. 24* A little rainy and quite windy. Only had 5 scholars. Sister Mary came home from Jonesville school.

*Nov. 25* Took the ballance of my wagon to Jim Whitlock's shop. Came by Benbow's oil mill and brought home a load of cakes—160. Went to Hamp[tonville] in the afternoon—recd. a letter from brother Wiley, a No. of the *Age.* Tolerbly cool.

*Nov. 26* A cold day. Heard Mr. Van Eaton preach James Martin's funeral at Aylesbury. His text was, "And as it is appointed unto men once to die, but after this the judgement." Heb 9 ch. 27 vs. The house was pretty well filled; one seat being so loaded with fat *galls* that it's legs broke, and down the fair ones went. I, as usual after the service, went home with Miss M. A. B.

*Nov. 27, 28, 29, 30* Cold and dry. School small.

*Dec. 1* Cold and cloudy in the forenoon, and I think it hailed a little. Pleasant in the afternoon and almost clear.

*Dec. 2* Cool. Went to Hamptonville—received the *Age,* No. 13 of Vol. VI, containing the governor's message etc; the *Blister & Critic,* No. 9 of Vol I. The *Blister* is quite neat in appearance, but filled with matter hardly worth reading—a few exceptions; the *Journal* No. 2 Vol I. The *Journal* is now published in neat pamphlet form monthly, of 64 pages. "All things human are imperfect,"—so the *Blister* & *Journal* are not as good as I think they might be, though there are a few good things in each. I also recd. a letter from brother William C., written from Shelby, Cleveland Co. at which place bro. W. is a[t] present employed.

*Dec. 3* Sunday, and a cold windy day. Staid at home all day! Mr. John D. & Miss Elizabeth Johnson came down and spent the afternoon with I & Mary, sis. John D. is a fine boy—destined to make a fine man.

*Dec. 4, 5, 6, 7* Very cold and windy. Clear & dry. School small; altogether the smallest one I ever taught.

*Dec. 8* Not quite so cold. Went to Aylesbury to hear the new preacher, Mr. Shelton, but he no come.

*Dec. 9* School closed. I taught 4 months in Dis. No. 31 for $72. Came by Hamptonville—recd. The *Age,* 2 letters, one from bro. Wiley, and the other from B. H. Day of New York—an advertisement of books. *Brother Johnathan* etc. I also received *Old Rip's Pop Gun*—a small paper published in Shelby, Cleveland Co. N. C.[11]

*Dec. 10 Sunday.* Cloudy but not very cold. Staid at home all day! and read, wrote, etc.

*Dec. 11* Rained some last night. Windy & tolerbly cool. Swaped my watch to Charles Benbow for a double barrel shot-gun, giving him about one dollar's worth of wagon felloe timber. This is the first gun I ever owned, except a pop, or squirt-gun, made of a reed, or an elder. I used to see more real fun shooting with my pop-gun at imaginary squirels than I see now with a real gun. Childhood is a pleasant time, but yet

"A day to childhood seems a year,
And years like passing ages."

*Dec. 12* Clear & cold. Went after my school money—$54. and on to Doweltown, where I bought a waistcoat for $1.60; also some indego [indigo] & copperos [copperas] for Ma. Alexander went with me to Sheek's & to D. Town.

*Dec. 13* A pleasant day after a cold morning. I and Caleb took a little hunt for rabbits etc. We started two and I shot at them as they run, but did them no particular harm that I know of, as they did not stop. Went to J. Couche's store and took my note, which was given last Apr. It called for $5.76, and the interest for 7 months & a few days—ten days—was .29 cents, making $6.05, but Couch only charged me $6.00. This paying interest is bad business.

*Dec. 14* A pleasant day. I, Pa and Alexander killed two hogs; one of them weighed 234 lbs., and the other 216, making 450 lbs. of pork. I

11. Apparently Thomasson received the inaugural issue of *Old Rip's Pop Gun.* The North Carolina Collection at the University of North Carolina at Chapel Hill has the issue dated December 23, 1854, which was vol. 1, no. 3, and the card catalog indicates that internal evidence in this issue points to December 9, 1854, as the date of the first number of the paper.

ing, as all preachers ought to do when they enter the sacred desk, the people crowded in, and when I took my seat, and cast my eyes round, I saw that a large congregation had assembled. As it was my first round on the circuit, I did not preach, but rather lectured on the utility of class meetings. To day I bound the "Songs for the People," and one vol. of the *Youth's Cabinet.*[1]

*Jan. 6* A cloudy, drizzly day, and old Christmas. Went to the office—received an enlarged No. of the *Blister,* tho' it's no better now, that I see, than it was when only half so large; and if the addition of a few more pages is the only improvement attached, it is only worth half as much now as it *use to was,* for it requires double the time to read over its big-worded-nonsense, that was sufficient to convince any one that it in its small size, contained nothing worth the attention of a—J[ack] A[ss]. *Old Rip's Pop Gun* blows off some pretty hard wads, which indicate that "Old Rip" is no "bug eater" so far as masticating tone is concerned. I read aloud, to night, some funny yarns in the *Age,* "The Judge's Big Shirt," dreaming over a bit of wed. cake too.

*Jan. 7 Sunday* Cloudy and rainy most of the day. Went to Hile Johnson's in the evening, and took John D. his news paper—the *Age*—and saw Miss E. John D. is a fine boy, and I predict that he will some day, be an honor to the Johnson family. He is now teaching school in Dis. No. 36, if I am not mistaken as to No., and getting $15 per month. That will do tolerbly well for this country, down east Teachers get from $20 to $25 for teaching Common School, and if a teacher does his duty he earns his money etc.

*Jan. 8* A cold, stormy day—a real northeaster. Staid in doors most of the day, and made out most of my last year's account. If my debts were paid I should not be worth more than 10.000. I intend, in the future, to spend less money in proportion to my income. This way of spending all one makes, and a little more, is a bad way. Economy has made many a man rich, and I intend trying it this year, and so I do.

1. Apparently Strong Thomasson bound together the separate issues of his newspapers with needle and heavy thread, just as he did the pages of his diary. The editor is uncertain whether "Songs for the People" is a publication or the hymns that Thomasson himself had written down.

*Jan. 9* Elisha Messick came over to swap horses with me to day, but we did not trade, I and Caleb cut and hauled wood. Cloudy, and rained a few drops. Read, in No. 3 of the *Journal,* "The Claims of Hydropathy," which is a truthful piece.

*Jan. 10* Cloudy, and rained some. Got up tolerbly early, and went to make on a fire, when a large stone, out of the back of the chimney, fell on my arm, just as I put it forth to draw forward a stick of wood that was lying against the back. The stone caught my arm on a stick of wood, and hurt it badly. Came near breaking it. I have not come as near fainting in many a day. Ma washed my arm in vinegar, and put on it a poultice of bruised cumfrey roots. I done no work to day, for my wounded arm pained me considerably all day. Wesley Pardue cut saw-logs for me all day, for which he charged me 37 1/2 cents. A. C. Johnson cut from dinner till night, and charged me 25 cents, as he boarded himself. I took my paint, and one gallon of flax seed oil, for which I paid $1 over to Whitlock's shop to paint my wagon. I hauled away a load of pine that some one, Wat Denny I'm told, cut on Pa's land.

I hope Mr. D. will take the hint and get pine on his own land in the future. Good pine is valuable property in this country, and those who wish to steal it, had better go some dark night, when they will not be so likely to be seen.

*Jan. 11* Still quite cloudy, and rained "a spoon full or two." My arm is a good deal better, tho' its swelled some yet, and is a little sore. Went out this afternoon, some half an hour before night, and shot 8 times at 2 squirrels, and only got one of them. I wasted more amunition than the worth of one squirrel. I do not understand shooting my double barrel gun so well yet. It requires practice to make perfect, even in shooting a shot gun.

*Jan. 12* Clear, and quite warm for the dead of winter. I think that I have seen corn planted on colder days that this has been. I, Caleb and Alexander took a short round after the squirrls; we hunted two or three hours, and killed three squirrels. The first one we found sprang on the ground, and ran a hundred yards or two, and sliped into a hole near the root of a little post oak. We, having an ax along, soon made a large aperture in the tree, and Mr Squirel ran out and took reffuge up among the branches of a small chestnut tree that stood near by. Squirelly now afforded a fair target, and I could not resist the temtation to take a pop at him. I took rest against a by-standing tree, and cut away, but few of the shot took affect, so I had the pleasure of another pop, and of then seeing him "come tumbling down." The next Alexander found on a large black oak, and took a shot at it before I got to the tree, as I had fired off "old betty" at a woodcock and had staid behind to reload. I came up tho' before Alexander was ready for another fire, so I leveled a way and down came the squirrel. The next Alex found on a large chestnut tree, and killed it the first pop. We now had three squirels, which we thought would do pretty well for so short a hunt; so we struck for home in order to get there in time to fix and go to church, to hear the new preacher. Soon Alex., Caleb, sisters Martha & Mary were on the road to Aylesbury. Soon after getting there we saw the preacher, Mr. Shelton, coming. He hitched his nag, and took his baggage, I think, into the house, then walked off to the spring in company with old Mr. T. Messick. In a short time they came back, stood and set in the yard for a short time, and then went into the house. We all followed the preacher and Mr. Messick's example, so far as entering the church was concerned; so we were all seated in the house in short order and ready to hear the new preacher. Mr. Shelton is a man, I would guess from his appearance, about 22 or 23 years old, has a fashionable *set* of whiskers, short, thick nose, tolerbly keen looking eyes, and a moderately high forhead, light hair, etc. In due time the preacher took his text,—"Wherefore, seeing we also are compassed about with so great a cloud of witnesses" etc. Heb. XII 1, and preached a short and tolerbly good sermon.

*Jan. 13* A little cloudy and windy. Took another squirel hunt with Caleb and Alexander. We found only two squirels and had to shoot at them a great many times before we got them. Went to the school house. A part of the Association met but did nothing more than appoint another meeting. Went to Hamptonville—received a letter from brothers W. W. & Wm C., the *Age* and some other trash.

*Jan. 14* A beautiful Sunday. Went over in the afternoon to see Miss M. A. B. Found her at home, wearing a red calico dress and looking sweet as usual.

*Jan. 15* Came home this morning about 10 o'clock. I paid Miss ——— a pretty lengthy visit, as I had not been over in 4 or 5 weeks. Took my boots to the shoemakers to get them mended. Making shoes is the last trade I should want, I think.

*Jan. 16* A pleasant day. I and Caleb cut and split wood. My poor goat got her leg broke, but I dont know how it was done.

*Jan. 17* A little hazy, but quite pleasant; fine weather for January. I hauled wood till 12 o'clock, and in the evening I & Caleb went to Couche's and Pa went with us as far as John Johnson's. John has been ill for several days.

*Jan. 18* Pleasant, and a little windy. Finished cutting saw logs for my office to be built at Union Academy.

*Jan. 19* Beautiful weather. Hung up the bacon to dry, cut a little wood, etc.

*Jan. 20* A little hazy and quite warm. Dont know that I ever saw as nice weather for business at this season of the year. Went to the office and got my paper—the *Age*—and Miss Mary A. Bell's—the *Dime.* The *Dime* is a neat little sheet, and this No. 5 contains an excellent story of a young man by the name of Paul Harrison.

*Jan. 21* A rainy Sunday. Read most of the day.

> "Of all the *days* that crown the *week,*
> Give *Sunday* unto me,"

for then I'm freed from the labor of the week, and have time, either to go to church or read good books.

*Jan. 22* Clear, and very cold. It rained, last night, a hard shower, and cleared off, and to day there has been a brisk cold wind blowing from the North all day. Done little more than set by the fire.

*Jan. 23* Cold but not so windy. Cut & hauled wood in the afternoon.

*Jan. 24* Still cold. I & Alexander went to James Vanhoy's. When we got there, no person being at home, Alex. got the key & unlocked the shop door and we went in. I got my boots etc. and was about ready to be off, when Alex. commenced hiding Jim's tools, leather, and every thing that

*Feb. 18 Sunday.* A pleasant day. Heard a Friend or Quaker, preach. He said some very good things, but it was a long time *between draws.* Went out to prayer meeting at night, heard one Tom Poindexter try to preach but he could not come it. Lots of people out but what they went for I know not.

*Feb. 19 Monday.* Cloudy & cool. I and the Dr. took down, brushed off, repriced, and put up lots of goods. Sold a few.

*Feb. 20 Tuesday.* Clear, cold & windy. Finished re-pricing goods, etc. Went down to Alexander Benbow's and assisted him in posting books till about 8 1/2 o'clock, after which Alexander prepared us a glass each of a first rate drink. He to[ok] 4 glasses, (tumblers) filled each one nearly full of water, then a spoonful of sugar, and two of Temmon Syrup (Teaspoon) It was good. That's all.

*Feb. 21* Clear & not so cold as yesterday. Sold $10.00 worth of goods, and recd in cash $2.20, and in trade 6 1/2 doz. eggs at 5 & 6 per doz, and 1 doz. rabbit skins at .15 per doz. Bought *Hitchcock's Geology* for $1.25. Geology is a very interesting study, I presume.

*Feb. 22* A pleasant day. Sold more goods for cash to day than I did yesterday. I find that I have but little time for study, and that I dislike.

*Feb. 23* Nothing of importance took place during the day that I am aware of. At night I went out to the debate, and took a part in speaking. The question was,—"Were the early settlers of this country justified in expelling the Indians from the same." I and Mr. John H. Kenyon, with a few others spoke on the affirmative and Mr. Kreel and others on the negative. The question was decided by a vote of the house and the negative gained it by a majority of one vote.

*Feb. 24 Saturday.* A little cloudy, not very warm. Sold a few goods.

*Feb. 25 Sunday.* Cloudy and quite cold, but notwithstanding the coolness of the North wind, I walked out about 2 miles to Masadonia to hear Mr. Shelton preach. His text was "What I say unto you, I say unto all. Watch." Mark XIII, 27. The sermon was a tolerbly good one, but the congregation was so restless it did not appear to produce much feeling among the people.

*Feb. 26* A cold day. Monday.

*Feb. 27* A *very* cold morning. Sold but few good as it was too cold for the people to go to the store! Mr. Bell came down from home, in company with Charles Benbow who brought me a letter from home, and such a letter! So short!

*Feb. 28* Last day of February 1855. Not quite so cold as yesterday. Sold but few goods.

*March 1 Thursday.* [No further entry]

*March 2 Friday.* A pleasant day. March comes in like a lamb. No important event takes place to day that I am aware of more than I received a new pair of shoes to day. I paid $2.10 for a hifr [heifer] skin; and .45 cents for half of a sheep skin for lining; and $1.25 for making, sole leather, thread, etc. My shoes have cost me about $2.50. Dear shoes! unless they are extra good ones. They are well lined, and I think they are well made. I hope so at least.

*March 3 Saturday.* Here I am in the store, and tolerbly well fixed for studying. To day I put up a book shelf on which to lay my books, set a clock etc. John P. Nicholson sat with me to night till about 9 o'clock, and we chatted over some of our experiences in love *matters*. This day has been a little cloudy but not very cold. Sold a few goods.

*March 4* Sunday, and a beautiful day it is. Clear and pleasant. I, with several others from East Bend, walked out to Prospect about 3 miles from town, to hear a gentleman lecture on temperance. At 10 o'clock a small congregation assembled and held a short prayer meeting, after which one Mr. Martin, a stranger to me, gave us a short talk of about 1 1/2 hour, after which Mr. J. H. Kinyon gave us one of about 2 hours, then Mr. Aquilla Spiers spoke a few minutes and sent the pledge of the Providence Temperance Society around for signatures. Mr J. H. Kinyon, John P. Nicholson, and myself with some 8 or 10 others signed it. There were several of the *Woods* ladies out but few of them signed the pledge. How strange it is that the ladies will not go forward and engage in this great reformation with might and main, and at once put down the liquor traffic when they might so easily do it. Our females generally are raised up in stupid ignorance, hence they are not aware of the extent of their influence, and of the good they might do were they to engage heartily in the temperance reform, and to tell all around them, by their actions, that they could not look upon drunkeness, or upon the use of spiritous liquor as a beverage in *any way,* with the least degree of allowance. Oh! that the female part of our community would rouse up, bid the enemy of our beloved state defiance, and enforce a liquor law among us as the ladies (be it said [to] their honor) of the far west have done in their country. Then would their mourning be turned into joy, their broken hearts healed, their tiers be wiped away, and peace and plenty would visit their homes, which have been for so many years comfortless, and not only so but the actual abodes of retchedness and want.

*March 5* A pleasant day. Great time for business, especially farming. Oh! that I had a little farm that I might have it well tilled, and a little cot that I might have it well filled. It strikes me that I should be one of the happiest of men if I had a little farm pleasantly situated in a healthy region, and about the center of it a vine covered cottage, and in front of that a beautiful flower garden with evergreen walks, etc.

*March 6* Cloudy but not cold.
Took in 3 bushels of Irish potatoes at $1 per bushel. Farmer can now get 75 cents for corn, $1 for wheat, 40 cents for oats and $1 for potatoes. I should think the present prices of produce ought to stimulate the farmer and induce them to exert every nerve in their farming operations.

*March 7 & 8 Wednesday & Thursday,* Pleasant weather. Nothing worthy of note.

*March 9 Friday.* This day, I guess, will long be remembered as the 'windy Friday.' I went up home, and I dont think that I ever travelled thro' as much wind in one day. I left East Bend about 8 o'clock in the morning, and got home about 3 P. M. I went on thro' Yadkinville, then took the left and went by Mr. Foot's buggy shop, but just before I reached Yadkinville, my brother Wiley overtaken. I had just been wishing that he would ride up or that I could overtake him. We travelled on together to the forks of the road between Yadkinville and Doweltown, when I turned to the left to go by Mr. Foot's and he kept the straight road for home. It is about 3 1/2 miles from Doweltown to the Wilkes road on which the above named shop is situated. Just before I reached the W. road, I narrowly escaped being killed by a falling tree. I was riding leisurely along through an old field that has so long been thrown out that it is now grown up in old field pines, and all at once I heard the noise of breaking timber and on looking to my right I saw a tolerbly large dead pine falling almost exactly towards me. Had not my horse taken a scare at the sudden breaking of the tree and made off with all speed, we should probably have been beaten dow[n] by the top of the tree, for with all the haste we could make, it fell into the road just behind us. Thus narrowly escaping I kept my eyes open whenever I thought there was danger near. The best plan is to "watch" at all times and be ready for in such an hour as we think not death may throw around us his icy arms, and in a moment of time usher us into eternity.

"Help me to watch and pray,
And on thyself rely,"

that I may ever be ready for the solemn change.

I called at Mr Foot's shop and paid $50 for Alex. Benbow towards a buggy that he bought there more than 12 months ago. I thought that people who are able to ride in $100 buggies usually paid for them when purchased. I find that all who do ride in fine buggies do not pay for them as they go. There is great danger of "living too fast," and most persons that buy on credit live in advance of their means.

On the road between Foot's shop and Hamptonville I passed through 2 or 3 severe blasts of wind, I had to hold my hat with one hand and with the other hold on to the saddle to keep from being blown away. Windy time! and no mistake. As I passed on by one Mr. Streetman's about 12 o'clock, a blast of wind hoisted off my hat and carried it about 50 yards across Mr. S's yard. A blank book, made of a sheet of cap paper, was in my hat, and the last I saw of it it was going over the top of Mr. S's kitchen. Where it stoped, I know not. A little girl brought my hat, but my blank book I've never seen nor heard of since. It may be flying over kitchens yet for ought I know. A kitchen is a great place, but I should take it as no particular honor to be blown over one.

Stoped as I went on and took a peep at Miss M. A. B. etc. etc.

*March 10* No so windy and quite a pleasant day. Took up my sale note of %5.42 given at the sale of Roughton & Gentry's goods one year &

one month ago to day. Went thro' Hamptonville after my papers, etc. Saw Miss Mary as I went on home hanging out clothes. I should think that the dirty articles of apparel ought to be washed and spread out to dry earlier in the week than Saturday. But this is "free America" where all have a right to *wash* when they please.

*March 11 Sunday.* Cloudy or smokey, but not cold. Went to Aylesbury to meeting, and then home with Miss M. A. B. Had quite a pleasant time so I staid till night, and till morning too. Miss Mary is a great *gall.*

*March 12* Cloudy. Thundered considerably this morning about day. Started at 9 o'clock for East Bend, and landed about 3 or 4 p.m. Came near going to sleep riding along.

*March 13* A cold damp day. Rained last night and this morning considerably. Sold but few goods.

*March 14* Not quite so cool.
I read, in the *W. A. Organ,* a piece headed "Characteristics of the American People" written by, or spoken by O. P. Baldwin. Mr. B. says, "No two people have so little idea of each other as those of the North & South." It is quite an interesting piece throughout.

*March 15* Rained very hard this morning.

*March 16 Friday.* Took my trunk off to John Spear's early this morning for him to take up to Thomas Benbow's as I expect to leave East Bend in a few days. Rained quite hard, just after I got from Spear's. Had I been a few minutes later I should have been treated to a shower bath.

*March 17 Saturday.* Still cloudy and rainy. Changed my notion in reference to going home. Mr. A. Benbow says he just can't give me up, and that he will almost double my wages if I'll stay, so now if I had my trunk back, I'd stay at least one month longer. Mr. A. B. agrees to have my trunk returned next week, and I agree to stay one month more.

*March 18 Sunday.* Still cloudy and rained on till evening, then commenced snowing and snowed for 2 or 3 hours quite fast.

*March 19 Monday.* Clear and tolerbly cold. Mr. J. Spear returns this morning and brings back my trunk. Mr. R. Bell heard I and Alex talking about my staying at E[ast] B[end] and told Mr. S. that I wanted my trunk back, so he returned it all right, Messr Bell & Spear in that particular, and I am oblidged to ye both.

*March 20* Tolerbly pleasant.

*March 21* Cloudy and quite cool. I've been reading "Paradice Lost" and find it quite interesting. John Milton thus describes Hell which he says is

"As far removed from God of Heaven and light
As from the center thrice to the utmost pole.
"Region of sorrow, doleful shades where peace
And rest can never dwell."

Milton also says that the flames of this dark abode are "Fed with ever burning sulphur unconsummed," and that this awful place was prepared "For those rebellious," meaning the Devil and his Angels, but the same will also include rebellious man unless he forakes his wicked course, for there are but the two parties, viz, God and His angels, the Devil and his angels, and all men that fight against God, fight in the ranks of the Devil, who will claim them as his *angels* in consideration of the valient service they did him while on earth, and will chain them down, forever to dwell, in the "dungeon horrible," prepared "For those rebellious." Oh God, forbid that I, the most unworthy of all created beings, should ever be found in the ranks of the Archenemy of Thee, and of fallen man, whom to save, Thou hast given thine only Son. For thy Son's sake have mercy upon me; guide me by thy Holy Spirit thro' life, and at last save me in Heaven.

*March 22* Yesterday I commented considerably on "Paradice Lost," after reading a few pages, but to day I've been studying Anatomy, Latin, moral science etc. to such an extent that I've only time to take a note of the day, that I with all that may come after me, may know that there was such a day as the Thursday the 22d of March 1855. Last night it snowed some as one could plainly see from the appearance of the bushes, fences, old logs, stones, etc. There was but little snow on the ground, as heat soon dissolves snow into its original eliment, which quickly disappears in the earth, or is evaporated and ascends again into the clouds to go through the same process of snow-making, and then fall in beautiful white flakes to "make glad the hearts of the children of men," and to protect the tender germ from the bitting frost, and give it a chance to take root that it may in due season spring up and bear flowers and fruit; or it (the rain drop) ascends to be condenced, then fall in April shower that it may invigorate vegatation in general. Noah Webster says "The rain makes the grass grow," but grass must have light and heat as well as rain. Air and light are necessary to the vegatable kingdom as well as to the animal. The leaves of plants answer to the lungs of animals and a plant could no more live without air than an animal. Light gives color, (which is as necessary to plants as it is to animals) to all things.

*March 23 Friday.* Weather moderate. Went out to the debate at night, and made two *thundering* speeches in favor of slavery.[3] The question was "Resolved that slavery should be tolerated in America." This question is calculated to produce warmness of feeling when argued by men, and when boys get hold of it, it produces no less the same phenomena. The affirmative gained by a majority of 2 or 3 votes.

*March 24* Saturday, and a day "long to be remembered" in consequence of the wind that blew. Dont think I ever saw it, or heard it blow harder. We, Dr. Benbow, Mr. John H. Kinyon, J. P. Nicholson, myself and others tried to raise a kite, but we could do but little at it.

*March 25 Sunday.* Not much wind but tolerbly cool.

*March 26 Monday.* Tolerbly pleasant day. Nothing of importance.

*March 27 Tuesday.* Windy and cold.

*March 28 Wednesday.* A cold morning for "Spring Time."
It is cold as winter; so cold that the vegatable kingdom seems not to be doing much. No flowers! no leaves! I have not yet seen so much as a swelled bud, that I remember, this season.

*March 29* Don't know that we had a colder night during the winter than was last night. The day is not quite so cold as yesterday, but its not very warm.

*March 30* A calm and beautiful morning. Borrowed the *Illuminated His-*

3. As the rest of Thomasson's diary makes clear, he was not in favor of slavery, but this was the position he argued in the debate.

*tory of North America* by John Frost, LLD. of Dr. E. Benbow, in which I read that North America instead of having been first discovered by Christopher Columbus, is supposed to have been visited by a band of Northmen about the year 1000.

The wars in which the Northman soon after engaged is supposed to have obliterated all remembrance of their discoveries, so we hear nothing more of the new world till some 500 years later when Christopher Columbus discovered some of the West India islands. The discovery seems to have been reserved for some mast[er?] mind, and Columbus was just the man to make it. Those Northmen who are supposed to have visited Martha's Vineyard and other places on the Atlantic Coast, intended to have planted colonies and taken possession of the territory, but in consiquence of losing a few of their men in a scrumage with the natives, they became discouraged and left the continent. Columbus was not the man who would back out for a triful, but would, if he failed in one intance, "try again." Would that I had a little of Columbus' energy or a little more of my own.

I heard the sweet and plaintive notes of a dove this morning, which is said, by old women at least, to be a true sign of the approach of spring, that blessed season when the April showers water the earth, and the flowers spring up, and all nature looks so inviting.

"I would not die in spring time," for then it is surely very pleasant to live. But would it not be pleasant then to die, and go up to heaven through the green leaves while the birds were sing[ing] their morning songs in praise to Him that ruleth all things well? To the Christian it is pleasant to live and enjoy all the blessings of this green earth; and to the Christian it is pleasant to die and go to Jesus.

*March 31* Saturday and last day of March. Tolerbly pleasant day.

*April 1 Sunday.* A little windy. Went to the Quarterly meeting at Masadonia, and heard the Rev. N. H. D. Wilson, the Elder, preach an excelent sermon to a large congregation. His text was "Ye are the salt of the earth, etc.," Mathew V, 13. After preaching the "Lord's Supper" was administered, and an invitation given to the members of other denominations to join in the commemoration of the death and suffering of our Savior, but the Baptist friends would not partake. I can't agree with the

Baptist in this particular, but its of no use to enter into discussion here. All men have a right to read the Bible and form their own opinions, and we should not persecute each other for mere opinion's sake.

*April 2* Cold and cloudy. Looks as tho' preparations were making to give us an April snow.

*April 3* Cool and cloudy, and after early morn it commenced snowing and hailing sure enough.

*April 4 Wednesday.* [No further entry]

*April 5 Thursday.* Rained very hard in the afternoon.

*April 6 Friday.* I and John P. Nicholson went up home from East Bend. Rained quite hard again just before night. I and Nicholson parted at Hamptonville, he turning toward his father's & I striking on to mine. It commenced raining soon after I left Town, and rained so fast I thought it best to take up, so I called in at Mr. Bell's and staid all night with *my little lark.* I found Miss Mary dressed and waiting for me, as I had written her that I should be there that night.

*April 7 Saturday.* I was off for home early, and got there, I spose about 9 o'clock. Found all up at home, and, I guessed, well as usual. I, Pa and brother Wiley & brother Caleb finished poleing[4] the garden.

*April 8 Easter Sunday.* A nice day indeed. Went out to Aylesbury to meeting. Pa gave a short discourse, after which we had a nice class meeting. I went home with Miss Mary, and enjoyed myself finely, in witness where of I refer to the time I staid, i.e. to the length of it.

*April 9 Monday.* Cloudy, and rained some in the afternoon, tho' not hard for I ploughed all the time about and did not get wet. I planted my May peas and Windsor beans to day in the East side of the garden. I had but 3 beans & 7 peas, and from them I hope to get in stock.

*April 10 Tuesday.* Still cloudy. I subscribed Five Dollars to Union Academy. Nicholson came and we put off for East Bend. We overtook brother Wiley, who started a little before we did, and all came on to E. B.

4. He is placing into the soil poles along which vegetable vines will climb.

together. We stoped at Jesse Vestol's and fed our nag 10 ears of corn for which we paid 10 cents. We landed at E. B. a little after 5 o'clock.

*April 11 Wednesday.* Clear and somewhat windy. The wind blew very hard just before day this morning. An Irish peddler stoped in to the store this morning who had table cloths for sale. I sold him a few articles and bought one of his table cloths for which I paid him one dollar. A. Benbow also bought one. This Irish peddler is a man of rather small stature, thin face, sharp nose, rather soft grey eyes, irregular teeth, dark brown hair and rather fairer skin than most Irishmen toat [tote]. He spoke English well; so well that I have my doubts as to his being a native of the "Green Isle," tho' he said he was and that he had only been in America 6 years. He is much opposed to Knownothingism and pretends to know all about it, how, where, and when it originated, who its author is etc etc, but yet he failed in being able to tell the true name! of the society known as Knownothings.[5] What strange animals men are! Know the author of a thing, the circumstances under which it originated, how, where, and when, and yet did not know the true name of this thing!! Ha! Ha!! It may be that Mr. Irishman knew some things, but I am disposed to doubt his knowing as much as he professed. His name was John Cassidy.

*April 12 Thursday,* Clear and cool. Dont feel so very well as I am very hoarse. I must try the cold bath to night.

*April 13* A little hazy, but warm. Tried the bath, and took off my whiskers last night, so I feel better to day.

*April 14 Sat.* A nice warm morning tho' a little smokey.
Went out to the debate last night and made 2 speeches on the affirmative side of the question which read as follows. "Is there more pleasure in pursuit than in possession." The negative gained by one or two votes.

*April 15 Sunday.* A little cloudy tho warm. Read in my new "Testament" one chapter, and also the notes and instructions by the Rev.

5. The Know Nothings, an anti-immigrant political party, was actually named the American Party. The party began as a secret organization, and its other name came into use because its members refused to answer questions about it, saying, "I know nothing." For a brief period the American Party showed great strength, but leaders could not keep its northern and southern wings together.

Justin Edwards, D. D. I like them well too, for their plainess. I also read a very good story entitled, "The Christian Traveler," in a new book I have of *Excellent Naratives.* I also read some in the *Age,* which has long been the best paper in the State, to my notion, and is daily or weekly growing better. It will soon be a perfect moddle of *newspaper virtue* if I may be allowed to *prophesy.* In the afternoon, at 3 or 4 o'clock, I went out to church and heard the Rev. John Webster, a Baptist minister, try to preach. His text was "Pray without ceasing." I Thes. V. 17.

*April 16 Monday.* Cloudy, and a little cool.

*April 17 Tuesday.* A little cloudy, but warm. The coffee, sugar, molasses, salt, etc, came to day about 12 o'clock. Coffee is worth 14 cents; sugar, a first rate article, 10 cents; molasses 50 cents; salt $1.50 per bushel.

*April 18 Wednesday.* Clear and quite warm.

*April 19* A hot day for April. Vegatation is a little late but it['s] growing green now, fast.

*April 20 & 21 F. and Sa.* Quite a pleasant morning, and the wind is blowing a brisk gale from the north.

*April 22 Sunday.* Quite a pleasant day. Went to Masadonia to class meeting, and to the Baptist church in the evening at 3 o'clock, and heard the Rev. Joseph R. Creel preach quite a good sermon from this text; "For the children of this world are wiser in their generation than the children of light." I do not remember that I ever heard this text used before. I gained a few new ideas from Mr. C's sermon.

*April 23 Monday.* Warm & dry. No news in particular.

*April 24* Clear and pleasant. Mr. R. Bell came down from home and brought good news relative to Union Academy and my people. The Academy dont pass into the hands of the Baptists, but is to continue on the union flag. Good news from the west!

*April 25 Wednesday.* Clear and pleasant. Tomorrow I expect to strike for home in company with Mr. James York. East Bend! Adieu.

*April 26* Left East Bend about eleven o'clock and landed home about nine in the night. Mr. York hauled me and my trunk up and charged me nothing. Stoped at Doweltown and give Hauser & Wilson my note for $9.50, and closed my book acct.

*April 27 Friday.* A little cool as the wind blew briskly from the North East. Helped Pa and Caleb grub in the swamp an hour or so. It is a dry time, very dry.

*April 28 Saturday,* "Warm and dry." Went to Hamptonville a foot—[and] back, received the *Age, Message,* and no letters. Came back and called in to see Miss Mary and staid till Sunday evening. Found Miss M. well and sweet as ever.

*April 29 Sunday.* Got up this morning at ______ dressed myself and walked out to breathe the fresh air.

After breakfast the Miss B's walked out into the garden to gather pinks, and I soon made tracks toward the same place. Gathered the flowers, and I and Miss M. retired to the shade of a small cherry tree, and talked over things in general, but nothing in particular. In the afternoon we walked out in search of flowers; found lots of honey suckles and a few dogwood blossoms. We then retired to the spring, which is shaded by the spreading branches of a Maple tree. Our fun over, we quit the spring for the house. Guess I had better change my subject lest I tell more than I wish all to know.

*April 30* And last day of April 1855. Went to plowing. Planted a little corn after dinner. Rained a nice shower.

*May 1 Tuesday.* Planted corn. The weather pleasant.

*May 2* Cloudy but no rain.

*May 3* Still cloudy. Finished planting corn. Borrowed of Col. N. Wilburn fifty dollars.

*May 4* Cloudy, cool, and rainy. Cut and hauled wood in the p. m.

*May 5 Saturday.* Clear and cool almost enough to frost. Ploughed the Irish potatoes whose tops are about 6 inches high. Finished planting corn a second time.

*May 6 Sunday.* Clear and cool. Went to the Flatrock Sunday School, I and Mr York. After School we called in to see the Miss B's. Mr Y. and

Miss J. retired to the garden to gather flowers and talk of ______ while I and Miss M. took our seats in the cool shade between the houses and talked of things in *particular.*

*May 7 Monday.* Went to ploughing the corn, sleepy as an owl in daytime, as I had company last night and did not get to bed till after daylight had peeped out from behind the curtain of night and bid darkness hide his gloomy features behind the huge mountains away toward the setting sun. I slept but a short nap, then rose and went to the field where the flower bushes of nature bloom, and where the birds flutter among the green branches and praise their Creator who feedeth them, in the sweetest of songs.

*May 8* Clear and cool. Went to hear the Rev. Shelton preach at Aylesbury. He preached quite a good, plain, short sermon from this text: "I will help thee." Isaiah 41 ch. & latter part of the 13 verse.

*May 9* Clear, windy, and almost cold. I've been plowing with my coat on all the day.

*May 10 Thursday.* Clear, and a few degrees warmer than yesterday. Rob[b]ed one stand of my bees, but it was not very rich. Paid John E. Grant $2.36 in full of an unjust account. He (Grant) charged me .75 cents for 1/2 day's work, & 37 1/2 cents for a few minute's work. I have a receipt in full, and intend to have no more dealings with John Grant, for a man can't deal with him and "live peaceably with all men."

*May 11 Friday.* Clear, calm and warm. There are too a few thunderheads to be seen away in the west, tho' there's but little appearance of rain. A light shower would be beneficial, I should think, just now, but He that sendeth the rain knows best, and will always order the workings of things in general for the good of those who love and fear him.

*May 12 Saturday.* Clear and warm. I & Clark bought a pint bottle of ink for which we paid 30 cents. I paid five sixths of the price and Clark one sixth. How much did each pay? I must shave, slick up etc, and go to see my darling.

*May 13 Sunday.* Clear, warm & pleasant. Staid last night at ______, went with Miss Mary to meeting, heard preaching, went home with

Miss M. and staid all night. Had a pleasant time, as I, Miss M., Mr. York and Miss J. took a walk to the spring. After the fun around the spring of pouring water on each other's heads was over, Mr. Y. and Miss J. retired to the wash bench while I and Miss M took shelter from the taning rays of the 'bright orb of day' beneath the spreading branches and green leaves of an Ash standing near the spring. So comfortably situated what could hinder our happiness?

*May 14* Still dry and warm, tho' there are some signs of rain visible.

*May 15 Tuesday,* Dry and cool. Took some of my timber to the factory for Mr. York to turn bed posts for me—two pair.

*May 16* Dry and windy. Sheared the sheep. Jo Cowles' wife has a young one!!!!! Nice hand. Ha! ha, ha, ha.

*May 17* Dry & warm. Some appearance of rain. Hauled rails to make the cross fence thro' the orchard, Hauled 60 at one load with old Med & Jenny. Unce Jo. Hauser came to see us yesterday, & left to day.

*May 18 Friday.* Cloudy & cool. Rained last night, also a light sprinkle to day. Hauled rails. Planted watermellon seed yesterday. in hills, made rich with ______.

*May 19* Hauled rails in the forenoon to make a cow-pen, and went to Town in the evening. Bought a hat for $1.15, pants pattern $1.62 1/2, one quire of paper .25. Received the *Age* and the *Pop-Gun.* Saw the skin of a large wild cat which is stuffed with bran and looks a good deal like it was yet alive. That cat was killed near Trap Hill in Wilkes County.

*May 20 Sunday.* Clear and pleasant. Went to Flat Rock to preaching; called in as we (I & Mr. York) went on at Mr. Bell's and walked out with the Miss Bells. The Rev. Green Brown preached old Mr. Burges' funeral from this text, "And as it is appointed unto man once to die, but after this the judgment." Heb. IX, 27, 28.

*May 21 Monday.* Warm and a little hazy. Commenced ploughing the corn the second time and planted the pea crop. Pea vines make great hay, and so they do.

*May 22* Clear and warm. Ate three ripe, or red strawberries, the first

*June 3 Sunday.* Clear & cool. Went to Sunday School, and then home with Miss B. as usual.

Received a letter from brother W. who is now boarding at Uncle Flem's in Guilford Co.

*June 4 Monday.* Cool and a little cloudy. Ploughed & hoed corn.

*June 5* Cloudy, and rained a few drops. Hoed corn etc, etc.

*June 6 Wednesday.* Nearly clear and tolerbly warm. Ploughed corn.

*June 7* Hoed corn, while Caleb ploughed. Rained a nice shower in the evening.

*June 8* Clear and cool. Hoed corn. Saw one of the wonders of the world, viz. a poem (or something else) written by Ann Benbow, on the death of her daughter-in-law, Caroline Benbow. One stanza will do for a specimen, and satisfy the reader.

"She decesed on the first month the 5 day
She fell a sleep in Jesus' armes,
And her spirit took its flight in the ralmes,
Who has said he would gather his lames
With his armes and cary them in his bosum."

Hem. Ha! Ha! Ha! I think Ann will have to screw up her machine, and try it again. The above stanza is letter for letter and word for word as it is in the original. See, what spelling. I hould hate for such *stuff* to be afloat in the neighborhood bearing my name, and so I should. Ann may practice medicine, preach, etc. but guess she'll never do much writing poetry.

*June 9 Saturday.* Mr. York and I took the Miss Bells, Mary and Jane, to the cherry orchard. We went to Wilburn's and got plenty of good ripe cherries, and had quite a *jolifycation* time. Miss Mary got sick before we got home, not from eating too many cherries, for she ate but few, but from some other cause, and that dampened my joy.

*June 10 Sunday.* A pleasant day. Went to Aylesbury, heard the Rev. Shelton preach quite a good sermon form the text, "After this I beheld and lo, a great multitude, which no man could number," etc, Rev VII, 9 and also the 10 verse. The house was crowded almost to *over-*

*flow*, or *underflow* one, for one bench came near falling thro' the floor. Pa preached after Shelton, and after we "got thro' with" dinner, from the text, "And they came over" etc. Mark V, and from the first to the 19 verse. Went home with Miss B. and back to night meeting. Shelton preached. His subject was "the prodigal Son," who, after his father had given him his portion of goods, took his journey to a far country, and after he had wasted his fortune, and began to suffer from want, he returned to his father's house, acknowledged his fault, and desired his father, not to call him son again, but to make him as one of his hired servants, etc. This is quite a good subject.

*June 11 Monday.* Clear and cool. Ploughed the corn. Yesterday my bees made out to swarm for the first time this year, which I had like to have forgotten to note.

*June 12* A little cloudy. Pleasant. Ploughed in the corn.

*June 13 Wednesday.* Clear and cool. The weather looks like being dry. Rob[b]ed one stand of my bees—the one nearest the garden—and found nearly as many young bees as honey.

*June 14* Tolerbly clear. A few little white clouds are all that are to be seen. Was asked to a barn raising to day at Thomas Benbow's but did not go for reasons best known to myself.[7]

*June 15 Friday.* Clear and warm. Went to the shop (Messick's) and had my buggy mended and old Med shod all round, as Mary and I are going to Senter to the Quarterly meeting next Sunday if nothing happens to prevent.

*June 16 Saturday.* Clear—not so much as a white cloud to be seen. I fear we shall have dry weather, and if the corn should be cut short what will become of the people? Corn is sald to be worth $1.50 per bushel in Tennissee! the greatest corn-raising state in the Union. Wheat, it is said, is worth over $2 in Charlotte! Our wheat looks good, and I trust it will turn out so. The corn looks tolerbly well, tho' its quite small for June.

7. Strong may have stayed away because he disapproved of the availability of liquor on such occasions. When he or his family sponsored a corn shucking or similar event, they did not provide liquor—and usually had few helpers.

Oats tolerbly—rain just now, I think, would help them, but the Lord knows best, and will sent us a shower when ever it should come.

*June 17* A hot day. I and Mary went to Senter to the Quarterly meeting, and heard the Rev. N. H. Wilson preach a missionary Sermon from the text, "Fear not: for behold" etc. Luke II ch. 10 & 11 verses. There were present a great many people, and the sermon was a tolerbly good one, but I dont think there was a large sum given to the Missionary cause. I gave one dime. May it aid in spreading the gospel among the heathern, and be, to me, as "bread cast upon the waters, which shall be gathered after many days." Amen. Went over in the afternoon to see Miss Mary A. B., found her at home and lively as usual. God bless her.

*June 18* Another hot day. Hoed corn. A light shower fell in the evening which was a great blessing to us, as well as to the vegetable kingdom.

*June 19* Quite warm. Alexander cut and I bound three doz. bundles of wheat—the first we cut this year. The wheat is well filled, and the heads are of medium length.

My bees swarmed. I & Alex. hived them and left the stand setting under the apple tree they settled upon as some contend that it is best to leave them to work at, or near the place where they settle. Another beautiful shower fell soon after the morning.

It is amusing, to see, with what care nature has provided "every creeping thing that creepeth upon the earth," a weapon of defence. She has not left the smallest insects destitute of means by which they can, in some degree, defend themselves, when attacked by their enemies. Some have sharp teeth, while others have stings, as bugs wasps, etc. Some do one thing to offend their persecutors and defend them selves, while others of different habits & natures resort to other means of self defence. The spider bites, the bee stings, and the grasshopper spits ambus. To day while I was setting under the sedars reading the *Spirit of the Age,* a grasshopper chanced to light on the leg of my pantaloons. I sliped my hand slily down and caught him. I took hm up holding the ends of his closed wings between the forefinger and thumb of my right hand, and his throat between the forefinger and thumb of my left hand. I turned him on his back, while thus confined, and several times saw him spit his ambus, rub his foremost l[e]g across his mouth and then my finger. This he continued to do, till, I suposed, his stock of ambus was exhausted, and then seemed to submit himself to mercy of his capturer. About this time I gave him a toss up, and he sailed off, thinking, perhaps, (if grass-hoppers ever think,) of what a narrow, but lucky, escape he had made.

*June 20* Clear and hot. Hoed corn in the forenoon, and ploughed in the after.

*June 21 Thursday.* Warm. Hoed corn in the forenoon and bound wheat in the after.

*June 22 Friday.* Cloudy. Cut & bound wheat. The wheat is low and short headed but very well filled. Rained a light shower in the evening.

*June 23 Saturday.* Bound wheat, tho' I am far from being well. Alex & I worked to[o] hard yesterday, so to day we are both sick. I gave out at 12 o'clock and went to bed. Pa gave me a wet-sheet pack. I staid in one hour & five minutes; came out wet with sweat, washed off in cold water and went to bed. Then with a warm rock at my feet and a dose of No. 6 in my stomach, I rested tolerbly thro' the night. Received a letter from bro Wiley, in which he writes—"Wheat (the new crop) is going at $1.75 per bushel."

*June 24 Sunday.* Some rain. Feel a little better but not well. Staid at home all day! as I was not able to go to meeting, or to see my *intended.*

*June 25 Monday.* Not exactly clear. I'm not well yet, but better. This world is nearly round, and quite funny in some respects. Ever since "The infernal Serpent deceived the Mother of mankind" there have been "evil doers" in our land. Some making short for their kindred spirits, and others mischief. Various are the principals, dispositions, etc. of men. Some laugh, while others weep. Some take the world fair and easy, while others "go mourning all their days." Seneca tells us of two Philosophers. One, he said, would laugh at the ways of the world, while the other wept over frail humanity.

*June 26 Tuesday.* Not able to work. Wrote brother Wiley a letter and took it to the office.

*June 27 Wed.* Warm and sunny. This is, indeed a beautiful day.

"Sweet day! so calm, so cool, so bright,
The bridal of the earth and air—
Sweet dews shall weep thy face tonight,
For thou must die."

"Thou must die" is written upon all things mortal. The grass, the flowers, yes, even the beautiful rose teaches us how to die. For a short time it flourishes and fills the garden with sweet perfume, thus teaching us how to live. Then

"When next the summer breeze comes by
And waves the bush the flower's dry."

*June 28 Thursday.* Quite warm. Hauled in our wheat.

*June 29 Friday.* Very warm. I ploughed in the forenoon, and Caleb in the afternoon. I, Pa & Mary went over to the Division room to meeting. The members of the "Silk stocking Church," as it is sometimes called, have a queer way of worshiping. The preacher said some good things, but his sermon did not appear to affect much, as there was but little weight about it.

*June 30* Saturday, and last day of June in the year 1855. This morning, I think, is rather warmer than yesterday. I & Caleb hoed corn.

*July 1 Sunday.* Came home from Mr. B's this morning about day. My bees swarmed about 10 o'clock, settled on a little peach tree near by, and I hived them without difficulty. Had a beautiful rain in the afternoon. Went over to see Miss Mary, but did not stay long.

*July 2 Monday.* A little cloudy, and not quite so warm as usual. Ploughed in the corn.

*July 3 Tuesday.* Rained moderately nearly all day. Mowed some in the morning. Went over to the factory in the evening and commenced a Prisim [prism].

*July 4* This is a great day. In some places it is celibrated in memory of the Declaration of Independence, and with us it is great on account of the refreshing showers of rain that fell on our parched earth. We have not had as nice a season in many a day.

Finished my Prisim, which is constructed of three pieces of glass, each 12 inches long, and 4 wide with end pieces of Walnut, and stuck together with white Lead, Bee's Wax. It is one of the best of the kind I ever saw. Went to Hamptonville after my papers etc.

*July 5 Thursday.* Mowed grass etc.

*July 6* Raked hay etc. Rained a beautiful shower about sun set.

*July 7 Saturday.* Spread and put up hay, in windrows, in the forenoon, and hauled and stacked it in the evening.

*July 8 Sunday.* Clear and hot. Went to Aylesbury but heard no preaching. It was class day, but the Leader, F. Reinhardt, is sick, so there was no meeting. Went home with Miss B. as usual.

*July 9* I and Caleb ploughed in the sandy bottom. Ploughed—laying by the corn. Ground point to go on my buggy.

*July 10* Very warm. Gave my buggy one coat of paint which helped considerbly. Ploughed in the evening after supper.

*July 11 Wednesday.* Gave my buggy a coat of varnish which helped it still more.

*July 12* Clear, with the exception of a few white clouds, and quite warm, Alex and I cut & bound oats. We cut and set ip [up] 71 doz., nailed up a gum and hived my bees which swarmed about 11 o'clock. The swarm was large, and we hived the [it] nicely notwithstanding they settled on the body of an apple tree some distance from the ground. Just after the "bright orb of day" had gone down behind the western hills, I histed the head of three of my beestands, two of which was, I think, as rich as I ever saw, and one having young bees almost to the head. The comb was white, and the honey beautiful. I guess from its appearance, that most of it was collected from the bloom of the sowerwood.

*July 13 Friday.* Went to Hamptonville to hear Candidates, (Col. R. C. Pryear, Whig and Knownothing; and Mr. Scales, Democrat and anti Knownothing of Rockingham,) make their great speeches. Scales spoke first for 2 1/4 hours, giving the Whigs and Knownothings *fits* as he tho't. Then Col. P. made a very plain, good talk of about 1 1/4 hrs. Mr. Scales replied, and Col. P replied, and they kept up an alternate replying till their hearers were, some of them at least, heartily tired of hearing them. Guess Col. P. will be elected.[8]

*July 14* Alex. cut, and I bound oats till about 11 o'clock. Went to Hamptonville in the evening and purchased license (marriage) for which I paid $1, and the C. C. C's [Clerk of County Court] fee .25. Received a letter for brother Wiley from Uncle Wm Houser, of Georgia. Uncle writes that crops look fine, etc. etc.

*July 15 Sunday.* Staid all day at Mr. B's. Rained some in the evening.

*July 16* Bound oats till the rain came and wet them. It was very warm till the shower.

*July 17* A hot day. Bound oats.

*July 18 Wednesday.* Clear and hot. Bound oats till 10 o'clock, then put on my Sunday and went over to Mr. Bell's and got married! Clark came down with his *gall* (Miss Mary Wilborne), her brother and cousin. We had quite a little wedding, as there were not a great many present. B. A. Johnson Esq. tied the matrimonial knot, which is to last for aye, and direct our lives, Mrs. Mary A. Thomasson's and mine, thro' the same channel.

*July 19* Still clear and hot. Brought my wife home to Pa's. Mr & Mrs. Bell, Mr. York & Miss Jane Bell, Alexander, Shadrack Bell & Caleb (brother) came on with us. We had in company four buggies only, but as we rolled on "*big as crout*" the dust rose in waves, sparkled in the rays of light for a moment and then returned, or descended to mingle again with its fellow dust. We might learn a lesson even from the coalescence of dust which are blown up, and for a time float upon the "dessirt air." It is dust, and unto dust shall it return. It rises, floats upon the wind for a moment, then falls and the first shower that comes converts it into mud.

We are dust, and unto dust shall we return. We rise—have our day—and then return to "old mother earth." If we would but open our eyes

8. Richard C. Puryear, a Whig from Surry County, had served in the North Carolina General Assembly and was currently the incumbent in Congress.

we might learn a useful lesson every day, and in time grow wise and good, live and die happy.[9]

*July 20* Clear and *hot.* Bound a few oats. Only had holyday one day and a half after getting married! This is Tax paying day. The tax payers in this Dis. meet at Jim Gree's [?] store this year.
Had quite a fine rain here, and a good deal of wind with some hail nearby. The corn appears to be doing finely.

*July 21 Saturday.* Quite warm. Took Mary to her father's. I and Mr. Bell went on to Town to the Taxpaying. I paid Pa's Tax for the year, 1853, which was $3.80. Received a letter from brother Wiley—a very good one too.

*July 22 Sunday.* A warm day. Rained some in the afternoon. Staid all day at Mr. Bell's with my sweet Mary.

*July 23* Came home and left Mary at her father's. Finished cutting oats.

*July 24 Tuesday.* Stacked a few oats. Rained lightly in the evening.

*July 25* Stacked oats in the forenoon. Went to see Mary in the evening.

*July 26 Thursday.* Went to Hamptonville—received a letter from brother Wiley, the *Age* and Esquire Cowles gave me a No. of the *Dollar Times,* a good weekly paper published in Cincinnati, Ohio. Brought Mary home, and in the evening we went over to Alex's. Some fine watermellons were gathered, and sliced, and we helped ourselves. These were the first ripe mellons I've seen this season. A fine shower of rain fell just before the close of day.

*July 27* Went to the shoe maker's shop, James Vanhoy's, to get a pair of shoes made for Mary. From the shop, I went to Wm Pardue's to see him about School. Saw Mr. Pardue, who said he wanted me to teach the school in Dis. No. __. Came home and got wet on the way as it rained quite hard, or fast.

Found Mary in bed with the sick head ache. Bathed her feet in warm

9. Perhaps nothing better illustrates the basic seriousness of Thomasson's personality than this reflection in the midst of the celebration of his wedding.

water, and gave her some Ditney tea, and a warm rock to her feet, covering her up nicely in bed. Soon she was sweating, and went to sleep. I & Alex made a cider trough of thick planks.

*July 28* Mary some better, but not well. Rained considerably in the afternoon.

*July 29 Sunday.* Warm day. Took Mary to her father's tho' she was hardly able to ride in the buggy.

*July 30* Mary not much better, if any. Alex Benbow came over. I went with him to Town, and he sold my note which yet calls for $20, with some interest, to Andrew Cowles. I shall have it to pay soon certain. Received a letter from my old friend, Thomas N. Hunter, which uses up the K[now] N[othing]'s in brown stile.

*July 31* And last day of July 1855. Clear, with the exception of a few white clouds, and quite warm. Mary, I trust, is some better as she did not complain of being so sore as here-to-fore. Made her a bottle of bitters, composed of wild cherry tree bark, Poplar root bark, and just enough of the "red eye" to preserve it.

*August 1 Wednesday.* Came home and took sister Mary to the burying of Fredrick Reinhardt at Flat Rock. There were present a large congregation of people, and many tears were shed for the departed. Brother Reinhardt was a fine citizen and an exemplary Christian. He is gone, and much he'll be missed both in the world and in the church.

*August 2 Thursday.* Election day. Went to Town and voted for Richard C. Praryer, Congres member. Left town soon after voting, so I saw but little of the fun, as some call drunken capers.[10]

*August 3 Friday.* Came home, and hauled 2 loads of dry-house wood, daubed the dry-house[11] etc., and then returned to Mr. B's.

*August 4 Saturday.* Went to Messers Elz. Messicks & Ira B. Messick's,

10. Thomasson's earlier prediction proved correct, as Puryear won reelection for one more term.

11. Thomasson refers to a house for drying fruit.

and made an engagement to teach the free schcol in Dis. No. 35. for $18 per month to commence on Monday the 13 inst. Brought Mary over home, in the evening, to stay a few days.

*August 5 Sunday.* A warm day. Went over to Alex's with Mr. Milton C. York & t[o] Mr. Loftin's, and ate mellons & peaches to my heart's content & my stomach's trouble. Came back home & brought Mary a watermellon and a basket of soft peaches, and spent the evening at home.

*August 6 Monday.* Helped fill the dryhouse with fruit, etc. About 2 o'clock it rained a beautiful shower.

*August 7 Tuesday.* Cloudy & rained moderately. Helped about drying fruit etc.

*August 8* Still cloudy and rainy. Old Mr. Bell came over this morning and I wrote for him, a letter to his brother Joseph, who lives in Illinois. We had, to day, a dinner of roasten-ears, boiled onions, cucumbers, musk-mellon, honey, bread, etc. etc., all vegetable, and the best, or at least the most wholesome meal, we could have had. Polly is looking on and laughing while I wright.

*August 9 Thursday.* Cloudy but not much rain. I & Mary went to Aylesbury to circuit preaching and heard brother Shelton preach a short and tolerbly good sermon from the text, "Let us hear the conclusion of the whole matter" etc Eccl. XII, 13 & 14. I & Mary went to Mr. Beall's. After eating a hearty dinner of vegetables, I went to Tow[n] and got my newspaper (the *Age*) and returned by James Vanhoy's, the shoemaker, and got Mary's shoes. Mr. V. charged me 75 cents for finding insoles and making the shoes which were lined and bound.

*August 10 Friday.* Came home and went to work. Read [a]n interesting article in the *Age* entitled "Coming Down" by Alice Cary, a good writer.[12] Had, this evening, quite a fine shower. Killed one stand of my bees. The

12. This article appeared in the issue for August 1, 1855 (vol. 6, no. 48). The story concerned a well-to-do couple who had everything but grew bored. Seeking amusement, they spent more but soon lived beyond their means and then had to give up their servants and luxuries. They experienced an awful "coming down" and many social humiliations but rediscovered love through helping each other. At the end of the story the husband concluded, "[W]hile we maintain honesty and self-respect, coming down is impossible."

gum was small, and but few bees in it, so I got but little honey. I do not like the driving system.

*August 11 Saturday.* Beat a few apples to make cider, and vinegar. Cloudy, and is raining moderately.

*August 12 Sunday.* I & Mary, Mr. & Mrs. Beall went to Flat Rock to meeting, and heard a Presbyterian, whose name I believe is Krider, preach, or read a sermon. His text was, "Examine yourselves," etc. II Cor. XIII, 5 verse. The Rev. Krider is of medium height, rather slim, and dark complected. He has a high forehead, dark eyes and a long nose. His eyes are set close together in his head, indicative of cunningness.

*August 13 Monday.* Peeled apples to dry, beat apples to make vinegar, etc. Just before sunset we had a very hard rain—the hardest, I think, we've had this season; there was also some hard thunder. Wesley Pardue says the lightening struck his chimney, ran down inside, and knocked him out of his chair, deafening him in one ear. The rain washed our fields some, but did no serious injury to our crops.

*August 14 Tuesday.* Quite warm. I & Caleb sprouted the turnip patch, gathered fruit, etc, etc.

*August 15* Cloudy. Rained some yesterday. Sawed off my cow's horns to keep her from killing the horses. Dried fruit, etc.

*August 16 Thursday.* Brought Mary over this morning. Clark came home in the evening from Wilkes Co.

*August 17* A swealtry day. I & Clark went to Couche's store, and I bought 8 yds. of bleached domestic at 12 1/2 cents per yd; 3 yds black calico 12 1/2 cents per yard; amounting to $1.37 1/2.

*August 18 Saturday.* Burned off the turnip patch, beat apples to make cider, etc. This day one month ago I was married. One month to day have I and Mary lived together and we have not had the "first fight" yet! Not so much as a quarrel!! May the love and peace which have made us happy thro' the first month of our married lives continue to the day when death shall part us for a season, follow us thro' every lane of our lives, and cheer us by day and by night.

*August 19 Sunday.* A beautiful day. I and Mary went to Union Grove (usually know[n] as Nigar Jack). The Rev. McCain was preaching the 8 o'clock sermon, when we got there. He took for his text a[ ]part of the 10th verse of the XXV chapter of Mathew: "And they that were ready went in with him to the marriage; and the door was shut." He made, as I thought, rather a poor out in the preaching line. At 11 o'clock the Rev. Q. Holton preached an excellent sermon from the text, "Go ye into all the world and preach the gospel to every creature." Mark XVI, 15 verse. At 3 o'clock a public collection was taken for the benefit of the superintendent, Rev. John Nailer. I guess the amount colected was small as the congregation was mostly composed of poor people who had but little to live, and was rather small for Sunday. After the collection Rev. Franklin Moss *tried* to preach from the text, "By faith Moses," etc. 24, 25, 26 verses of the XI chapter of Hebrews. After meeting broke for the day, I and Mary went to Mr. S. Hicks', 4 or 5 miles south east of the campground, and staid all night. Mr. Hicks is a curriosity, and no mistake.

*August 20* A pleasant day. I & Mary, Mary & Rebeckah Hicks, and Mr. Thomas Hicks, came to the campmeeting. Rev. McCain was preaching the 11 o'clock sermon, when we got there, from the first Psalm. He made but little better out than yester day, if any. We had no dinner with us to day so we suffered some with hunger. Mr. Holton preached the 3 o'clock sermon from the text, "Behold, a sower went forth to sow," etc. on to the end of the 8 verse. Mathew XIII chapter. The sermon was good. Holton can preach if he is ugly. Before meeting broke, and while the good people were bawling and squalling, and kicking up a considerable dust, Mary and I left for "home, sweet home."

*August 21 Tuesday.* A rainy day. I and Caleb hauled one load of wood, which as about all the work we done out of doors. Took Mary to see her pa, who has something the matter with his under lip, which has caused it to swell very much and is exceedingly painful.

*August 22 Wednesday.* Rained some in the morning. I came home from Mr. Beall's, and I and Caleb went to hauling Dry-house wood. We hauled three loads. I killed Pa's old stand of bees. It was two years old, and had old black comb enough, but not quite so much honey. One wretched bee, by way of retaliation, stung me just under the left eye.

*August 23* My face is considerably fuller on the left side than usual, owing to that sting I got last night. Fixed bows on my wagon, which are made of small hickory poles split once and shaved to the right thickness prepatory to going to the camp meeting at Temple Hill, which is to commence on the 24th inst.

*August 24 Friday.* The day was spent in putting things in readyness for the campmeeting, etc. Rained a light shower in the afternoon.

*August 25* Quite a warm morning, as there is but little air stirring. We rolled off for the campmeeting, and got there about 2 o'clock. We found but few people assembled; a few more came in about night. Our first work after we got there, was to establish ourselves, for a few days in our tent, which was erected of small pine logs hewed on 2 sides and notched up. The size of it was about 12 by 16 feet—quite a good tent. At the sound of the horn for the 3 o'clock sermon, the people gathered to the stand, and the Rev. M. T. Wood, a young man from Randaulph [Randolph] Co. N C., preached quite a good sermon from the text, "Work out your own salvation with fear and trembling, for it is God" etc Philippians 2 chap. a part of the 12 with the 13 verse. After services Clark, Alexander & I fixed up a table at the side of our tent, and over it we erected a small brush arber. At night Pa preached, his text was the first verse of the VI chapter of 2 Cor. "We then", etc.

After preaching we went to our tent and soon were sleeping soundly, but alas! our rest was soon broken, and sleep driven from our eyes by the noisey rain drops which came down in torrents making a tremendous clatter upon the roof of our tent, and by "the dreadful thunder" which roared long and loud.

*August 26 Sunday.* Things are a little wet. Thomas Nailor preached at 8. His text was "All scripture is given by inspiration of God" etc. II Timothy III. 16.

At 11 Pa preached. His text was "There was a certain rich man" etc. from the 19 verse to the end of the chapter. Luke XVI. At 3 the Rev. Mr. Braim, the preacher in charge, preached. The text was "Yes, doubtless"' etc. Philippians III. 8. Mr. Braim preaches mildly, and uses but little energy.

At night Mr. Wood preached another good sermon. He preaches with energy. His text was "We have sinned," etc. Judges X. 15. To night Mr. York and another young man presented themselves as mourners in the altar, but neither of them professed.

*August 27 Monday.* The Rev. T. Nailor preached again at 8. The text, from which he preached quite a good sermon, was, "For this purpose the Son of God has manifested," etc. First epistle of John III and latter part of the 8 verse.

At 11 Pa preached again. His text was, "The last enemy that shall be destroyed is death." I Cor. XV 26. He preached the funeral of Hiram Reding, a young man who died last winter. At 3 Mr. Wood preached another excellent sermon. His text was, "Because there is wrath," etc. Job 36 ch. 18 verse. Before night it rained and wet the straw and seats under the arber, so the congregation assembled in the church for night meeting. There was no preaching, but the congregation sung, a few prayed, and many shouted aloud the praise of Israel's God.

*August 28* After being dismissed at the stand, we loaded up our dry goods and rolled out for home. Among all the campmeetings that ever I attended I've no recollection of such a one as this. Notwithstanding the many gospels warning sermons, there was not the first soul converted during the meeting!

*August 29* Clear and warm.

*August 30 Thursday.* Commenced pulling fodder in the evening.

*August 31 Friday.* Sowed the turnip seed, worked the road, etc.

*September 1 Saturday.* Went to muster, and of all the musters that ever I saw, this was the poorest.

*Sept. 2 Sunday.* Staid all day at Mr. Beall's, and read the Bible, an interesting piece in a book entitled *Excellent Narratives,* headed "It's all for the best." What more need a man have to contribute to his happiness than good books to read, and

"A pretty, little, loving wife,
To chase dull care away?"

*Sept. 3 Monday.* Clear and warm. Pulled fodder.

*Sept. 4 Tuesday,* Thrashed our wheat this afternoon.

Ann Eliza Martin's funeral to day at Aylesbury, but it rained too hard for the people to assemble or the preacher to come without being in danger of a "shower bath," (which by the way would, no doubt, do some of our unwashed neighbors a vast amount of good, as the large drops of pure cold water might remove the scales and dirt which have for years been accumulating on their filthy persons, open the pores of the skin and cool the fever occasioned by filthyness of person), so I suppose, for I was not there, the funeral was not attended to.

*Oct. 1* Came home early and went to ploughing, but soon tired of that and went to pulling pea vines. This I found to be rather worse than ploughing, but still I pulled on till most night.

*Oct. 2 Tuesday.* Hauled rails, ploughed, etc.

*Oct. 3 Wed.* Ploughed. Went to Mr. Beall's to see my darling wife, Mary.

*Oct. 4 Thursday.* Commenced sowing wheat by sowing and harrowing in one bushel in the orchard field. I turned over the soil with a twisting plough, soaked the wheat in Blue Stone, or Blue vitrol sowed it down while wet and then harrowed the land over twice. If we would raise wheat, in this country where the soil is so very thin, and where the crop is subject to so many diseases, we *must* take pains, and the more the better.

*Oct. 5* Sowed 1/2 bushel of wheat, ploughed, hauled up the pea vines I & Caleb pulled on Monday, etc.

*Oct. 6* Saturday, Quite a rainy morning. We have had nice weather for business this week, and now at the close of the working period—six days—He who makes all things work together for good to those that love Him, is sending His rain to water his earth, on which we live and prepare the soil for the seed to be sown next week. "Surely God is good to Israel." I went to Couche's store and purchased with bee's wax the paper on which I now write, 1/2 lb. of black pepper, a blue water bucket.

*Oct. 7 Sunday.* This morning there was quite a frost, followed by a pleasant sunny day. I and Mary went to our cousin William Martin's a distance of about four miles from Mr. Beall's. We found Martin at home, and a queer man he is. He talks but little, and that little in a low soft tone of voice. Martin is a close, hardworking, money-loving kind of a man, has but one eye, or can see with but one, and appears never to have done a great deal of harm any way. Sunday evening I and Mary came on to Pa's.

*Oct. 8 Monday.* Another small frost and a nice day. Ploughed.

*Oct. 9* Not much frost, if any, and a few degrees warmer than Sunday. Ploughed to day also.

*Oct. 10 Wednesday.* A pleasant day.

*Oct. 11* Finished breaking up the orchard field—all of it that I shall sow in wheat—the south side only. Hauled hay, or weeds for there are more weeds than grass in all that we hauled this afternoon. I and Caleb went to old Tom Benbow's to a corn husking, but it commenced raining soon after we got there, so we shucked but few ears of corn. We went in, as we could not shuck corn and was anxious to do something and assisted of emptying the plates and dishes of their delicious contents. That done, we went in and took seats in the sitting room, chatted with Dr. Benbow, who arrived just before it set in to raining, for some half hour, and then struck out thro' the dark for home. It was so dark that Caleb, to keep the path, had to hold to the skirt of my coat, while I led the way. We arrived safe home about 9 o'clock.

*Oct. 12 Friday.* Finished hauling hay. Quite cool, as the weather changed last night.

*Oct. 13* Split and hauled wood and pine, and in the evening late took Mollie to her pa's.

*Oct. 14 Sunday.* To day Pa preached Fredrick Reinhardt's funeral at Aylesbury. His text was "Honor thy father and mother," Ephesians VI, and a part of the second verse. As large a congregation I never saw at Aylesbury as was in attendence.[14]

*Oct. 15* But little frost. Quite a pleasant day. Commenced gathering corn. Sold my wagon, which cost me $57, to Thomas Jennings for $56![15]

14. The entry for August 1, 1855, described Fredrick Reinhardt as "a fine citizen and an exemplary Christian" and predicted that he would be "much . . . missed both in the world and in the church."

15. This, like most of Thomasson's efforts to make a profit, proved disappointing.

*Oct. 16* Pulled corn in forenoon, and hauled that which we pulled yesterday in th afternoon, and at night shucked it out.

*Oct. 17* Hauled corn in the forenoon, and four loads of manure in the afternoon, and night went to Mr. B's as Mary is staying there this week. Found them all well, and Mary standing in the door anxiously looking out for me.

*Oct. 18 Thursday.* Clear and quite pleasant. Hauled manure on the wheat ground.

*Oct. 19 Friday.* Somewhat cloudy. Went to the General Muster at Y[adkin]ville. There were eleven companies in parade. Clark mustered to day for the first time.[16] Col. Spiers was the chief commanding officer. Mustering appears to me to be a childish practice.

*Oct. 20 Sat.* Nearly clear. I & Clark put up the corn which grew on the field round the house. I can't write to day, so I'll quit.

*Oct. 21 Sunday.* Quite a pleasant morning, as the sun shines but dimly thro' the thin white clouds which hide from view the bright blue sky. "Of all the months that crown the year" Oct. is surely the most pleasant. In this month the corn is gathered, the wheat sown and the fruit, potatoes etc. laid away for winter; in this month the trees shed their yellow, red and various colored leaves, reminding us that we too must soon be striped of our all on earth, and, if worthy, be clothed with brighter and unfading garments; in this month Nature looks beautiful but solemn; in this month the vegetable kingdom lays at the feet of men her ample store of provision and bids them "slay and eat," but their depraved appetites are not to be satisfied with mere vegetables, which, in our opinion, is the only natural food and therefore the best ever tasted by man, so they attract the faithful ox, innocent lamb, pet pig or some other member of the Animal Kingdom, drench themselves in blood and make a near approach toward canableism. O! that we had a few Casper Housers and Luthers to reform our taste, and thus bring about, or establis[h] a purely vegetable diet throughout the world.[17]

16. Clark reached the age of twenty-one on September 10.

17. The reader will notice that although Thomasson favored a vegetarian diet and praised his meatless meals, he raised meat for his family and probably ate some meat.

*Oct. 22 Monday.* Rained some early in the morning. Sowed wheat in the evening.

*Oct. 23* Finished sowing wheat in the orchard field. Cloudy but no rain.

*Oct. 24 Wed.* Rained most of the day. Mended my shoes, cut wood, etc. I & Mollie staid last night at Alexander's, and I read aloud the story of Ellin & Robin, a poem by Thomas MaCeller of Philadelphia. It's a very touching piece, and contains some excellent advice to young ladies. The following stanza should be laid away and frequently remembered by all husband seekers.

"Beware of him whose speech is smooth,
The mother spake her daughter,
The deepest depths are ever found,
Where flows the smoothest water."

Be advised, young ladies, and "look before you leap."

*Oct. 25 Thursday.* Thanksgiving day.[18] Meeting at Aylesbury. Pa, Cal, & Martha are gone. I & Caleb put up 3 of Pa's hogs, in a lot, for porkers, and cut and hauled 2 loads of wood, all in the forenoon! Jennings was to have been after my wagon last week, or the first of this week, but he has not come yet.

*Oct 26 Friday.* Pleasant weather. Sowed wheat.

*Oct. 27 Saturday.* A beautiful day, Can any month in the year beat old October for nice days and pleasant weather? Ploughed most of the day. Went to Town and to Mr. Beall's late in the evening.

*Oct. 28 Sunday.* Still pleasant. Staid at Mr. B's all day, and read the *Weekly Message.*

*Oct. 29 Monday.* Sowed wheat. Rev. Thomas Nicholson came up to look at my wagon but he did not trade for it.

*Oct. 30 Tuesday.* Ploughed till 2 o'clock then went to Mr. Beall's corn husking. We got thro' about 11 o'clock at night, as there were but few hands and a tolerbly large pile of corn to shuck.

18. Although a day of thanksgiving following the gathering of crops was traditional, the United States had no common day for the festival until 1863.

*Oct. 31* Wednesday, and last day of October. Came home and brought Mary with me. Sowed wheat. Mr. Beall helped me plough to day.

*November 1 Thursday.* I and Mr. B. finished sowing wheat in the "house field" to day. Nov. come in with a warm day, and will perhaps go out with a cold one.

*Nov. 2* Cloudy most of the day, and rained a light shower in the evening. Sowed wheat.

*Nov. 3 Saturday.* Still warm and cloudy, and a little rain in the morning. I & Clark hauled the boards on my wagon which Wesley Pardue made for Pa last winter. We brought 644. At night Alex & Jacob Johnson, Caleb & myself went hunting and caught two opossums, but they were not large.

*Nov. 4 Sunday.* Cloudy early in the day, but the sun shone out about 9 or 10 o'clock. Unusally warm for Nov. This is a pleasant day, indeed, and while I write Mary is sitting near with the *Bible* & *Sunday Book,* and at my right elbow is a white pitcher of red and white Winter Roses—emblem of love and purity, and just before me lies that great "Store house of the English Language"—*Webster's American Dictionary.* Took Ma and Mary to Mr. Beall's in the evening.

*Nov. 5 Monday.* Cloudy and raining. Worked in the blacksmith shop with Wiley Messick, Scenior.

*Nov. 6* Still rainy. Tried my hand at coopering, and made quite a rough vessel of the species of the tub tribe, and kitchen name of which is Kuler.

At night the dogs, Sanco and Juno, kept up almost a incesent barking, and all the inmates of the "house of Thomasson" were wondering what could give the dogs so much trouble. One said "there must be a strange dog about" while another supposed "some michief brewing" this dark and dreary night, and enquired from whence the trouble of the dogs seemed to proceed. The watcher at the door replied, "They are barking towards the shop." "They are barking at Strong's Kuler(?)," said the youngest of the company seated round the "log fire", but by no means the most stupid one of the family. The sentenal then left his post and retired to bed while the mother and children united in a hearty laught at the idea of "Strong's Kuler(?)" occasioning so much trouble and noise among the dogs.

*Nov. 7 Wednesday.* Still cloudy and rainy, and no prospect, that I see, of fair weather.

In finest mist the rain comes down
Making the day quite gloomy;
But I must leave and go to "Town,"
And make the *room* more *roomy*.

*Nov. 8* Rained all day. I staid at Mr Beall's.

*Nov. 9* Cleared off at last after 4 or 5 days raining. Finished gathering corn, and hauled half the day.

*Nov. 10 Saturday.* Finished hauling corn in the forenoon, and hauled wood in the evening—4 loads from Mrs. Craff's new ground, and 2 from our woods. Went over to Mr. Beall's late in the evening, and found my Mary busily engaged ironing the clothes.

*Nov. 11 Sunday.* Cloudy and warm. Rained a little about the turn of the day. Heard Mr. Van Eton preach the funeral of Mrs. Martin's child, at Aylesbury, from the following text "Is it well with thee? Is it well with thy husband? Is it well with the child? And she answered, It is well."

It rained some about the turn of the day. I & Mary came over home in the evening.

*Nov. 12 Monday.* Ploughed in the "second bottom" with the two-horse plough. Cloudy. Finished husking out the corn at night.

*Nov. 13* Ploughed and sowed wheat. Exceedingly warm for Nov.

*Nov. 14 Wednesday.* Nearly clear, and quite warm. Finished sowing wheat. Guess I've sowed about 8 bushels. Some I harrowed in, after breaking the land with the one-horse turning plough; and some I sowed down and ploughed in with the above named plough, and some with the Scooter. About half of the "Second bottom" I turned with tbe big plough, sowed the wheat and brushed it in with a large pine brush, and the other half of said bottom, I sowed down and turned it under with the last mentioned plough, and then brushed it with the said pine brush. Now which of the wheat will do the best, that which was put in with the plough, brush, or harrow?

*Nov. 15* A nice warm day. I and Clark hauled a sawlog to Rev. T. Nicholson's and brought back a small load of flooring plank.
In the evening we hauled shed and smoakhouse timber.

*Nov. 16* Quite foggy early in the morning. Hauled another log to the mill in the morning, and shed and smokehouse timber, fire-wood, rock, etc. in the afternoon.

*Nov. 17 Sat.* Foggy and warm as yesterday. Took 3 bushels of corn to Benbow's beating mill to be beat up cob and all. Went to Hamptonville and bought of Jo. Cowles, Scen. [Senior] 4 yds bleached domestic, 15 cents per yd; one wash pan 25; one lard kan $1.25; 2 papers of pine [?] 18 3/4; Almanac for 1856 .05; one hand-saw file .15; 9 yards of bed ticking 33 1/3 cents per yd. Recd no news

*Nov. 18 Sunday.* I and Mary went to the Quaker's monthly meeting at Hunting Creek Church. The congregation was not very large. The Rev. Martin, a Quaker minister, talked interestingly on the peculiararities of their church, giving the reason why they discard Baptism, the Sacraments, etc.

*Nov. 19, 20, 21, 22* Cool and dry, tho' cloudy part of the time. I & Pa hauled sawlogs to Rev. T. Nicholson's mill, a distance of about 6 miles.

*Nov. 23* Cool and cloudy.

*Nov. 24 Saturday.* Nothing as I remember of importance.

*Nov. 25 Sun.* Cloudy and rained moderately. Staid at Mr. B's all day.

*Nov. 26* I & Clark cut a tree to make boards to cover my house.[19] Steven & Henry Godfrey came after my Black-smith tools, and took them away.

*Nov. 27* Hauled & rived some of my board timber.

*Nov. 28* I & Pa took a log to the mill in the forenoon, and I & Clark sawed board timber in the evening.

19. Thomasson is referring to the house that he and Mary will occupy as their separate residence. At this point they were still living with his or her parents.

*Nov. 29* A little cloudy and quite cool. I & Clark went to Yadkinville, bought of Houser & Wilson; 40 lbs of nails, a few screws etc. I bought an ax for $1.25. This is the first ax I ever bought, and it proved so far to be very good.

*Nov. 30* Quite cold early. Pa & Clark went to the sawmill, and I rove boards, boalted timber [?] etc.

*Dec. 1 Sat.* Cloudy and looks like it might snow. Holed timber[20] and rove boards.

*Dec. 2 Sunday.* I & Mary went to Mt. Moriah to meeting. The Rev. T. Nicholson & Pa preached the funeral of two of Thomas Lindly's children.

*Dec. 3, 4, 5, 6, 7, 8* Nice weather for business—cool and dry. Worked on my house. Bought of the widow Hutchins, one sash plane, one bench crew and one hoop saw, all for $1.

*Dec. 9 Sunday.* Cold and rained in the morning quite hard. The wind also blew strongly. Staid at Mr Beall's all day.

*Dec. 10* Very cold and windy. Worked some on my house but it was rather cool for me.

*Dec. 11 Tuesday.* Cold but not windy as yesterday. Worked on my house. This is a cold night—too cold to write.[21]

*Dec. 12* Staid last night Mr. Beall's with my darling Mary. Came home early this morning and finished facing the windows and doors of my "log cabbin" and commenced "chinkin" the wall.[22]

*Dec. 13* Finished the work I commenced yesterday, and made a two-light slip shash for the small window of my house. This is my "first born" in the shash line, and is not very well jointed, but it's a bettter shash than some John E. Grant made, for this same cabbin while I taught school in it, and charged me ten cents a light for.

20. The meaning in this and the previous day's reference is obscure, but perhaps Thomasson is referring to some method of joining wood akin to mortise and tenon construction.

21. This entry suggests how poorly the houses of yeoman farmers were heated and insulated.

22. This process—filling in the cracks between logs with mud—had to be repeated almost every year.

*Dec. 14* Cloudy and warm, in the morning, but turned cool and rained considerably in the evening. Mr. Abnew Davis, the book distributor, staid with us last night, and this morning I bought of him *Sears' Pictorial Description of the U. S.* Price $2.50. I & Mr. Alexander Hutson daubed my house, laid the hearth, and wouled up on each side of the fire place. A pretty fair day's work. Caleb (bro) and Calvin Johnson (hired) assisted us some. This is a gloomy night, and

"The rain comes down with all its might
Darkening the face of the sullen night."

*Dec. 15 Saturday.* Rained moderately all day. Made one door shutter for my house. Went over to Mr. Beall's in the evening, and staid all night.

*Dec. 16 Sunday.* Still cloudy and rained some. I & Mary came home in the morning, and examined my new book. I like it tolerbly well and regret that it is not better bound. Book binders, like some of other trades, greatly slight their work at times. They surely forget the old saying, "Whatever is worth doing at all is worth doing well," and throw aside the book ere it's half bound. Such hands ought to be dismissed, and others employed that would do honest work.

In the evening I and Mary walked out to see our new home. The rough walls and fire place afforded Eliza & Caleb (they being with [us]) a good deal of merriment.

*Dec. 17 Monday.* I rose this morning about 4 o'clock, as usual and wrote this page before day. Not exactly clear, but quite pleasant. Dressed off and put railings in my bedstead. Scowered my house, etc. Went to Mr. Beall's at night to make ready for moving the next day.

*Dec. 18* I write to night in my own house, as I and Mary moved to ourselves to day. We have set up some of what little furniture we have, and lined the rough, clay stained walls of our house with pictures, newspapers, advertisements etc. in order to hide them as much as possible from view, and to make our little house look cheerful and enviting.

*Dec. 19 Wednesday.* Cloudy and cool. Made the shutter for the East door of my house, hung it, and in the evening I & Caleb hauled three loads of wood and a few sticks of pine to my house. I rose this morning about 3 'clock.

*Dec. 20 Thursday.* Made a 4 light sash for the East window of my house, put glass in it and in the sash I made previously and placed them in the window.

*Dec. 21 Friday.* A very frosty morning. Fixed up a fire board, water shelf, etc. in the forenoon, and in the p. m. I went to J. Couche's store and bought one set of knives & forks—price .80 cts; two small dishes—10 cts. each; one pair of shears—30 cts; one saltsellar—10 cts; 2 lbs. coffee, for Pa—16 2/3 per lb; 2 lbs sugar—12 1/2 cts. per lb; one box of small tax—.05 cts.

*Dec. 22 Saturday.* Cloudy, and rained moderately most of the day. I & Pa went to Leonard Messick's to make a bedcord, and as we walked along he told me the following story of a murder case in Missouri.

There were on a certain day an old man and a doctor in Jefferson City. The old man received $900 to be paid at the Land Office. The Dr. received a rifle gun to be left where he expected to stay that night. The old man and the Dr. both ready to leave the city they started off together; the old man with the $900 dollars, and the Dr. with the gun. They had not traveled far however when the old man as he was an old hunter, thought it best that he should carry the gun; and being opposed to carrying an empty gun, he took the gun and loaded it. Now the old man had the money and gun, too, and thus they traveled on for some distance. By and by they came to another road and here the old man gave to Dr the $900 and requested him to leave it at the Land Office, and took a different road from the one the Dr[.] was going to travel. The Dr. had not gone far when he heard a gun fire, and soon another fire, but no doubt thought nothing strange as that was a country in which game was plenty, and knowing that many lived by hunting, so he rode on to the place where he intended putting up that night. The old man did not get home. Early the next day his friends set out in search of him. It snowed the day the old man and the Dr. left the city, and when the old man was found the following morning he was lying dead on the cold snow having been pierced through by a rifle ball. The same gun the old man thought it proper that he should carry instead of the Dr., and had loaded with his own hands was now lying at his side empty, and about there were other tracks than those of the horse the old man was riding. The strange track was followed, and strange enough it led to where the

Dr. staid all night. Now the neighbors began to suspicion the Dr. The citizens of Jefferson knew that the old man & Dr[.] left the city together, and that the Dr. knew that the old man had the $900, and as the same rifle the Dr. received in the city was lying at the old man's side, and as there were the print of a horse's feet leading from w[h]ere the old man was killed to the house the Dr. lodged in that night, Public Opinion said, at once, "The Dr. must be the murder[er]." So the officers of the law were seen in pursuit of the Dr., and when they found him, poor man, he was sitting by his own fireside counting the $900 which the old man had received in the city the day before. This was evidence enough. The Dr. was taken and lodged in prison to await his trial. The day of trial rolled round. Witnesses from the city swore that "The old man and the Dr. left the city together, the old man having $900, and the Dr. a rifle gun." This same gun, and a powder flask, found with the gun, at the old man's sid[e], having the initials of the Dr.[']s name and profession carved upon it, were brought into court. The family, with whom the Dr. staid on the night the old man was killed, was questioned as to who staid there that night. "The Dr." "No one (save the family) but the Dr?" "No other person." Circumstancial evidence good against the Dr. so far, and when the officers who took the Dr. gave in their evidence to the effect that the Dr. was setting at his own fireside counting the old man's $900 when they enetered his house, his fate was sealed. He was tried, condemned to hang, and the day of exicution appointed. The day rolled round and early in the morning a large company of the Dr[.]'s friends and acquaintances assembled to see the man who had for years filled his place as a respectable physician honorably, and who all thought to be an honest, law abiding man, but now under the sentence of death for the awful crime of *murder.* Among others that assembled to witness the execution, was a free spoken, inquisitive kind of a man who wished to know the cause of every effect. This man soon after he arrived at the place where justice was pretended to be dealt out to all men according to their dues, and before the hour appointed for the exicution of the condemmed arrived, he entered the clerk's office and while interogating that officer on various subjects, he happened to see the identical powder flask which had been found with the gun at the old man's side, it having been left in court from some cause or other, and instantly sung out as follows: "Clerk what have you got my powder flask here for?" "Ah,["] replied the C., "I wish it was yours, instead of the unfortunat Dr. ______'s" "It is mine,["] said the owner of the flask, "and I'll swear it. Here are the first two letters of my name, and next two—M.D. are the initials of my partner's name in the flask." "You see, Clerk, neither of us was able to buy a flask, so we bought this one in co[mpany], and I carved these very letters with my own knife. The flask is *ours,* and I'll swear it." Here the scale was about to turn in the Dr.'s favor. Public opinion began to loose the chains he had so tightly bound the Dr. with, and commenced winding them about the bodies of the owners of the powder flask, who were soon arrested and imprisoned. The Dr[.]'s life was prolonged six days; the witnesses recalled, and the trial renewed. The force of Public Opinion being turned in a different channel[,] hope for the Dr[.]'s aquittal revived, his old friends returned with greater interest in his favor, and the witnesses were examined more closely. The family with whom the Dr. staid the night the murder was committed was again questioned as to who staid with them that night. "Are you certain no person staid in your house on the night the murder was comitted but the Dr?" "Let me see. Yes Mr. ______ a young man staid there too, and it was late in the night before he arrived." This "Mr. ______ a young man" was the partner in the flask of whose name the M. D. carved on the flask were the initials. Circumstancial evidence growing so plain in favor of the Dr. and against M. D., and he being so overwhelmed with guilt confessed guilty, and gave the history of the whole affair. He knowing that the old man had received $900 in the city, waylaid him and shot him with his own rifle, which carried a smaller ball than the one the old man was carrying, (this the examiners in the case found out on the last trial) and then shot off the gun the old man was carrying, which made the second report heard by the Dr[.], and threw it down by his dead body, searched for money but found none, (as the old man had given the money to the Dr. at the forks of the road,) mounted his horse and rode rapidly away.

*Dec. 23 Sunday.* A little hazy, but quite a pleasant day. I & Mary staid at home till evening and then walked out to Pa's, but we soon returned and spent the later part as we had the morning—in reading, etc.

*Dec. 24* I am not in the habit of telling my dreams, much less recording them, but I had such a singular diverting one last night, I guess I must give it a place in my Diary. It is as follows. Mollie, her sister Jane, my sister Mary, Mr. York and myself all went to meeting. Mollie and the two girls had preceeded Mr. York and myself, and as we went on (we went by Thomas Benbow's house and oil mill and, just up on the hill not far, stood a very old meeting house). Mr. York asked me if I had "religion." I told "I hoped so." Then he asked me if "a man could not get to heaven by being an honest, upright man without a change of heart." I told him "there is no other way than by faith in the Lord Jesus Christ." Then he told me that he had religion. When we arrived at the old church before mentioned the congregation had assembled, but no preacher had come yet. Soon the class Leader came, and enquired of me why I had not been there sooner. I told him I was there soon enough. "But, why have you not been here to meeting before?" asked the C. L. Then "*Thinks-I-to-myself*" I have lived not far from here about two years and this is the first time I've been here! My answer to the C. L. I do not remember. By and by Uncle Gunn came and took his seat in the congregation. Some person rose up in the pulpit and commenced talking. I do not know whether he read out a text or not—he said but few words anyway, so few that a remark he made at the close was considered his sermon. It was "*Put in Armstrong's shirt tail!*" The congregation "snorted right out." I looked down, (I was sitting some distance from the speaker, *flat on the ground*) and seeing a large portion of the tail of my shirt exposed to view, I felt some what embarressed, but however, I joined in the general laugh which was kept up by being now and then renewed by a witty remark from some one in the congregation concerning the *sermon,* my shirt tail etc. By and by Uncle Gunn said "Let's quit." He rose up, gave out a hymn, and handed the book to me with a request to "pitch and carry the tune." I was a little *bothered,* but however I started up "Auld Lank Sine." We sung the hymn, Pa pronounced the "benediction" and every man was off for home. Guess I made as much haste as any for my shirt tail was still hanging out.

*Dec. 25 Tuesday.* Christmas and I think, as rainy a day as ever I saw. Brother Wiley landed home early. I suppose he staid last night at "bridge Isaac Long's." He has been teaching school in Rockingham Co. and this is his first visit since Easter. This day 12 months ago, I, a single man dined at Uncle James Browning's in Forsyth Co., and to day I and my wife ate Christmas dinner at Pa's. How things have changed since last Christmas day which seems as yesterday! Time flies swiftly and

"The more we live more brief appear
Life's succeeding stages."

*Dec. 26* Clear and very cold. Mr. & Mrs. Beall came over to see Mollie & I for the first time, dined with us and returned home about night.

*Dec. 27 Thursday.* A little cloudy and cold. I and Pa killed our hogs—4 of them. Alex and brother Wiley helped us. Think I have about 300 lbs of pork.

*Dec. 28* Raining and sleeting; and so it continued till about 12 o'clock, when it turned warmer, as usual after a sleet, and the ice melted to the great joy, perhaps, of the vegetable Kingdom. Fixing up our cupboard, (the old bookcase) and cutting out 2 pair, and closing one, of shoes for myself completed the labors of the day.

*Dec. 29 Saturday.* Still sleeting. Yesterday and to-day will count for winter—especially in the South. I suppose a few hundred miles north the weather is, on an average colder the winter through, than any of *our* coldest days. Sat within doors most of the day, as it was to[o] disagreeable to be out, and quite agreeable to be in warming distance of a good fire. It is a great thing to have a good, warm house to sheald us from the blasts of winter. How thankful we should be for our houses, warm clothes and all the comforts of life. Many there are in the world who are compelled to beg their bread from door to door, and to seek shelter from the stormy blast beneath another's roof. Yea, even "the Son of Man had not where to lay his head."

*Dec. 30* Sunday, Cold and hazy. Brother Wiley came out and sat with us till bout noon, and we read and talked, and had quite a pleasant time. When shall we pass another forenoon so pleasantly?

"Oft shall glowing hope retire,
Oft shall wearied love expire."

I fear, ere we are again thus situated.

In the evening I and Mollie went to her Pa's for the first time since we moved to ourselves.

*Dec. 31* Monday—and last day of the year 1855, as well as of December. Cloudy and very cold. Looks "for all the world" like it would end the dieing year with a snowstorm. Went to H[ampton]ville, and bought the following articles;—one "dishpan" .50; one set of plates 30; one set tea cups & saucers .40; and a few other little notions. Moved over Mollies chickens—six—five hens, and a "tender young pullet." I hauled one load of wood to begin the New Year with, and thus completed the labors of the day, month, and the year 1855.

# Diary 1856

*January 1.* No snow yet, but there is, this morning, a large circle around the moon which is said to be a sign of falling weather. I rose early, and wrote for a speciman [specimen] No. of each *The Water cure Journal* and *Life Illustrated*—New York. And *The Country Gentleman* and *The Cultivator*—Albany, N. Y.

*Jan. 2 Wednesday.* Cold, cloudy and sleeting. Shadrack Beall came over yesterday evening late while I and Mollie were at Pa's making our *Sossingers,*[1] and this morning carried away the hogs, (a sow and 4 pigs, nice ones too) which I sold to his pa. Made a pair of shoes for myself, with Pa's assistance, he doing the lasting and triming, and I the peging. Pa also worked some on a pair for himself.

*Jan. 3* Rained all day yesterday, last night, and is still raining this morning. The above was written early this morning, perhaps before day. It cleared off about 12 o'clock quite pleasant, but its now turning cold very fast. I don't know that I ever saw the serface of the earth softer and fuller of water. The creek is past foarding.

*Jan. 4* Quite cold, and not exactly clear. Worked the road, or walked over it and talked about what ought to be done, the weather, etc. I believe it was a given up point, among all hands, the road *needed* a great deal of work, and that it would snow to-night. We finished the road about one o'clock p.m, to the great joy of all hands, and *scattered off* to our "respective places of abode." In the evening I hauled 2 loads of wood for myself, and 1 for Pa.

1. Thomasson was making a kind of sausage.

*Jan. 5 Sat.* Snowed last night sure enough, and to-day also till about 12 o'clock. Dont think I ever saw it snow faster. It is a real North East snow storm. This is the first snow for the winter of '55–6, and the deepest one we've had "for lo! these many" winters. Did little else than make fires and set by them, as the snow has rendered out-door work disagreeable. I guess the snow is, on an average, about a foot deep.

*Jan. 6 Sunday.* Clear, and, but for the snow, the most pleasant day we've had for a time. Read the XXVI ch. of "The Acts"—St. Paul's speech, before King Agrippa, in defence of himself. I also read the XXV ch., and—lots more.

*Jan. 7 M[on].* Took a small search for rabbits but found none. The depth of the snow kept them *in bed* last night, and as they "hide in secret places" they are not easily found to-day, tracks being scarce. Clark and his co. caught 14 this forenoon notwithstanding. I hunted birds, but was only lucky enough to scare them up, and see them fly away. Dont think I was *cut out* for a hunter, or if so, I was screwed up to[o] tight, for I am entirely "to[o] slow on triger." Just before night it snowed again very fast for a few minutes, perhaps 10 or 15.

*Jan. 8 T[ues].* Cold and a little cloudy. Snow melted but slowly. Went out early and tracked a rabbit to his bed, but he was so completely hid under the sedge and snow that I could only see where he passed in and out. I let him set and went to the house after my gun, Jo. Beall, and the little dog—Juno, expecting to fire upon Mr. Rabbit and kill him in his nest, or if I should miss, and he jumped up[,] to give him a fine race. I guessed at his whereabouts, blazed away, but saw no rabbit. "*Thinks-I-to-myself*" he may not be here, so to be certain I gave the sedge and snow a few flirts with the toe of my shoe, and out jumped the rabbit and away he ran. I, Jo. and the little dog took after him in full speed, but he soon left us far in the rear if the snow was deep, and our legs the longest. We soon lost sight of him and his tracks too, and went home to warm our toes, and break our fast. After starting said rabbit twice more, and as often shooting at him, Sanco, Pa's dog, caught him. I shot one in his bed—the first, and only one I've killed this snow. Had better luck to-day. Who knows but I'll yet be a Nimrod?

*Jan. 9 Wednesday.* Coldest day we've had, I think, in four years. I and Mollie went to her Pa's in the evening.

*Jan. 10* Still very cold. Mollie was very sick last night—had pain in her sides and back, and sick stomach.

I and Mr. Beall went to D[owthart]'s Cotton Factory to get lining and pading, but did not get any of either. We took a quantity of tangled yarn to make into bedcords on the shears: we are to have one third for making. Jane & Catharine rode home in the buggy, while Mr. B. walked behind to see that all went safe, and I took it on foot by Hamptonville, and bought 2 1/2 yards, lacking 1/8, of flannel for Mollie, 50 cts. per yd; 1 pitcher 45; ginger 6; black thread 10.

Came home in the evening tho' Mollie was hardly able.

*Jan. 11* Cold and cloudy. Mollie quite unwell yet, tho' she rested better last night than night before. Cut some wood in the forenoon, and "toward the edge of the day["] I hauled two loads, one for myself, and the other for Pa. One week to night since the snow fell, and it's melted but very little.

Received a "Lottery paper" the other day in which I find the following good advice:

"Treat ladies' favor with respect,
Good will of woman ne'er neglect,
No man ever slighted woman yet,
But found good cause for sharp regret."

*Jan. 12 Saturday.* Snowed last night, and to-day, sure enough. The snow is, now, nearly knee-deep to a man! This the deepest snow we've had for many years. In fact, I don't recollect its equal in the snow line, and am in my 27 year. It quit snowing about the turn of the day, and hailed and sleeted till night, and then toped off with a sprinkle of snow. The ice on the snow is about half, or three quarters of an inch thick, and will bear the weight of a small boy, or a dog, but gives way at every step of a man, and he sinks almost to his knees in snow, and if he's not careful in lifting his foot, for another step, he'll bark his shin, or spot it blue and black with the sharp ice. Its very difficult, and somewhat dangerous for man or beast to travil just now, especially if they leave the "beaten track." Each and every crime has it's penalty annexed, and sometimes the latter so far exceeds the former, that we are not aware of our guilt until the evil consequences fall flat upon our heads. Such seems, now, to be the case with the crime of discovery, (the discovery of, and treading out new paths) the penalty falling on the legs instead of the head being the only difference.

*Jan. 13 Sunday.* Quite a nice day over head, with plenty of snow and ice under foot. Staid at home, and spent the day in reading, etc.

*Jan. 14 M.* Clear excep a few scattering clouds. Helped Pa kill three hogs, got wood, & hung the west door shutters of my cotage. Read, to night, the "happy Typo"—Mr. Gettyphat Take, who said; "The happiest day I ever spent was one time when I had[ ]not but one shirt and a pair of pants to put on, had spent all my money and gone hungry for forty hours."

A happy man is he,
Who thus can fast and be,
Always in good temper
Both summer and winter.

*Jan. 15 T.* Clear & cold. Made a 6 light shash, for window in the south end of my cottage, this afternoon, and put in the glass to-night. Mrs. N.[2] says this "Planet is in a very unlovely condition." So it is, in many respects. First. It is, just now, in an "unlovely condition" to thos who do not love snow and ice; and 2d, to those who hate briar patches, mudholes, snakes, spiders, frogs and lizards; and thirdly, to those who have an antipathy to distilleries, butcheries, tobacco fields, etc. etc. All these we have, and many other "unlovely" things, and yet we are commanded to love! for "love is the fulfilling the law," but we are not to love these things by any means, for they are all the fruit of sin. But we are to love the earth for what it contains that is lovely. We are to love truth and justice, our neighbors as ourselves, and God with all our hearts.

"In the cool grey dawn of morning,
When the swallows dart and sing,"

who can help loving this old grassy Planet? notwithstanding its polution.

*Jan. 16* Clear and cold. Snow melted some, tho' it is yet deep. "The

2. Mrs. N. is probably someone Thomasson has been reading.

fouls of the air" must suffer for food, though it takes but little to supply their wants, and that little they can steal from the farmer, who has enough, and to spare.

To the fowls of the air,
Which are depending always,
Upon their neighbors to raise,
The wheat, the oats, and the rye,
And when they have all dovour'd,
What nature in the wilds shower'd,
To the farmer's barn they haste,
And there eat, scatter and waste,
All they can, early and late.
How proud would be the farmer,
If some kind friend, or charmer,
Would teach those birds a lesson,
And learn them how to reason,
To reap, and sow, in season.

I have, to night, been reading Byron and Moore, and see the effects above. After looking over the above one might presume

"This world is not so bad a world,
As some would like to make it."

But let me tell you, generous reader, I did not write the above with a view of altering "this world," or any other "world," in any shape, form, or fashion. So let it pass for what it's worth and no more.

*Jan. 17 Th.* A clear, nice day. I and Mollie went to her Pa's in the evening, and in crossing the little branch that runs down from old Mr. Godfrey's spring five or six spokes, in the right fore-wheel of my buggy, broke off at the hub. The banks of the branch were high, and hard frozen, and the spokes being frozen too made them break so easily. I thought it was the axil-tree I heard break, and did not find out the spokes were broken, altho' I could see no crack in the axil, till I and Mr B. started to H'ville on the following [morning].

I and Mollie will have to stay at home now as our riding machine is broken down. It will, perhaps, be well enough to for us to "bide at home," for of all places on this broad snowey earth

"There's no place like home."

What would Sir John Franklin[3] and his ill fated crew have given, when they were perishing amid the icebergs, to have been at "home?" All this world, I dare say, had they have had it in their power. If a man has a home he had better content himself, and stay there—even if its a poor one its better than none.

*Jan. 18 F.* Still clear and fine. I, Mr. B. and Shadrack made 18 bed cords, and 2 or 3 pair of plough lines. We made them of No 10 cotton yarn from Dowthart's factory that was to[o] badly tangled to weave.

*Jan. 19 Sa.* A fine day. The snow fell one and two weeks ago to day, and I think, at least half of it is here yet, and for ought I know will be for many a day to come.

*Jan. 20 Sunday.* Cold and snowy, and the wind is whistleing mournfully around the corners, and over the snow covered roof of our cottage. This is the third snow we've had in a short time. We have been well blessed with snow this winter, so far. Read, in the latter part of B[ook]. V. of *Paradise Lost,* Milton's account of the cause, and "rebellion in Heaven." The "Omnipotent" decreed that to his Son "should bow all knees in Heaven, and shall confess him Lord." Satan, thinking this to[o] much homage to pay to one whom he considered rather his inferior, and being, thro' fear, unwilling to discus the subject in the "camp" which extend

"By living streams among the trees of life," fled to
"the quarters of the north,"
"and with lies
Drew after him the third part of Heaven's host", pretending to make preparation to receive the king. But when Satan reached his "royal seat," "he assembled all his train, addessed them "about the great reception of their king,

Thither to come, and with caluminious art
Of counterfieted truth thus held their ears:"

See Book V, page 158, for Satan's speech.

*Jan. 21 M.* Very cold northwest wind, with now and then fragments of snow and wind cloud moving over, and shuting out from earth and

3. Sir John Franklin (1786–1847), who pioneered English exploration in the Arctic, died trying to find the Northwest Passage.

its frozen vegetable production the warm rays of the sun. What cruel clouds! thus do deprive the innocent plants of their means of living. Went over to Buck Shoal Factory and bought 5 lbs of waste cotton, for 5 cts per lb. of which I and Mollie paded a comfort, and it is, indeed, comfortable to sleep under these cold nights. I, Pa and Caleb built a sheep shelter, by driving one fork down in the bend of the fence, letting one end of the rails rest on it, and the other end on the fence, and thickly covering over the top and north side with pine brush. The first lamb in my "little flock" of sheep, for this season came last night. It's black and quite peart.

*Jan. 22 T.* Still very cold, but not quite so windy. Cut a little firewood, etc.

*Jan. 23 W.* Clear & cold. Snow melts quite slowly. Cut and hauled firewood: It's about all the outdoor work one can do this weather. Received "a few lines" from brother Wiley, who writes that he and Uncle Flem caught about 40 rabbits this snow. G[u]ess they have "fir skins" for sale, and *bacon* too, perhaps. If so, they might, with propriety, stick up

Bacon and fir
For sale here.

*Jan. 24 Th.* Clear and cold. Made a small book-case.

*Jan. 25* Clear and windy. Spent the day in reading, fetching water for Mollie to wash, etc. Read to night, in the *U. S. Magazine* "The White-Stone Canoe"—an Indian Tradition in verse, of which the following are extracts.

"Where parted spirits meet again
Beyond the reach of pain and woe."

"We have no war nor bloodshed here."

"And virtues pleasant path pursue."

Resolved to hope and labor on
With better prospects in his view."

*Jan. 26 Sa.* Cloudy and commenced snowing about the middle of the day, and snowed on till next morning. This is the fourth "great snow"

we've had in about a month! The winter of '55–6 will long be remembered as the "snowy winter."

*Jan. 27 S.* Quit snowing about, a little after day-break, but is still cloudy. Read the X & XI chs. of "The Acts of the Apostles." Not many months since I conversed with a member of the Quaker Church, who told me the reason why their church do not believe in baptism is, because it was used (they say) as a tipe of the "Holy Ghost" which was to, and which *did* des[c]end on the "Apostles," Acts II, 4, and that it (Baptism) was there done away, as the scripture was then fulfilled so far as to baptism. Now if this had been the case would not the "Apostles" have known it? And knowing this, would Peter have commanded the "multitude," (when it cried out, "Men and brethren what shall we do?") to "repent and be baptised every one of you" etc? Would he not have left out the word "baptised" if the gift of the "Holy Ghost" had taken the place of baptism? It looks reasonable to us that, had this been the case, Peter, taught of the "Holy Ghost" as he was, would have been conscious of the act, and that he would not any more have commanded sinners to "be baptised," but merely "Repent, have faith in Jesus Christ and ye shall receive the gift of the Holy Ghost." But we see that the apostles continued to baptise such as believed, (Acts XIII, 12.) and that even some were baptised after they had received the "gift of the Holy Ghost," (Acts X, 44 and 48 verses.)

We are of opinion that the Apostles, notwithstanding some of them "were unlearned and ignorant men," knew as much about this matter as the most learned of our Quaker friends. We therefore choose to differ with them on this point, and walk in the way taught us so plainly both by the precept and example of the Apostles.

Night. "The day is past and gone
The evening shades appear,"

and I again essay to write a few lines in my Diary.

Mollie and I, in ease,
Have spent the day in re[a]ding,
Save what was spent in writing,
And in—*eating our peas.*

In the days of the "Apostles" there reigned, in Jerusalem, a wicked king whose name was Herod Agrippa. This wicked king "stretched forth his hands["] to vex certain of the church. "And he killed James the brother of John with the sword.["] And because it pleased the Jews, for him thus to persicute the disiples of Jesus, he took Peter also. And having taken him he lodged him in prison to await his trial. Now Peter was chained and the prison guarded by men who were appointed for that purpose. On the night previous to the day on which Peter was to be tried, an "angel of the Lord" came to him in the prison, "and smote Peter on the side, and raised him up, saying arise up quickly." Then Peter rose up "And his chains fell off from his hands." And Peter having dressed himself the angel led him out of the prison, and out of the city, and then left him. Peter at first thought he saw a vision, but when he was come to himself, he said, ["]Now I know of a surety, that the Lord hath sent his angel, and hath delivered me out of the hand of Herod, and from all the expectation of the people of the Jews."

Now Peter goes to the hous[e] of Mary and knocks at the gate, and Rhoda comes out to see who is there "And when she knew Peter's voice, she opened not the gate for gladness, but ran in, and told how Peter stood at the gate." Those in the house were loth to believe Rhoda's report, but "when they had opened the door and saw him (Peter) they were astonished." Then Peter told them "how the Lord had brought him out of the prison," and after commanding them that they should "show these things unto James, and to the brethren,["] he departed, and went into another place.

*Jan. 28 M.* Clear and fine. Snow some 8 or 10 [inches] generaly, and in places knee deep.

Read Geology, which being rightly understood, and the Bible correctly translated, would not disagree.

*Jan. 29 T.* Much as yesterday, tho' if any differance, a little colder, I believe, as the snow seemed to melt more slowly.

*Jan. 30 W.* Cold and windy. I and Mollie went over to Pa's and took dinner. Mollie helped Eliza quilt, and I shelled corn.

*Jan. 31* And last day of January. Not so cold, quite as yesterday. Cut and hauled two small loads of wood in the forenoon, and commenced making a sleigh in the evening.

*February 1 F.* Quite pleasant. Snow melting fast. Finished my sleigh, and at night I and Mollie went over to Pa's and (among hands) shelled out 9 1/2 bushels of my corn.

*Feb. 2 Sa.* Pleasant, but a little hazy—fixing to snow. Made a grainary and put up the corn we shelled last night, hauled two loads of wood in the evening, and at night shelled out and put up the ballance (10 bushels) of my corn. I have 19 bushels of corn stored away,—enough, I hope, to bread Mollie and me till next corn harvest, and some to spare.

*Feb. 3 S.* Snowing sure enough! This is the fi[f]th snow from the north east, besides one or two little flurries from the south west, in but little over five weeks.

Cleared off about 12 p. m. and I and Mollie to a ride in our *new* sleigh over to her pa's.

*Feb. 4 M.* Clear and very cold. Went to Town, received a specimen No. of each, the *W[ater] C[ure] Journal, P Journal* and *Life Illustrated.* On my return from Town I came by Lindly's shop to get my buggy wheel mended. Dont think I've come as near freezing this winter as I did to day.

*Feb. 5 T.* Still cold. I and M. came home, and had a slick ride too.

*Feb. 6 W.* A little hazy and not quite so cold. I and Mollie commenced our first pot of homany, and run it thro' the lye boil, washed it out, and set it to cook till morning. Sold my goats to W. R. Johnson at $1 a head. What will 2 goats come to at $1 a piece? Ans $2.00.

*Feb. 7* Cloudy and rained quite fast till about 8 or 9 A. M. It also sleeted considerably.

*Feb. 8 F.* Still cloudy, and warm compaired with some days we've had this winter. Quite foggy, or misty, most of the day, and *actualy* rained some. It may be that we have had our last snow for this winter, and I don't care if we have as the ground is now covered, in places, five snows deep. The snow is melting slowly, so I guess it will "get off" some time between this and August.

Hauled five loads of wood for Alex. Hutson in the forenoon, and he helped m[e] chop 2 or 3 hours in the evening.

Read aloud "Narative XXII" in P.N. The "Aged Sailor" was no doubt, converted in answer to the prayers of his parents, who had long since been dead, and whose spirits, perhaps, had been his guardian angels for many years.

Who can estimate the value of a pious parent's prayers? Of more worth are they than all the gold that ever has been, or ever will be extracted from the mines.

*Feb. 9 Sa.* Cold again. Snow melted but little. Hauled 2 loads of wood, made a board shelter over my wagon etc.

Gave Mollie her first lesson in Grammar. She learned and recited three lesson[s]! They were of course not very long, but I am well pleased with her start. Think she'll be a grammarian some day.

*Feb. 10 S.* A clear pleasant day over head, but somewhat sloppy under foot as the snow is melting a little. Shad. B[eall] & Jane came by from Benbow's quilting, and staid with us till after dinner. This is Jane's first visit since we moved.

It is not customary for us to go, or send to the office after news on the Sabbath, but Clark went to day, as we were all *news hungry,* and brought out *The Times* and two nos. of *The Country Gentleman* and one of *The Cultivator. The C. G. & C* are Agricultural papers published in Albany N. Y. Rather to[o] far north to be of much interest to a southeran planter, tho' they are good works.

*Feb. 11 M.* Cold, cloudy and snowed a little just after the turn of the day. There are said to be 600 varieties of snow flakes; the few that fell to day, I think, were round, and about the size of a pea.

*Feb. 12 T.* Cold and windy. Staid within doors most of the day. Only ventured out to gather in wood and water, and to feed [the livestock]. This must be a "foretaste" of March.

*Feb. 13 W.* Cold but not quite so windy. Snowed a small sprinkle last night, and is cloudy yet to day.

Thrashed oats for seed, or to sow. Oat sowing time will soon be here, and the "face of the *earth*" is yet covered with snow.

*Feb. 14 Th.* Rather colder, I believe, than yesterday, and a little cloudy this afternoon.

This is Valentine's day—a day sacred to St. Valentine. Webster says, "It was a very old notion, alluded to by Shakspear, that on this day birds begin to couple." From this notion, I guess, orrigionated the practice of sending valentines.

The following rule for weighing by measure I found in the *Times* Vol. I, No. 5.

Take a tape measure with feet and inches marked upon it, pass it rou[n]d the animal just behind the shoulder blades, and note the circumference. Then measure from the shoulder blade to the rear of the buttock, and note the length. To obtain the superficial feet multiply the circumference by the length, then to ascertain the weight of the animal, if it is less than one foot in girth, multiply superficial feet by 8.

If less than 3 and more than 1 by 11.

| | | | | | |
|---|---|---|---|---|---|
| " | 5 | " | 3 | " | 16. |
| " | 7 | " | 5 | " | 23. |
| " | 9 | " | 7 | " | 32. |
| " | 11 | " | 9 | " | 43. |

*Example.*—Suppose the girth of a hog to be 4 feet, the length 3 feet, then 4 × 3 = 12, and 12 × 16 = 192 lbs. the weight of the hog after it is dressed.

The above rule, I think, is worthy of preservation. I should have proved it tho' before writing in my Diary.

*Feb. 15 F.* A little cloudy, and many degrees the warmest day we've had in a time. Split and hauled five loads of wood this forenoon, and carried water for Mollie to wash this evening.

The snow is getting off to day with a *double rush;* and I'm glad that it is, for I never was so tired of the "feathery stuff." Guess this *protracted snow* has been a vast deal of service to the wheat in protecting it from the many severe frosts we've had of late, and in keeping the soil warm and moist, and giving it a fair chance to get good root hold.

*Feb. 16 Sa.* Warm, and windy. Snow melted fast. I & Pa hung up our bacon. I mended Mollie's shoes, laid in a supply of wood and pine to last over Sunday, and thus about "the going down of the sun" completed the labors of the day. After eating my third meal for the day of bread and fruit, I read about ten pages in the first vol. of the "Reformation," a chapter in the Testament, (Acts XXIV) and taking a *wash down* with tepid water applied with spunge, I drew on my *long shirt,* (which is made of cotton domestic, and extends from my shoulders to my ankles—the very idea to sleep in—) and went to bed.

*Feb. 17 S.* Clear, cold and very windy. Staid within doors, and spent the day in reading, writing, etc.

What a sad, and awful thing it is to spend our golden moments in idle *chit-chat,* when there are so many good books we might read and be eternaly benefited there-by. There is seldom any anything gained by visiting, so I must visit less and read more.

*Feb. 18 M.* Cold and windy. The flood gates of the "four Winds of Heaven" must surely be wide open. Went to H[ampton]ville—received a letter from bro. W., and some news for Clarkie.

*Feb. 19 T.* Calm, and not very cold. Had our first mess of ground hommony to-day. I took it to Benbow's mill this morning, and Mollie had it ready and on the table at dinner!

Corn is perhaps the most valuable grain raised in this country, from the fact that it yealds more in proportion to the seed, and that there can be made of it such a variety of good, holesome diet. It's good in hommony boiled in or ground. It's good in mush, which is perhaps the best thing that can be eaten for supper. It's good in the various kinds of bread into which it is made from the ash, and "*Old Virginia hoe-cake*" up to the most improved modes of preparing it. It is also said that a table spoon full of the meal will sustain a usor for 24 hours without other food. Not only is corn good, for man; almost every insect, fowl, or animal from the ant to the elephant, will eat it in some shape or other. Corn and iron stand about on the same footing with respect to other grains and mettles.

*Feb. 20 W.* Cloudy and warm. Worked in the shop on W. Messick's Jr. shash.

*Feb. 21 Th.* A pleasant day? No. Quite cool, cloudy and rained a small sprinkle. Finished M's sash, and commenced his bookcase.

*Feb. 22 F.* Clear and pleasant. Moved 2 stands of my bees. Finished M's book-case, etc. Sold James Windsor my filly timber, and his son carried away a portion (32 ft.) of it to day. He gives 2 cts. a foot for 2 inch timber. Mine is of that thickness.

*Feb. 23 Sa.* A blustry day, but not cold. Made 2 picture frames—the first I ever made—of poplar.
Messick came after his sash, and took them away.

*Feb. 24 S.* Wind a little cool, but tolerbly pleasant day. I and Mollie walked over to Alex's and staid an hour or two; then we went to Pa's a few minutes. Nothing strange, or wonderful happened that I wot of.

*Feb. 25 M.* Went to H[ampton]ville—received 2 nos. of the *Criterian* and one of each the *Mother's Magazine* and *Merry's Museum*—all tolerbly good of their kind.

Got my buggy-wheel from T's shop, and I and Mr. B. put the tire on. I then came home. Mrs. B. came with me, and staid "till next day."

*Feb. 26 T.* Fine weather. Hauled manure on a small lot of wheat which I did not manure in the fall. I must notice which does the best Spring or fall manuring.[4] Mr. B. came after Mrs. B. and they left for home just before night.

*Feb. 27 W.* Cloudy, and a little rainy. Staid within doors most of the day.

*Feb. 28 Th.* Went to Rev. T. A. Nicholson's mill with a load of wheat. Found it hard pulling as the road was very muddy. Read in the *Criterian* a review of W. Irving's life of Washington, in three volumes. I must have them. Just as I was reading of the surrender at Trenton, the poverty of the army, and the wealthy Quakers advancing the "sinews of war", etc. I drove by the Quaker church, where a small band of "Friends," as they call themselves, meet twice a week to hold their silent meetings. Their house begins to show the marks of Time, and unless its repared it will soon "go the way of all the earth." Just at the end of the church is the burying ground, inclosed by a rail fence, where nothing marks the resting places of the dead but rough stones which time has marked and colored. Some of them have been there for many years filling the place with solemnity. I felt solemn as I gazed on them, and repeated, to myself, the following lines;—

"Perhaps in this neglected spot is laid
Some heart once pregnant with selistial fire
Some hand the rod of empire might have swa'd
Or waked to extacy the living lyre."

However, I tarried not, but drove on to mill, got my wheat ground, and struck for home, at which place I landed just before the "bright Orb of

4. The reader will notice that Thomasson was an inquiring and experimenting farmer.

day" went down behind the Pacific's waves, and darkness took possession.

*Feb. 29 F.* And last day of February. Pleasant with signs for rain. Cut and hauled wood, painted my buggy-wheel, etc.

*March 1 S.* Cloudy, rather cooler than yesterday, and raining. Muster day. Went, but no muster. Elected 2 officers, who treated to something most of the voters had a dreadful hankiring after. How can I describe the scene? Imagin[e] 50 half famished hogs in a lot, throw them an ear of corn, see the rush and bustle—the stronger over powering the weaker—and all together making a hideous noise, and you will have a faint idea of the scuffle, to day, around the tin cup that contained the liquor. How disgusting! "Oh! that *our* people would consider," and abandon the use of rum, tobacco, and coffee.

If the money that is anualy expended in the trafic of rum, tobacco, & coffee, was added to the School fund what a great blessing it would bring upon our people, where as it now only adds to the cup of their misery and wretchedness. How long—O how long will the people continue to "lay up for themselves wrath against the day of wrath?"

*March 2 S.* Clear and windy, but not very cold.

I and Mollie went to her Pa's in the evening. Mollie had not been over in four weeks!

*March 3 M.* Quite cold early, and not very warm any part of the day. Went to H[ampton]ville after the news, but did not get much. There seems to be some chance, just now, for a war between U. S. & England. If England could not manage America in 1776 I think she had better not undertake it in 1856, or in any succeeding year while the Union last, and I hope that will remain unbroken to the end of time.

I and Mollie came home.

*March 4 T.* Quite windy. So was yesterday. Hauled wood at Pa's in the forenoon, and commenced ploughing—breaking oat ground—in the orchard in the p.m.

*March 5 W.* Clear and windy. One might tell that this is March, by the wind's blowing every day, even if he had not an Almanac. Commenced sowing oats, broke up my potatoe patches, etc.

*March 6 Th.* Still windy. Sowed oats in the evening.

*March 7 F.* Calm! and pleasant. Planted a few potatoes, helped Mollie wash,[5] Sowed oats, etc,

*March 8 Sa.* A little cloudy this morning, windy and warm. Finished planting my potatoes, ploughed Ma's garden, hauled two loads of wood and quit work for this week.

*March 9 S.* Clear and pleasant in the forenoon, but the evening was quite blustry and cloudy. I & Mollie went to Aylesbury to hear the Rev. James Minish preach but we were rather late owing to our not knowing the appointed hour, and the sermon was closing up when we got there, it being about twelve o'clock. We heard enough tho' to enable us to say, at a rough guess, that so far as the sermon was concerned we did not lose much. Guess all that was said was well intended, as we believe J. M. to be a good man. If he had been educated he might have been of great use to the church, but as it is he can only exert quite a limited influence. "Knowledge is power," and the good or evil a man can do is limited to the extent of his knowledge, but wo! to that man who acquires learning and uses it in a evil way. He had better never acquired it, for it will prove a "savor of death unto death."

*March 10 M.* Windy, and quite cold. Ploughed, sowed oats, etc.

*March 11 T.* Cloudy, calm, and not quite as cold as yesterday. Burned some brush in the orchard, ploughed, etc. Clark bought a young horse (Pete) of the Widow M. Wilborn.

*March 12 W.* A little cloudy and not very cold. I & Pa hauled rails (300) to rebuild some of the old fences.

*March 13 Th.* Pa planted his potatoes. Cold, cloudy and commenced snowing about 2 o'clock. It snowed very fast about "the close of the day" and soon covered the ground. I & Clark hauled wood in the evening, 3 loads for Pa and 2 for myself.

*March 14* Quite clear and pleasant. The little snow that was on the

5. Thomasson helps his wife wash clothes on several occasions, almost always when she is ill or pregnant. At this time Mollie is carrying their first child (who will be born on June 29).

ground this morning soon returned to its original element which soon disappeared. Went to mill, built a factory! (an ash hopper) of rock, sticks, boards, and straw. This is a factory, or hopper in which to make lye, by filling it full of ashes from the hearth, and pouring up water and letting it drain thro' them.

I also straitened up some shackeled[6] fence, ploughed a little, etc, etc. I am reading the *Reformation* at the rate of ten pages per night. Read to night of Luther's students burning the theses of Tetzel, an affair which Luther hated very much at the time. Luther had a great many enemies, but he did not fear them. He was made of quite the stuff to cut his way thro' the world, and leave "land marks" by which others could follow.

"Who will be free themselves must strike the blow," no doubt, thought Luther on many an occasion, and he kept striking "blows" until he blowed things up. Do you ask what he exploded? Read the *Reformation.*

*March 15 Sa.* Clear, (save a few scattering clouds which occasionaly passed between us and the sun catching for a moment his rays of light, and hiding from our view his bright face,) and quite warm. Mollie and I went to her Pa's to a "choping," wood cutting, or clearing, as you like. The hands were but few, and tho' they worked well they made but a small rent in the forest.

*March 16 S.* Tolerbly pleasant, a little cloudy, windy in the p.m. I & M came home about the middle of the day, and I spent the evening in reading the *Reformation, Testament* and Mr. Wesley's sermon on the restoration of all things.

*March 17 M.* Cool and cloudy. Received a letter from brother W., also a No. of the *Independent* from N. Y.—a very large, and tolerbly good paper, I think.

*March 18 T.* A cold east rain—the Easter storm perhaps. Half-soled my shoes, etc. Been married 8 months to day.

*March 19 W.* Cleared off quite pleasant. Ploughed, etc. Shot a blue hawk, a real chicken catcher.

*March 20* A fine day. Ploughed some, fixed our spring path, etc.

6. ramshackled

*March 21 "Good Friday."* Finished sowing oats. Cloudy, a little cool, and commenced raining about 10 o'clock. It always rains either before, at, or after Easter, to the great annoyance of the "little folks," who like then to be paying and receiving visits, dying eggs etc.

*March 22 Sa.* A rainy day.

*March 23 Easter Sunday.* Clear, most of the day, and cool. I & Mollie walked out to Pa's in the evening; this was all of our Easter visiting. I never did, nor do I ever expect to see the necessity of ones being always "on the pad." Home is home if never so poor, and if "There's no place like home" why not stay at it? Necessary visits excepted.

*March 24 M.* Hard wind.

*March 25 T.* Clear, and still windy. March is making us feel his presence in his latter days.

*March 26 W.* Quite a pleasant day. I & Mollie went to her grand pa's in the evening to warp a piece of cloth. This is the first cloth Mollie has attempted to make since we were married, and I hope it will be the last, save one, she'll ever be compelled to make. Mollie is not able to make cloth, and trust I shall, by the blessings of heaven, soon be able to buy our clothes. The future will tell whether I am or not.

*March 27 Th.* Another snow! Cold and windy. The dark clouds chasing each other in almost every direction gave us two very hasty showers of snow. Dont know that I ever saw it snow faster for the time. The ground was soon whitened, but the snow did not lay long.

*March 28 F.* A cold day for March. Hard wind. Vegetation, tho' slow in drawing on its green coat, is now receiving a sevear check. March seems determined that she shall wear her gay winter dress while he reigns. He's so conservative! He wont give up the fassions of winter.

*March 29 Sa.* Clear most of the day, and quite pleasant. How changeable the weather is! and yet, some people seem to be almost as changeable. One day is fair and pleasant, the sun shines warm, and all nature seems to smile and cheer up. The next, cold winds sweep by, dark clouds rise up and tell us that "winter is not yet gone," but, by and by there appears to each a "silver lining," they pass by, the warmth of the sun again reaches the earth, and all rejoice.

*March 30 S.* Somewhat cloudy, snowed *just a little.* Read, besides several chapters in the *Testament,* Mr. Wesley's sermon on "The Danger of Riches." Text.—"They that will be rich" etc 1 Tim. VI, 9. The sermon is good, and I hope that I shall be the better for having read it. It is, doubtless, very dangerous to be rich, or even to desire riches. We should not "lay up" for ourselves "treasures upon earth," but "having food and raiment" we should "be therewith content."

This it is difficult to do, since we are never satisfied with gain; but we should "watch and pray" and keep ourselves under subjection.

*March 31 M.* Snowed till near the middle of the day, but the ground being wet from last night's rain the snow melted nearly as fast as it fell. Cut firewood in the evening—some for myself, and some for Pa—and Caleb hauled.

*April 1 T.* April comes in like a lion, for there is, this morning, a devouring frost. Clear and cool. I & Mollie planted our beet seeds, beans, cucumber seeds. For the beans and cucumber seeds, I ploughed the ground twice, then run two furrows about 2 feet apart, and raked them out with a hoe. Then filled them about half full of ashes, and raked dirt on the ashes till the trenches were full. Then I opened a small trench with the corner of the hoe, droped in the seeds, and covered them about 2 inches deep with the hoe. Time will show whether this is a good plan or not.

*April 2 W.* A cool rainy day. Bed[d]ed out my sweetpotatoes, ploughed some, etc.

*April 4 F.* tolerbly pleasant. Went to "work the road" but the hands not coming in the overseer (Wiley Messick Jr.) postponed till another day.

*April 5 Sa.* Cool and a little cloudy. Worked the road, helped Alex. roll logs, etc.

*April 6 S.* Quite a nice day. Mr. & Mrs. Beall came over to see us.

*April 7 M.* Nice weather.

*April 8 T.* The finest day, I believe, we have had this year. Planted our onions, cut wood, ploughed some, etc.

*April 9 W.* Nice as yesterday. Ploughed Clark's horse, (Peter). He does well for a beginner.

*April 10 Th.* Smokey, cloudy & warm. Ploughed half the day. Killed three snakes—Guine Mockison [Guinea moccasin], a highland Mockison and a black racer—the first I've seen this season, and if I meet no more I shant be sorry. They had better not, however "cross my path." If they do—"wake snakes"! But perhaps I had better remember "What ere be the danger by night or by day" *Frogs, spiders* or *snakes* "get out of the way."

*April 11 F.* Quite warm & dry. It was a little cloudy yesterday evening, and rained a little. Commenced listing for corn.

*April 12 Sa.* Listed, hauled wood, etc. Clear and windy.

*April 13 S.* Quite a fine, warm day.

*April 14 M.* Commenced planting corn.

*April 15 T.* Dry and warm.

*April 16 W.* Rained some about the middle of the day. We have had quite a spell of dry weather.

*April 17. Th.* Rained a considerable shower last night, and thundered very hard. Cloudy, and rained a light sprinkle this forenoon. Planted corn. I lay-off, Eliza (sis.) drops and Caleb covers, so we go right ahead.

*April 18 F.* Windy. I & Caleb listed. Mollie and I have been married nine months to day, and no increase in the family yet!! If Mollie sees this *spect* she'll pull my ears, but howsomever—

Nine months have passed and gone,
And we are yet alone,
But how long we thus shall be,
In the future we'll see.

*April 19 Sa.* I & Caleb finished listing. I & Mollie went to her Pa's in the evening, as we had not been there in 5 weeks. Soon after we got there it commenced raining, and rained quite fast for some time—perhaps an hour.

*April 20 S.* Quite clear "at early day," After a bit the wind blew up from the north, dark clouds came over, and it was quite cool & blustry. I & Mollie came home about the "close of day."

*April 21 M.* Windy, cloudy and cold. Winter has not entirely let go yet. Planted corn. This is Mollie's birthday, but I did not think of it till it was too late to have any thing like a *kick-up* on the occasion. Guess we'll have a feast of honey and mush, at supper; Mollie will be 19 years old, and we'll all go to bed without a dance.

*April 22 T.* Quite pleasant. Planted corn, cut briers, elders, sprouts etc on the branch banks. Planted my ground peas,
A little hazy about sun set, and looks favorable for rain. Hope it will come soon.

*April 23 W.* Warm, and quite rainy. Great season, I guess, for the wheat and oats, which look well.

*April 24 & 25 Th. & F.* Nothing of importance.

*April 26 Sa.* I & Mollie went to Eagle Mills and bought a bolt (30 yds) of domestic cloth at 9 cts per yd.; one large dish 35; 1/2 lb. putty, 2 1/2 yds. calico, etc, etc, to the amount of $3.40 which was all the change I had.

*April 27 S.* Quite a nice day, I & Mollie went to Pa's and staid till after dinner, and in the evening we went over to see old Mr. Beall.

*April 28 M.* Finished planting corn to day, just 2 weeks from the day we commenced. Cloudy & rained a little. Planted my sweet potatoes.

*April 29 & 30 T. & W.* Warm with signes for rain. This has been, so far, quite a dry spring. Wheat (the growing crop) looks well for all and the corn is coming up nicely. If crops should come in good this year, and I hope they will, there will be a vast quanity of grain in the country. Wheat is now ready at $1 per bu. and corn at $1/2.

*May 1 Th.* Cloudy with a cool north east wind. In the evening we had a very hasty shower—a real washing rain. Clark went to Nicholson's mill with wheat. He got back with the flour just before the rain.

*May 2 F.* Cloudy & cool yet this A.M. Cleared off about 9 or 10 o'clock, and the sun shown out quite warm. Went to Aylesbury to meeting. Uncle Gunn was there and preached from a text he had used there before. In the evening I took the Class Book to B. H. Johnson Esq's where Uncle Gunn went for dinner, to get it revised, and succeeded. I also bought a new Class Book, of Uncle Gunn, for 15 cents, containing a portion of the Discipline.

*May 3 Sa.* Windy, but warm. I & Pa hauled rails in the forenoon. Uncle Gunn was to have preached at Pa's to night but did not come. It seems that he has disappointed the congregation he promised *himself* to meet in order to fill an appointment that some one else made for him!

*May 4 S.* Quite a pleasant morning. I & Mollie went to her pa's, and she staid there while I went on to Mt. Mariah to meeting. I heard the Rev. B. Clegg preach a nice, excellent sermon from the text "Now we believe not because of thy saying; for we have heard *him* ourselves, and know that this is indeed the Christ, the Saviour of the world." John IV, 42. It clouded up and commenced raining about noon, and rained on till night.

*May 5 M.* Still rainy & cool. I & M came home. Helped build a cross fence, in the evening, between the pasture and Alexanders corn field.

*May 6 T.* A very rainy day, and tolerbly cool. Guess this is the "sheep killer" which usualy comes in May. This day I've spent, as I would like to spend a great many, in reading and writing.

*May 7 W.* Cloudy & cool. Worked for Pa, hughing [hewing] timber to build a stable.

*May 8 Th.* Took a sawlog to D. Johnson's saw-mill, in Iredel Co., and brought back a load of studing etc for the stable.

*May 9 F.* Hauled rails for Pa in the forenoon, and helped Mollie wash in the evening.

*May 10 Sa.* Quite cool, for May, and windy. Wheat is growing finely, but the corn looks white, and grows but slowly, if at all.

*May 11 S.* Clear, and a few degrees warmer than yesterday.

*May 12 M.* A nice, warm day. Commenced ploughing the corn, or the ground, rather, for the benefit of the corn.

*May 13 T.* Looks favorable for rain. I & Caleb replanted corn—most of our whole crop to-day.

*May 14 W.* Nice weather. I & Pa sheared our sheep. It is well for us that the sheep are so timed [timid], and easy to overcome, for all we some times blame them for being so fearful of dogs, and for not defending themselves with more courage when attacted by them. If a sheep had the pluck of a goat it would be difficult to unfleece.

*May 15 Th.* Warm, rained a little this morning. Put up 3 shoats (small ones) and of them I expect to make our next year's pork, bacon, etc.

*May 16 F.* Ploughed. Rained in the evening.

*May 17 Sa.* Cloudy, and rained some. Went to mill in the morning, and to the creek in the evening. I had better luck at the mill than at the creek, for I only caught 3 fishes, and one of them was so small that I threw it back into the water. Fishing is a poor paying business in this country.

*May 18 S.* Been cloudy all day, and cool too. Since *dinner* it commenced thundering, the clouds hasten up from the north east, and now, while I write,

"The rain comes down with all its might,"

and from all that I can see, tho' I don't profess to be a "Weather Prophet"—Stop. Let me poetize my thoughts just here.

And from all that I can see,
Though I don't profess to be,
"A Weather Prophet,"
Guess we'll have plenty of rain,
To aid in making the grain,
For store and basket.

Now the wheat looks green and fresh,
The corn a little whiteish,
Owing to the cold.
Should the rain to[o] often come,

Wheat will be "as good as none"—
Can't be bought with gold.

Then "perchance" some one will learn
That many labour hard to earn
Their bread and butter,
And be taught with pride to look
After the farm, as to work
Than starve is better.

*May 19 M.* Cloudy in the forepart of the day, and rained some little. I helped Mollie wash our clothes, and then wash the wool. Wool washing is not a very pleasant business.

*May 20 T.* Ploughed. Pa planted his cotton and Oregon Peas.

*May 21 W.* Clear and cool. Ploughed, hoed corn, etc.

*May 22 Th.* Still cool. Fine weather for wheat. Ploughed, replanted corn, etc.

*May 23 F.* Clear, and warm. Ploughed and hoed our swamp corn.

*May 24 Sa.* Clear and quite warm. Went to Senter to the Quarterly meeting—the third one for this year. Heard Rev. W. Barringer preach quite a good sermon from the text.—

I also sat in Q. Conference for the first time. There was but little business done. Isaac Jester applied for, but did not obtain, licens to preach. And why? Because the Conference thought it best that he should wait awhile, and acquaint himelf with the Discipline, as he said that he had read it but very little, before undertaking to preach. Jester seems to despise "the day of small things." I was present when he applied for licens to teach Common School. He then could not tell how many letters there are in the English Alphabet. Now if Webster's Spelling Book, and the Methodist Discipline were each a hugh volume, I dare say Jester would know more about them than he does.

*May 25 S.* Went again to Senter, and heard Rev. W. B. preach another fine sermon from the 1, 2 & 3 verses of the 40 Psalm. A little shower of rain, near the close of the sermon, scared the people (the ladies in particular) unto the house, their carages. When the few drops of rain had fallen, the people returned to their seats, and Mr. B. finished his sermon.

*May 26 M.* Hoed corn. The weather fine,—rather dry for corn & oats.

*May 27 T.* Hoed corn again to-day.

*May 28 W.* Helped wash and scour. Commenced cross plowing, planting peas, etc.

*May 29 Th.* Dry and pleasant. Plowed. Pa and Caroline went to the Exam. at Jonesville.

*May 30 F.* Went to meeting. Uncle Gunn preached quite a plain sermon from the parable of the field, Mathew XIII, 24 to 30. After preaching J. F. Reinhardt was had up in the church for swearing, and was expelled. His mother, and wife of our late C[lass] L[eader?], bro, F. R., was also tried in refference to an evil report against her, but was acquited.

*May 31* Sa., and last day of May. Clear and quite cool. Went to the Cherry Orchard; gathered as many as we wanted to eat, and some to bring home. Eliza & Caleb went with me. The cherries were not fairly ripe, but they were May cherries anyway.

*June 1 S.* Quite a nice warm day. Staid at home, and read & wrote.

*June 2 M.* Quite warm. I & C[aleb] plowed. Late in the evening we had quite a nice little shower, not enough to wet the ground much, but enough to lay the dust, and do a vast deal of good.

*June 3 T.* Warm. Plowed. Had our first mess of peas & potatoes for dinner. Rained a fine shower, and I and Caleb got pretty well soaked. We thought the cloud had passed by, and it had, but it gathered, wheeled, and came rushing back and gave us a real shower bath. It, however, was soon over and we plowed on.

*June 4 W.* Had another light shower.

*June 5 Th.* Nice and warm. Great time to clean the corn. Our dinner of bread, fruit and potatoes is just over. It was good enough for a king and his lords to sit down to.

*June 6 F.* Hoed corn. Rained in the evening quite a nice shower.

*June 7 Sa.* I & Caleb hauled 2 loads of wood—one for me, and one for Pa, set out a few cabbage plants, etc.

*June 8 S.* A nice, warm day. Meeting at Aylesbury. I was there. Pa preached. His text was the 5 & 6 verses of the V chapter of Luke. The sermon was quite a plain one, and that is the right kind of sermons for the common people. After preaching we had a short class-meeting, and all left for home.

*June 9 M.* Hoed corn part of the day. Another great washing rain fell about 12 o'clock. Such a rain we've not had for days.

*June 10 T.* Rainy.

*June 11 W.* Still it rains. Made some picture frames.

*June 12 Th.* Very warm. Rained a little at night. Hauled rails and put up some 8 pannels of fence that was washed down.

*June 13 F.* Not so warm as yesterday. Wed [weeded] out Pa's sweet potatoes, hauled up some hay, etc.

*June 14 S.* Muster day—and we had the nicest muster, and the shortest one we've had since I've belonged to the company.

After the parade Miles Cowles, Esq. made a political speech. I suppose this was his *first born,* and I think it came a little before its time. We need not be surprised tho' for the doctors tell us that such things happen frequently.

*June 15 S.* A nice day. Went to Flatrock to meeting, and heard the Rev. Z. Adams preach a tolerbly good sermon from the text—"By faith" etc. Heb. XI, 24, 25, 26.

*June 16 M.* Nice weather. Plowed—commenced laying by the corn crop to day. Went to Couche's store in the evening and bought a gallon and a half pot for .75 cts.

*June 17 T.* Rained *lots* to day, but we commenced cutting wheat between showers. Some of our wheat is, I think, as good as I ever saw, and some of it has the rust. Corn looks nice.

*June 18 W.* Went to Town, but did not get anything for myself. Rained a few light showers to day too. Bought 1 gall. mollasses of Mr. Beall for ———.

I & Clark & Caleb cut, bound and shocked about 14 doz. [bundles] of wheat. We were at it but few minutes. I made 2 picture frames for Clark, and thus ended the business of the day.

*June 19 Th.* Hoed corn 1/2 the day, and plowed 1/2 the day. Awful hot, sure.

*June 20 F.* Clear and hot. Plowed, helped Mollie wash, cut & bond wheat, etc.

*June 21 Sa.* Very hot. Plowed till noon. Rob[b]ed one stand of my bees this morning, but it was not very rich. They now have room to work, and I want them to fill up with sowerwood honey. Went over to Mr. Beall's in the evening, and staid all night. I engaged to teach the District School in [District] No. 30 for $16 per month. School to commence on Monday the 14 day of July.

*June 22 S.* Another warmday. I & bro. J. D. Johnson hel[d] our first class meeting to-day. There were but five members present. In the evening it rained, and the wind blew, and bent the corn considerably.

*June 23, 24, 25, 26, 27 M. T. W. Th. F.* Very warm. Cut and bound wheat, I believe, more or less every day. Nothing of very great importance has taken place this week, on Pa's farm, that I now remember.

*June 28 Sa.* Still hot. Went to the store (N's) and bought

| | | |
|---|---|---|
| one little wooden bucket - - | .30 | |
| one lb brown sugar | .12 1/2 | |
| " gro. shirt buttons - - - | .10 | |
| Two boxes matches | .05 | .57 1/2 |

Hoed corn (3 rows)—and hauled in 80 doz. of wheat, hilled up some of my sweet-potatoes, etc.

*June 29 S.* Warm, warm, warm. So little air in motion that the leaves can scarcely be seen to move. Joseph L. Thomasson born.

*June 30 M.* Last day of June. Warm as yesterday but for the soft, cool breeze that occasionly fans our sweating brows. Finished cuting wheat.

We have this year, 245 doz. [bundles] of wheat, good, I guess, for 60 bushels.

*July 1 T.* Clear and hot.

*July 2 W.* Hoed corn, plowed, etc.

*July 3 Th.* Plowed, notwithstanding the hot weather.

*July 4* Great day throughout the U. S. in general—the people celebrating the Declaration of Independence—and a joyful time on our farm in particular, as to day we finish plowing our crop. May the people of the U. S. long remain free and independant—if, indeed, they are free—,[7] and may our crop flourish and bring forth abundant harvest. Amen.

*July 5 Sa.* Finished hauling in wheat. Rained a nice shower in the evening. I saw it and felt it too.

*July 6 S.* A little cool early, but quite warm later in the day. Staid at home *all* day. I always loved home, and was never satisfied any where else long at a time, and now that the *home chain* has another and a strong *link* (my boy) attached to it, and as it binds closer in proportion as it increases in length guess I shall have to stay *about.*

*July 7 M.* One of the hot[t]est days we've had this season. Had a nice shower in the evening.

*July 8 T.* A rainy day. I was taken very sick last night, and don't feel well yet to-day. Pa took Caroline to H. Barker's where she is to board and go to school in Jonesville. Mr. B. lives about one mile east, I think it is, of J'ville. Lit with a vol. of *Capt. R. B. Marcy's Exploration of the Red River.* I find the work interesting as far as I have read. He describes some fine farming lands. Think I would like pretty well to live either on *Sweetwater, Otter,* or *Cache Creek,* where Capt. M. says the water is good, the land rich, and portions of it well timbered, with any amount of grass for stock. Dont know if I shant visit this "good country" some day.

*July 9 W.* A little cloudy part of the day, and warm. Dont feel very stout.

7. Again Thomasson alludes to his concerns about slavery and the status of women.

*July 10 Th.* Pleasant. I & Clark cut and bound oats part of the day. I scraped down the hill in which my sweet potatoe slips are planted, etc.

*July 11 F.* A cool east wind, and cloudy part of the day. Cut some fire-wood, etc.

*July 12 Sa.* Clear and would be very warm were it not for the cool breeze that is stirring. Hauled one load of wood, hoed corn, etc.

*July 13 S.* Clear and quite warm. I & Mollie are sitting in the south end of our cabin; the window is up to admit the pleasant south wind which comes in gently through the green leaves of the peach-tree that stands just in front of the window. Our boy is sleeping on a pillow in a chair just at my left hand, and the dogie is also quite [quiet] just now. I've been reading this morning, in the *Testament,* and I can't see, for the life of me, how the Baptists can conceive that Baptism is essential to Salvation. Where is their scripturial proof? Perhaps this is it, "He that believeth and is baptized, shall be saved." Well, if this be it which baptism is here meant, John's or Christ's? John had a baptism for he baptized with water. Christ also had a baptism distinct from John's for He baptised not with water, but with the "Holy Ghost and with fire." Now it seems plain to me that Christ's baptism is the one here alluded to, for the latter clause of the verse I have quoted above says, "but he that believeth not shall be damned." Mark XVI, 16. Now a man may be baptised with water and yet never believe, but unless he believes he will never be baptised with the "Holy Ghost," hence Christ, instead of saying, "He that believeth not *and is not baptised* shall be damned," only says, "He that believeth not shall be damned." Then if Christ here means his own baptism, water baptism is not essential to salvation. "But one thing is needful," Luke X, 42. "Love the Lord thy God with all thy heart, soul, mind and strength, and thy neighbor as thyself." "This do, and thou shalt be saved."

*July 14 M.* Clear and *hot.* Commenced school in Dis. No. thirty. Had 11 scholars the first day![8]

*July 15 T.* Still hot, and not much air stirring. Had 12 Scholars to-day.

8. Eleven seems a scandalously small number to Thomasson.

*July 16 W.* Dry weather, and warm. Corn begins to need rain. Had 15 scholars to-day.

*July 17 Th.* Rained a little. The little bull I bought of W. Messick, Jr. last winter died to-day, and I & Caleb took off his hide after I came from school. I suppose he had the "Distemper."

*July 18 F.* Cloudy most of the day, and rained some. Had 16 scholars, both yesterday & to-day.

We have progressed finely this week. I have no a b c scholars! and I believe this is the only school I ever commenced without at least a few of the A B C class.

*July 19 Sa.* Hutson finished cutting our oats; Clark & Caleb bound them.

*July 20 S.* Mollie and I took our boy and went to Pa's for the first time since he (the boy) was born.

*July 21, 22, 23, 24, 25 M. T. W. Th. F.* Things moved on as usual. Very dry, but not quite so hot as last week. School geting on finely. From 15 to 17 students out each day. Teaching school has its disagreeables, but I dont know that it has any more than most other occupations, and perhaps not as many as some, so I dont know if I shant take it for a life business yet. One can make out to live at it, and if he does his duty he is laying up "treasure" in a better world.

*July 26 S.* Clear and hot. Made some medicine for myself. Settled with Martha and Alexander, and came out in their debt $2.05. I & Mollie went to her pa's in the evening.

*July 27 S.* Warm weather and no rain yet. The corn begins to look badly. Went to class meeting. But few of the members out. I & M came home in the cool of the evening.

*July 28 M.* Hot & dry. Had 19 scholars.

*July 29 T.* The wind blew some to day else it would have been swealtry. Had 21 brats today. The old rooster crowed last evening after sun-down, and this morning the smoke settled to the ground; these are said to be signs of rain, but we've had none yet. "All signs fail in a dry time." A few more such dry, hot days, should they come, will, I fear, do the work (sad work) for our corn, potatoes, etc. I do hope, and think it will rain soon.

*July 30 W.* Very dry, and quite hot.

*July 31* Th. and last day of July. No rain yet. Corn begins to look badly, especialy about the heat of the day. The grass in the yard is dead, and if the dry weather holds on a few days the pastures will be grassless, the corn and potatoe crops short.

*August 1 F.* Hot. Quite a rain went just north of us to-day, and it even droped a little here.

*August 2 S.* Clear, hot and still. A "Two-days meeting" commences at Aylesbury.

*August 3 S.* Cloudy in the evening, but no rain. The congregation, to-day, was tolerbly large. Rev. T. A. Nicholson preached at 11 o'clock, and Rev. W. L. Van Eaton at 3. After Van. finished his sermon 8 seekers of religion came forward to be prayed for, and two of them (sisters, Eliza and Catharine) professed. Some 2 or 3 others also obtained pardon at the night meeting. I came home about night, and went to my school on Monday though the meeting continued part of the day.

*August 4, 5, 6 M. T. W.* Hot & dry. The Flux is raging considerably. Several children have died in this, and Iredell Co. lately.

*August 7 Th.* Election day. Dismissed school, but did not get to go to the Election.

*August 8 F.* Taught school.

*August 9 S.* Took our bee's-wax (5 lbs-9 oz) to Madison's store and sold it at .20 per lb. 5-9 × 20 = $1.11 1/4. I gave $4 for 2 bee stands 3 years ago. The first year I got, say one dollars worth of honey and wax; the second about $2, and this year $3, making in the 3 years $6 which is $2 over paying for themselves, and I now have as many as I bought.

*August 10 S.* Went to class meeting. More out than usual, as the Rev. James Minish was to have preached, but he did not come. I & bro. J. D. held class meeting.

*August 11 M.* Very dry and warm.

*August 12* Thrashed our wheat, but I did not help. Rained! a nice little shower abo[u]t 12 o'clock. We have had a long drought. The corn is well nigh dried up, but this little shower makes it look a little fresh.

*August 13 W.* Had a fine rain last night, and some more to-day. Hope it will continue coming till the "dry land" is well soaked once more. Think it will help the corn yet, or the potatoes at least.

*August 14 Th.* Brother Wiley came up from Oak Ridge.

*August 15 F.* A little cloudy, and pleasant. Wiley staid with us last night for the first time in his life.

*August 16 Sa.* Helped to clean wheat, stack straw, etc.

*August 17 S.* A little cool and hazy. Read Book VI of *Paradise Lost.* Strange that there should be an uproar in "Heaven, the seat of bliss," and "that Angel should with Angel war"! These Angels were formed equal in their creation, and had often met "in festivals of joy and love." But one of them became wicked, and drew after him "A third part of Heaven's host." and instead of feasting, and "Hymning the Eternal Father,"

"Now storming fury rose,
And clamor such as heard in
Heaven till now was never."

The wicked angel was cast out of Heaven, but he is still fighting against God, and he sometimes makes a great deal of noise on Earth, and seems almost ready to and would if he could, devour the "little flock" that dares to show him fight.

*August 18 M.* Cloudy, and rained a good deal in the evening. School rather small.

*August 19 T.* Still cloudy part of the day and rained a fine shower. As it rained yesterday, last night and to-day the ground is now tolerbly well soaked once more. Fear the rain is too late to do the corn much good, but think it will help the potatoes.

*August 20 W. 21 Th. 22 F.* Cool. School about as usual. Received $2, the amount of a note I had on W. Snipes. I handed 3 notes on the boys Andrew, James & Write Snipes to brother Wiley more than a year ago.

He has collected 2 of them, and there is one of $4.00 yet due. Paid W. Louis, yesterday one bushel of wheat for a days cradling. Good wages too, for a bushel of wheat is now worth $1.

*August 23* A little hazy. Saturday. Cut and hauled a load of wood, picked up and hauled a load of apples to make vinegar. Not well.

*August 24 S.* Rained some. Class day but I am not well enough to attend—got the Flux.

*August 25 M.* Not well, but taught school till night and then dismissed till Wednesday morning.

*August 26 T. 27 W.* Lay by these two days. Not well enough to go to preaching on Tuesday, or to teach school on Wednesday. Never had the Flux before and it uses me roughly.

*August 28 Th.* Went out to school, tho' I was hardly able to walk.

*August 29 F.* Taught school tho' I feel worse to-day than I did yesterday.

*August 30 Sa.* Not much, if any better.

*August 31 S.* And last day of Aug. Still unwell. Have taken *lots* of things, and don't know that I have been benefited much by any. It seems that the disease has got fast hold on me, and that it is determined to have it's way.

This is a stormy day. A real "North Easter" is raging furiously this evening. C[amp] Meeting at T[rap] Hill, Pa, Ma, E, C, & Wiley are gone. I & Mollie stay with Aunt.

*Sept. 1 M.* A stormy day. The "Northeaster," spoken of yesterday, still continues. Guess seamen and campers faired but *midlin* last night. Our corn is blowed down, or into such a slant that gathering fodder will be rather a low business this fall. Small boys can gather now without the trouble of bending the stalk. The wind and rain abated about 12 o'clock, and the sun shown out for a few minutes, but another set of clouds rolled up and put a stop to his shining on our part of the "moral vineyard," and thus the day closed, cool & cloudy. Not able to teach school, so I sent Caleb over to dismiss till Wed. morning.

*Sept. 2 T.* Clear and pleasant. Felt some better, but not well yet. Cut a small load of wood in the morning, and in the evening I went to Madison's store and bought 1 quire of paper (this is a sheet of it) for 20 cents. I sold A. Hutson a mutton yesterday for $1.50, and should have set it down under yesterday's date, but I did not think of it till after I had commenced writing for to-day.

*Sept. 3* Cooler than yesterday, and cloudy. Taught school to-day. Had 8 scholars.

*Sept. 4 Th.* A little cloudy; but pleasant. Only 8 scholars again to-day. Caroline came home from school to stay. She had only been there 2 months, so she says, and was not satisfied; so she left.

*Sept. 5 F.* Cloudy, and drizzled a little once. Only had 5 little brats to-day. Dismissed for 2 weeks. If my school don't increase after fodder time is over, I shall think the people in this District are not very anxious to send their children to school at all.

*Sept. 6 Sa.* A pleasant day. Sold my wagon to Mr. Thirstin Redin for $57.50. He is to be after it next Tuesday. I sold the same wagon last fall to Thomas Jennings for $56., but he did not come after it, neither did he pay for it, so our trade was broken. I fear that Readin will not come, and that this bargain will be no more lasting than the other. I paid A. Hutson 30 cents, the ballance due him, so we (I & H) are now even.
I commenced gathering fodder to-day.

*Sept. 7 S.* A beautiful morning. The sun rose in seemingly unusual splendor, gilding the long line of clouds streched across the eastern horizon. Where shall we turn our eyes to behold a more beautiful and splendid scene than the rising of the sun? It is not only a beautiful scene, but it is a time of rejoicing,

"The shifting clouds
Assembled gay, a richly gorgeous train,
In all their pomp attend his *rising* throne."

The lark—the humblest of birds—soar high up in the sky to greet with a song of welcome the first rays of the morning; while the other feathered tribes enliven and vocalize mountain, hill and valley.

*Sept. 8 M.* Warm and a little cloudy. Cut tops.[9]

9. He refers to cutting the tops of the corn for fodder.

*Sept. 9 T.* Still warm as summer. Cut tops. We (I & Caleb) cut in the forenoon, and tie up in the evening.

This day completes my 27 year.

"Thus far the Lord hath led me on,
Thus far His power prolong my days."

Sold my wagon sure enough! but I only got $55 for it!! It was worth about 65 dollars. Mr. Redin came according to promise and paid me the cash for it.

*Sept. 10 W.* Still hot & dry. Helped to butcher a beef that Pa bought of Elzy Messick. He gave $9 for the beef, and it only weighed about 45 lbs. to the quarter. He gave at least $2 too much. In the evening I went to Town and paid off the "Benbow note." I paid A. Benbow $30, and Cowles $23.08 cents. I bought a 6 qt. tin bucket for .35 cts. a stra[i]ner for .20 cts. a pocket glass for .10 cts. & a pound of sulphur for .10 cts.

*Sept. 11 Th.* A little cloudy and very warm. Cut tops A. M. and tied them up P. M.

*Sept. 12 F.* A rainy day. Pulled a few bundles of fodder, etc, etc, etc.

*Sept. 13 Sa.* Cloudy early in the morning, but soon cleared off quite warm. Wiley went to his school, (at Smith Grove) and I took Mollie to her pa's and then went on to Yadkinville to get my School License renewed. There were thirteen applied for, but only twelve obtained, certificates. One Mr. Monroe Nicks was rather to[o] ______. Well, the fact is he was not sober, so he was dismissed minus a "paper." I took up my note which H. & W. had. The principal was $9.50, and it had amounted to something over $10. I received the balance due ($2) on Williamson's school acct.

I left Yadkinville soon after we were dismissed (about 5) and came on to Mr. Beall's where I had left Mollie.

*Sept. 14 S.* Clear and warm. Went to Aylesbury and h[e]ard the Rev. James Minish preach, or try to preach. His text was "The men of Nineveh" etc, Math. XII, 41. Jimmy told us all about Jona[h], the storm at sea and the whale, and then wound up with an exertation, leaving the preaching place (between the account of Jona[h], the storm, whale, etc; and his exertation) all blank. As a general rule we do not expect much from a new hand, taking it for granted that they have as yet received but little.

*Sept. 15 M.* Clear and warm. Took our wool (mine & Pa's) to the machine at Eagle Mills. Bought 2 stone crocks for 25 cts., 6 small readers for 75 cts. Amount $1.00 Took up my store due bill $2.87 at Madisons. Pulled fodder in the evening. etc, etc.

*Sept. 16 T.* Clear and warm. Helped Mollie scald and scour, cut tops, pulled fodder, etc.

Clark landed with his books. I bought *Dow's Works, The Prince of the House of David.* I have not examined either of them much yet.

*Sept. 17 W.* Still clear and hot. Cut tops in the forenoon, and hauled hay in the evening.

*Sept. 18 Th.* Would be very warm but for the wind that is blowing from the south west. Hauled in top-fodder in the morning, and then cut tops. Loaned Alexander Johnson $4. I find Dow to be quite interesting. O! that there were more Dows, Dows, Dows.

*Sept. 19 F.* A little hazy but warm. Hauled in fodder early in the morning, and then went to Eagle Mills after our wool rolls. I paid for carding 6 1/4 cts per pound and something for greasing. There being 17 lbs of rolls, I paid $1.20.[10] I also bought five yards of sheeting 10 cts per yd.—.50. Finished cutting tops. Alexander & Martha commenced moving to their new home.

*Sept. 20 Sa.* Somewhat cloudy. Had a fine rain last night. The more I read of Dow the better I like him, though I think he ought not to have ridden his poor horse so hard. When it comes to riding fifty five miles in a day, preaching five times and meeting two classes, its doing business in double quick time.

*Sept. 21 S.* Cloudy most of the day. I & Mollie went to Flat Rock to meeting. Billy Garner was up giving out his hymns when we got there. I heard him pray, and then left the congregation as I chose not to

10. Thomasson's multiplication is not faulty, as this total includes "something for greasing."

he[a]r him talk about Baptism and other churches. "The preacher is to preach the gospel," says Webster; but Billy takes it out in fighting other churches. One Samuel Smith preached the second sermon, and I think he did tolerbly well. We left before the third sermon was commenced as it look likely for rain.

*Sept. 22 M.* Commenced school again after an intermission of two weeks. Had only 8 scholars.

*Sept. 23 T.* Cloudy, rained some in the morning, quite cool. We had fire in the School house to day for the first time this fall. Should not be surprised to see frost about in places in the morning. Only 8 scholars again to day!

*Sept. 24 W.* Clear and cool. Frost this morning sure enough. The sweet-potatoe tops, pea vines, etc, are bitten considerably. Commenced reading the *Vicar of Wakefield* and read 12 chapters. It's a great work of the kind, and should be read by "all the world and the rest of mankind." O! that we could always remember that

"Man wants but little here below
Nor wants that little long,"

and deal with our neighbors as for eternity. The family of *Wakefield* was a curiosity, but there are many such. Many such *wives* and *daughters,* and not a few *Moseses,* who have paid dearly for their *specks.*

*Sept. 25 Th.* School small!

*Sept. 26 F.* School smaller!! Nice weather, and children enough in the District to make a good school, but their parents are too negligent to send them out. How can such parents expect their children to rise up in a coming day and call them blessed?

*Sept. 27 Sa.* A pleasant day. Finished hauling in our top-fodder, got up some wood and pine, put up a quanity of hay, etc.

*Sept. 28 S.* Warm, and a little cloudy. Went to classmeeting. Not many of the members out. Had no *stir,* but I trust there was some good affect.

*Sept. 29 M.* Cloudy, and rained a nice shower. Jodie's [three-month] birthday. Mollie is gone to see her ma, and I am left alone.

*Sept. 30 T.* Rain and hail. Quite cool. School small. Mollie came home. Her ma has a daughter. It was born yesterday.

*Oct. 1 W.* Windy and cool. "Jack Frost" paid us another visit, and left his *mark* this time. Look for him round again to-night.
This District, No. 30, numbers 91 children, and out of that number 10 were at school to-day. Why do the people not send to school? Their children must be educated, or grow up in ignorance. The State offers to give them an education, on conditions that they use the means she has provided, but they won't comply. They choose rather to travel on in the beaten track of their know nothing fathers and run the risk of becoming an easy prey of designing Tricksters. "Oh! that our *people* would consider."

*Oct. 2 Th.* Somewhat cloudy, but warmer than yesterday. Had ten scholars.

*Oct. 3 F.* Hot, almost, as summer.

*Oct. 4 S.* Still warm, and quite clear. Bought my first sack of salt of J. Madison & Son for $3.50, It measured out 3 bushels and a half. Been keeping house upwards of nine months, and living on borrowed salt. "The borrower is a slave to the lender," so I concluded to go to them that sold and buy for myself, as I am, and of right ought to be free and indipendant.

*Oct. 5 S.* A little cloudy, but quite a pleasant day. Went to a Protesant Methodist meeting which was held at their new stand near Jim Green's. Preachers, McCain, Q. Holton and B. Weasner. McCain preached the 11 o'clock sermon. His text was "Behold, I stand at the door and knock," etc. Rev. III, 20. He preached, I think, at least two hours, and some good I trust was done. After a short intermission Holton preached quite a good sermon from the 18, 19 & 20 verses of the VI chapter of Hebrews. The congregation was not very large for a "big meeting,["] and a new place. At night Weasner preached. His text was, "Say ye to the righteous it shall be well with him." His sermon was short, but tolerbly good. Some of the would-be-gentlemen showed themselves. Holton, in his exertation after the sermon, gave them some broad sides, which cooled them down somewhat. What a pitty it is that young men will treat the

Gospel with such contempt, and suffer themselves to be led captive, by the devil, at his will.[11]

*Oct. 6 M.* Quite warm. Only 9 scholars to-day. The more I read of Dow the better I like his works. Every person should read them.

*Oct. 7* Clear, calm and warm. Had only 8 scholars. Read a portion of Dow's tour through Virginia. "The world will love its own," and as Dow was not of this world he had to bear percicution, and that not a little. Christ said to his deciples, "If they percicute me, they will percicute you also," and Dow found the same to be true, even in his day. The wicked world will percicute the Christian. It ever has, and ever will till the Great Author of the universe shall "create all things new."

*Oct. 8 W. 9 Th.* Ma's birthday—she being 49 years old to-day. School small.

*Oct. 10 F. 11 Sa.* This has been a warm week for Oct. Butchered a beef. Sold it nearly all out at .03 cts. and .03 1/2 per pound. It weighed 208 lbs only.

*Oct. 12 S.* We went to Shilo to hear W. L. Van Eaton preach, but he did not come. Pa and E[lzy] M[essick?] done the speaking.

*Oct. 13 M.* Clear, or nearly so, and pleasant. We have been blessed with fine weather for business this fall. School small. The moon was in eclipse, in the early part of the night.

*Oct. 14 T.* Commenced raining this morning about day, and rained on, *and on.* The clouds and wind came fron the North East, and it is quite cold. Had ten scholars to-day.

*Oct. 15 W.* Still cold and rainy. This is a real "North Easter." Would not like to be on sea about this time; nor any other. Only six [students] out to-day. North Carolina has a large family of boys and girls, and she has them divided off in lots. As she can not see after them all her self, she has appointed a great many overseers, to attend to the bringing up of her sons and daughters. One she gives $1500 a year; others 2 1/2 per

11. Evidently some young men in the audience were more interested in gaining the attention of young women than in listening to the sermon.

cent of all the money that passes thro' their hands; but the majority get nothing but the *honor* of the office; so they dont attend to the lots assigned to their care—they dont keep the fountain heads of the stream open, so it loses a vast deal of its strength among *rubbish.*[12]

*Oct. 16 Th.* Still rainy, and cold.

*Oct. 17 F.* Rained till about noon; cleared off in the evening, and the sun shone in through the roof, windows, and cracks of our school house for the first time since Monday.

*Oct. 18 Sa.* Rose and wrote this morning before day. Cut two loads of wood, and went to the store (M's) and to the shoe shop in the forenoon, and in the evening I hauled my wood, moved out the old clock, etc. I bought, at the store, one water bucket for the school house,—.30 cts.; 2 1/2 yds. pla[i]d to make Jodie a Sunday coat, .37 1/2 cts. per yd.—.94 cts.; two lead pencils for .05; four awls for .05. Whole amount $1.34. Half soled one of my shoes to-night, and the old clock has just struck 9.

*Oct. 19 S.* Clear and quite pleasant. Staid at home, and read Dow, P. H. D. [*Prince of the House of David*], and so forth. "Swear not at all."

*Oct. 20 M.* Pleasant. One of my hogs—one that I had up fattening—died last night. All things that live have to die, and some think all things that have lived will live again—live eternally. Whether that "*all*" in the scripture has refferance only to the human family, or to all living things, I know not.

*Oct. 21 T.* Clear and pleasant. Only five scholars yesterday and today. Shadrack Beall set in yesterday morning to work for me awhile. Went to M's store to-night, and I bought goods for Shadrack to the amount of three dollars and twelve cents, I also bought two boxes of matches for five cents.

*Oct. 22 W.* Warm. Paid Ann Benbow two dollars for her services when Jodie was born. Only five at school.

12. In the state's common-school system, he complains, the committeemen at the local level receive no pay, and without financial incentive many of these individuals do little to promote the schools.

*Oct. 23 Th.* So warm that we did not build a fire in the School house. Caleb and Shadrack finished getting up our corn.

*Oct. 24 F.* Cloudy and cool. Only taught school till twelve o'clock, as A. H. Thomasson and lady called at pa's, and I was sent for to come home and see them. They were our near neighbors while we lived in Forsyth, and I was *very* glad to see them. They are out in search of a home; having sold out in Forsyth.

*Oct. 25 Sa.* Still cool and cloudy. Mr. & Mrs. Thomasson left early this morning for Jo. Weasner's in Iredell Co. Commence sowing wheat. Rather a late beginning, but I hope to make a good drop. On one occasion, if no more, those that came up at the eleventh hour received as much as those that commence early in the day;[13] and I trust I shall make a fine crop of wheat, though I sow it at the eleventh hour. I & M. went to her pa's, late.

*Oct. 26 S.* Cleared off warm. Went over to see Alex. and Martha for the first time since they moved. Found them up, and living at a nice place, but their land is not a very rich one. I & M came home in the evening. Mr. Beall has a patch of the finest turnips I've seen in many a day; and, I'm told that he never could have any luck in the line of turnip raising before.

*Oct. 27 M.* A rainy day.

*Oct. 28 T.* Clear and pleasant. School small.

*Oct. 29 W.* Jodie's birthday—he being 4 months old to-day. Prepared the *goodies,* invited the hands, etc. and tried to have a corn husking, but no one came to help us.[14]

*Oct. 30 Th.* School small.

*Oct. 31* F. and last day of Oct. Rev. J. W. Crews and lady from Forsyth (more of our old neighbors) called at pa's, and staid all night. They also, are out land hunting.

Land! Land! All want land.

13. He refers to Jesus' parable about the landowner who hired laborers, at different times of the day, to work in his vineyard. See Matt. 20:1–16.

14. His "*goodies*" undoubtedly did not include alcohol.

*Nov. 1 Sa.* Clear and pleasant.

*Nov. 2 S.* Staid at home one more Sunday, and read my books.

*Nov. 3 M.* And as rainy a day, I think, as I ever saw. Had 8 scholars at School.

*Nov. 4 T.* Election day, and as rainy a time, I think, as I ever saw. I dismissed school to go to the election, but it rained so I did not go; I shucked corn.

*Nov. 5 W.* Clear and cold, the wind blowing very hard from the N. W.

*Nov. 6 Th.* Coldest morning we've had. School about as usual.

*Nov. 7 F.* Weather moderated. Had 9 scholars. Cribed our corn. Dont think I've got more than ten bushels to my shear [share]. It was so dry last summer our corn crop failed.

*Nov. 8 Sa.* Another rainy day. Made Jodie a cradle, but it is rather a rough specimen. It is a flared box of half dressed pine plank nailed together, and set on rockers of White Poplar. Such a cradle!

*Nov. 9 S.* Windy and cold. Staid at home and read.

*Nov. 10 M., 11 T., 12 W., 13 Th.* Cold frosty mornings, but nice weather for sowing wheat. School quite small yet.

*Nov. 14 F., 15 Sa.* Finished sowing wheat. I've sowed 4 3/4 bushels. Brothers, Wiley and Clark came in from their schools this evening. Bought Clark's tools, and sold him my gun.

*Nov. 16 S.* A nice day. Took M. to her pa's, but did not stay long.

*Nov. 17 18* Very cold, frosty mornings, but pleasant evenings. School small.

*Nov. 19 W., 20 Th.* Weather moderates some. School about as usual.

*Nov. 21 F.* A *very* rainy day, but not very cold.

*Nov. 22 Sa.* Clear and warm. Cut and hauled wood.

*Nov. 23 S.* A very rainy day, but not cold. I & M went to Pa's.

*Nov. 24 M.* Cleared off warm. Had fifteen scholars to day; three new ones came out.

*Nov. 25 T.* Another very rainy day. Had 19 or 20 scholars!

*Nov. 26 W.* Cleared off very warm for Nov. Had 18 Scholars to-day. Free schools are generaly like a wasp, big at both ends and little in the middle. I am reading the *Lives of the Presidents.* What a queer animal man is! While he lives [h]is errors are kept bright by frequent handling, but after his death they are burried with his body, and his good deeds are gathered up to be handed down to future generations. And this is well enough, in one sense, as the children should be taught [to] reverence the memory of their fathers and imitate them in their greatness; but is so much exageration fair?

*Nov. 27 Th., 28 F.* Rained one of these days.

*Nov. 29 Sa.* Jodie's birthday. Drudged about as I usually do on Sat. The wind blew quite hard in the evening.

*Nov. 30 S.* Clear and pleasant. Heard the Rev. Jerymiah Johnson preach. His text was, "To him that overcometh will I grant to sit with me in my throne," etc. Rev. III, 21. The sermon, I thought, was a tolerbly fair one, tho' it contained at least two remarks that, perhaps, should not have been made. In speaking of religion he said that if it only benefits us in this life it would not be worth seeking for. This he called back, and said that religion is better than the world anyway. And, in speaking of the bible he said that it was the brightest treasure that ever illumined our world. Aught he not to have excepted the Author who "gave himself for us, that we might live."? I think so. I and M. came round by her pa's, took dinner and then came on home; and when we got here we found one of our fat hogs crippled.

*Dec. 1 M.* Quite a frosty morning. School as usual. Butchered our cripple[d] hog, which had badly hurt one of its hind-legs.

*Dec. 2* A very rainy day.

*Dec. 3, 4, 5* Cold, frosty mornings. School about so. Rev. J. Johnson preached at Pa's on Thursday night. His text was "Ye believe in God, believe also in me," etc. John XIV, 1, 2, 3. The congregation was small, and the sermon short. H. B. Godfrey paid me ten dollars, the amount of one of his notes.

*Dec. 6 Sa.* Cloudy part of the day and quite cool. Went to Eagle Mills.

| | |
|---|---|
| Bought Mollie a black woollen shawl— | 2.75 |
| Nine quires of paper | 1.00 |
| Sundries | 1.07 |
| | 4.82 |

I also bought, as I went on, of Madison, 15 3/4 yds. of quilt

| | |
|---|---|
| lining— | 97 |
| two pounds of sugar | 28 |
| one & a half yds calico | 18 |
| | 1.43 |

*Dec. 7 S.* Cloudy part of the day and quite cool. Snowed a little, and but little. Staid at home. Mrs. Charity Hicks came by to see us, and staid an hour or so.

*Dec. 8 M.* Coldest day we've had this winter.

*Dec. 9 T.* Quite cold. Had 20 scholars.

*Dec. 10 W.* Cold. The frost has not melted in the shade this week, and where the red gullies spew up ice it remains all day. Cold weather this.

*Dec. 11 T.* Rained lots last night, and cleared off warm this morning. Sold Alex. my oxen (a match to the one that died last summer) for $4.00, and bought of him a note on Alfred Johnson of $8.85.

*Dec. 12 F.* Quite frosty this morning.

*Dec. 13 Sa.* Not exactly clear, but several degrees warmer than last Monday. Cut and hauled some wood, went to mill, etc. Old Med has lost the use of her lips. They seem to be swelled and hang down quite limber.

*Dec. 14 S.* Rained last night and the forenoon lots. Staid at home. Quite windy in the evening.

*Dec. 15 M.* Cold and cloudy. Had 18 scholars, and trouble enough. The girls *will not study.*

*Dec. 16 T., 17 W., 18 Th., 19 F.* This has been a very cold week. School tolerbly full. Dismissed this evening (Friday) till the first Monday in January 1857. If the house is not repaired previous to that time the school is out. I've taught four months and fourteen days, so the Com. owes me about $75.

*Dec. 20 Sa.* The trees, fences and the ground are all glazed over with ice. This is the first sleet we've had this winter, and this has been a very rainy day. Not a fit time for out of door work, so I shelled corn, fixed the clock, etc. And now that I have but two lines I'll fill them.

*Dec. 21 S.* Cleared off cold. I & Mollie went to her pa's. We found the road very rough.

*Dec. 22 M.* Went to H[ampton]ville, and bought an Almanac for 1857, and a collar for Jenny. I paid 10 cents for the Almanac, $1.00 for the collar. I also bought me a cap for which I paid $1.25. We got home about 12 o'clock, and in the evening Pa & I hauled railes, wood, etc.

*Dec. 23 T.* Tolerbly clear, windy and very cold. Several degrees the coldest day we've had this winter. Mr. B. brought M's cow over, and at night she [the cow] got out and went back.

*Dec. 24 W.* Not as cold as Tuesday. I & Pa upset some fence in the forenoon; in the evening I cut, split and hauled wood.

*Dec. 25 Th.* Clear and pleasant. Brought Mollie's cow back. We ate Christmas-dinner with Ma.

*Dec. 26 F.* Fine and warm. I & Alex. built a sheep stable, etc. etc.

*Dec. 27 Sa.* Clear, and very warm for winter. Helped fix up the School house, cut wood, etc.

*Dec. 28 S.* Rained a little last night, and some since day. A little foggy, but nearly clear, and quite warm. Staid at home and read Josephus.[15]

*Dec. 29 M.* Clear and pleasant. Jodie's birthday, he being six months old to-day. He is sitting up in the cradle, before the fire, so sleepy he can hardly hold his peepers open. Sweet boy, he is gone to sleep. I & Caleb went to B's mill with Alex.'s bull and cart, cut fire wood, etc,

15. He refers to Flavius Josephus (37?–100?), author of a history of the Jews.

*Dec. 30 T.* Cloudy part of the day, but not much cold. Went to Madison's store and bought the following articles.—Calico for Mollie a sack, 2 1/2 yds; one butterplate .15; one tin cup 10;

| | |
|---|---|
| The Calico 12 1/2 per yd . . . . . . . . | 30 |
| The butterplate . . . . . . . . . . . . | 15 |
| The tin cup . . . . . . . . . . . . . . | 10 |
| | ·55 |

*Dec. 31 W.* Last day of the month, and of the year. Shelled out my corn last night. I've only got *six bushels!* except a few nubbins. Bought a Shangiah rooster of Alex, for which I am to give him two quires of paper (letter paper) worth 40 cts. Bought 100 boards of Alex. for 30 cents. Hauled wood, built cow shelters, etc. Cloudy and cold. Commenced snowing about 2 or 3 o'clock in the evening, and continued till sometime in the night.

# Diary 1857

*Jan. 1 Th.* The ground is covered with snow about 2 inches deep. Cloudy and looks much like it would snow again soon.

*Jan. 2 F.* Still cloudy.

*Jan. 3 Sa.* Clear, windy and cold. Went to the shoemaker's shop, (John G. Johnson's) and got Mollie's shoes half-soled. Bought, at Madisons, one water bucket, and one box of tax [tacks]; but those articles with their prices I've set in my "Cost of living."[1]

*Jan. 4 S.* Clear and cold. Spent the day in reading at home. It is much more pleasant to me to stay at home and read than to spend the Sabbath in unnessary visiting. Time is precious, and the day is coming when we will have to give an account of the manner in which we spend it.

"Arm me with jealous care,
As in thy sight to live;
And, O thy servant Lord, prepare,
A strict account to give."

*Jan. 5 M.* Cold and cloudy. Commenced school again after having two weeks holy-day—had 13 scholars.

*Jan. 6 T.* Cold. Pa and Alex. killed their hogs. School "so, so."

*Jan. 7 W.* This evening one week ago the snow fell, and there is *lots* of it on the earth yet. Had twelve scholars to-day.

1. Despite the detail with which Thomasson recorded purchases and expenditures in his diary, he evidently kept a separate book of financial records. It probably was a more detailed financial record, but it has not survived.

*Jan. 8 Th., 9 F.* Nothing worthy of note that I remember.

*Jan. 10 Sa.* Cool and cloudy, and rained some. I and Mollie went to her pa's.

*Jan. 11 S.* A very cold day. The wind blew and it snowed some. I & M. came home.

*Jan. 12 M.* Clear and cold. Had fifteen scholars.

*Jan. 13 T.* Clear and pleasant after the morning which was stinging cold. School closed. I've taught five months, or 105 days. I taught 21 days for a month, at $16 per month. Don't think that No. 30 will ever get my services as cheap again.

*Jan. 14 W.* Cold, cloudy and looks favorable for snow. Cut and hauled three loads of wood, took home the loom, etc.

*Jan. 15 Th.* Very cold and windy. Hung up our bacon, etc. Went after my shoes, and bought at Madison's, some thread and spice to the amount of 25 cts.

*Jan. 16 F.* Clear, cold, and windy. Went over to Mr. B[eall]'s to see about matters and things.

*Jan. 17 Sa.* Cloudy. Rained and hailed some little. Pa cut out my new yarn coat. I repaired the buggy shelter, cut some fire-wood, etc.

*" 18 S.* A stormy day. We had several snows last winter, but none, I dont think, to compair with this. The wind blows hard from the north east, and the snow falls very fast.

The snow! the snow!
It whirls about.
The snow! the snow!
It fills the rout.
How can a traveler see his way,
And pursue his journey such a day?

*Jan. 19 M.* A very cold, windy day. The snow is light, and the wind has drifted it up, in places, belly deep to a horse. I think if the snow had have lain as it fell it would have been 12 or 15 inches deep generaly, and perhaps deeper. This will long be remembered as the "big snow," unless there comes a biger one.

> The wind! the wind!
> It whirls the snow.
> The wind! the wind!
> What makes it blow?
> O that larger were our piles of wood,
> And all our fires were very good.

*Jan. 20 T.* A little cloudy, and not quite so cold as yesterday. Hauled two loads of wood.

*Jan. 21 W.* Hunted half the day and only caught one rabbit!

*Jan. 22 Th.* Clear and very cold. We (Alex., Hudson, Caleb and myself) hunted again to-day, and killed three rabbits and three squirles. The snow has melted but very little yet. Think its good to hang on till spring.

*Jan. 23 F.* A little cloudy, and very cold.

*Jan. 24 Sa.* Cloudy part of the day and snowed a little. Cut and hauled one load of wood. The snow has not melted much yet.

*Jan. 25 S.* A nice day—the earth covered in snow, and the sun shining warm. The snow melts slowly the morning. I've had no news in many a day, so I know not what "all the world, and the rest of mankind" are about these snowy times.

*Jan. 26 M.* Cloudy, but not very cold. Went to Town and received a little news, etc.

*Jan. 27 T.* Cloudy and rained considerably in the evening. Cut and hauled two loads of wood, finished my bridle, etc.

*Jan. 28 W.* A rainy day.

*Jan. 29 Th.* The warmest day we've had in some time. Went to Grant's mill twice.

*Jan. 30 F.* Commenced snowing early, and snowed, and snowed. Cut and hauled two loads of wood.

*Jan. 31 Sa.* Cloudy but not very cold. Snow melted considerbly.

*Feb. 1 S.* Quite foggy early in the morning. The sun shoan out about 8 or 9 o'clock. Think we shall have a pleasant day. Finished reading Frost's life of General Jackson this morning. Old Andrew was a great man, and a great general. He defended New Orleans as bravely perhaps as any man on earth would have done. He spoke the words, "Our country must and shall be defended," and then confirmed them by his actions. He also said, "We will enjoy our liberty or die in the last ditch." Let every lover of freedom set his "pegs" as firmly, and then stand up to them as bravely as did General Andrew Jackson.

*Feb. 2 M.* Quite a fine day over head. Went to Town and bought Jodie a cloak. Received two Nos. of the *Age,* (the first) and paid the postage (13 cts) in advance for the year. The *Age* is a good and cheap paper, and I wish to take it for a time. The snow melted very fast.

*Feb. 3 T.* A little cloudy. Cut and hauled one load of wood, commenced a tongue for Pa's waggon, etc. Cleared off in the evening quite pleasant.

*Feb. 4 W.* Clear, and the warmest day we've had for some time. Went over to our intended home, and I and Shade daubed and scored the cabin in style.

*Feb. 5 Th.* Cloudy most of the day and tolerbly warm. Made a meal tub or box.

*Feb. 6 F.* Still warm and cloudy, and rained some. Alexander and I cut the sills and plate for Pa's waggon shed, etc.

*Feb. 7 Sa.* Still warm, cloudy and rained some. Went to the Institute, met the trustees and told them I was willing to exchange the deed for lot No. 4, made to me and W. Messick jr. by the trustees of Union Academy Association for the note we gave them for said lot, but they would not accept the proposal so there was no definite action taken, but another meeting appointed for the 7th of March next. I put some hard questions to the board of Trustees, which is composed of the following *honorable* and *very wis* (in their own eyes) would be gentlemen, viz. A. W. Martin, *Esq.* Bennett Windsor, *Esq;* E. T. Burgis, Esq; Mort. Jacks, drunkard, (the other two John Reese and Dr. Parks were not present) which they did not answer either because they could not or would not. And as the reader is inquisitive enough to ask what those questions were I'll state on which will do for a sample.

A W. Martin, Esq. said (if not in these words to this account) that the present Association and Union Academy Association were one and the same thing, and that there has been no change affected! Then I asked him why there was such a dust kicked up in the spring of '55, the union desolved, etc. etc. when there was no change to be affected? The Esq. got ashy and said that I had thrown out some insinuations against the Baptist as a denomination, (when I had not opened my mouth about the "Baptist as a denomination" and he knew it.) and said so far as he was concerned I could pay for the lot or let it alone, signifying that if I did not pay for it they (the Trustees) would resort to law. Now the Esq. is like the fabled wolf which said to the lamb "Why do you make the stream of which I drink so full of mud," and when the lamb cleared ltself of that charge the wolf said "I've been told you spoke ill of me last spring." "Indeed" said the lamb "I was not then born." "Well, if it was not you" said the wolf "it was one of your race, and as it is all one to me you shall pay for it."

So he forthwith tore the lamb limb from limb. Now if I did not throw out "insinuations against the Baptist as a denomination" perhaps some one else had, and as it was all one to Esq. M. he determined I should pay for it.

*Feb. 8 S.* Still warm, cloudy and quite rainy in the evening.

*Feb. 9 M.* Clear, windy and cold.

*Feb. 10 T.* Moved over to our new home, on the Burton farm.

*Feb. 11 W.* A pleasant day. Went over to Pa's and made some laders to haul hay on, and loaded on the hay.

*Feb. 12 Th.* A nice day. Went over after my load of hay in the morning, and in the evening I split a few rails, built a pig pen, etc.

*Feb. 13 F.* Cloudy, and rained some. Went to Pa's and brought over a load of plunder.

*Feb. 14 Sa.* Warm. Brought our cows over. Moving is no small job even if one has but little to move.

*Feb. 15 S.* Quite a pleasant day. Staid at home and read, wrote etc till late in the evening, then I & M. took a walk to her pa's.

*Feb. 16 M.* Cloudy. Rained last night. Went to the old place after my hogs but did not get them. I & Caleb brought over a few articles. In the evening I cut timber to build a shed on the south side of our cabin.

*Feb. 17 T.* Cloudy and warm.

*Feb. 18 W.* Clear and hot. This has been the warmest day we've had this year. Hauled over a load of straw, and hauled up some of my shed timber, three loads of fire-wood.

*Feb. 19 Th.* Tooled my hogs (6 in no) over. Hogs love corn, and with a few nubbins I think I could tool mine to ______.

*Feb. 20 F.* Hewed some of my shed timber, cut and split a few rails, etc. etc, etc, etc, etc, and so on.

*Feb. 21 Sa.* I and Mr. B. finished hewing the timber for the shed, framed and raised it all to-day! We have had fine warm weather this week. Very warm for Feb.

> The frogs hollow, the birds sing
> The ticks crawl, the fleas bite
> And ______ what a time we have.

*Feb. 22 S.* A nice pleasant day. Pa and Ma came over to see us for the first time since we moved. Pa and I walked around my farm. Pa thinks I will not be satisfied here, for he says that I can not make anything, the land is so poor.

*Feb. 23 M.* Had a little frost this morning, for the first in some time. The day was clear and warm. Hauled a load of plank from my black smith shop, and weatherboarded up most of our shed.

*Feb. 24 T.* Went after my school money ($80.)

*Feb. 25 W.* Rained considerbly. Covered part of my shed, cut down a tree, in Mr. B's new ground, to make board to finish covering it. We sawed the timber after night.

*Feb. 26 Th.* Clear and pleasant. Went to Madison's store, and bought some iron, Mollie a calico dress, etc. I also went to H[ampton]ville and bought iron, nails, etc.

*Feb. 27 F.* Clear and pleasant. Went up on the river after my corn that Clark had engaged for me. I bought sixteen bushels of James Harris at 60 cents per bushel. I also bought half bushel of potatoes for 25 cents.

*Feb. 28 Sa.* Cloudy. Staid last night at Martin Redding's. Went over to Harris' early this morning, loaded up my corn and struck for home. I also loaded in ten bushels of oats as I came on at the widow Willburn's. Ps had been up and bought them at 30 cents per bushel. Sixteen bushels of corn in the ears and ten bushels of oats made me a tolerbly good load. It was a little after night had set in when I got home. Sister Mary came up to see us to-day for the first time since she left for Smith Grove.

This page I would like to fill
With something, very well,
But what, and where to find it
I'm sure I can not tell.

*March 1 S.* Cool. I and Mollie went over to Pa's on a visit for the first time since we moved. Had a pleasant time.

*March 2 M.* Very cold and windy. This is March in earnest. Rove a few boards, but sat by the fire most of the day. Dont know that I ever heard the wind blow harder than it blew last night.

*March 3 T.* Calm and not quite so cold. Went to mill, finished covering our shed, etc. etc. etc. Tom Messick cut and split rails for me today. He cut the timber and split chestnut rails.

*March 4 W.* A tolerbly pleasant day. Went to court; paid my taxes ($3.24) for the years 1854 & 1855; bought a side of sole leather for $1.65. (paid 30 cts. per lb.)

*March 5 Th.* A dark, rainy day.

*March 6 F.* Clear, cold and windy. Commenced sowing oats.

*March 7 Sa.* Cold and windy. Took Pa's waggon home, shelled one bushel and a half of corn. Caleb went to Town and brought me the *Age* and a letter from Dr. Hunter.

*March 8 S.* A little cloudy, and quite cold. Looks like it might snow.

*March 9 M.* Quite cold.

*March 10 Tu.* Went to mill with a load of corn, and had it ground up cob & all.

*March 11 W.* Cold. Hailed, rained and snowed.

*March 12 Th.* A nice day. Pa cut his leg badly. I cut and split a few rails, went over to see Pa, cut some wood for him, went to Madison's store and bought a pair of drawing cha[i]ns for 65 cts.

*March 13 F.* Cold and cloudy.

*March 14 Sa.* Clear and pleasant over head. The snow that fell yesterday evening and last night is six or 7 inches deep. Went to muster. The roll was called and dismissed. No muster. Paid for my lot at the Inst. $35. Would be glad to sell for $40.

*March 15 S.* Quite a pleasant day. I and M. went to Pa's.

*March 16 M.* Nice. Hauled 2 loads of wood for Pa, brought over a load of hay, moved my bus etc. Snow about all gone.

*March 17 T.* Clear and pleasant. Took up Pa's land note that J. L. Roughton, Exc. of S. Gentrys will, held against him. I paid $3.60 on it, (the note). In the evening I planted my potatoes, one half bushel.

*March 18 W.* Cloudy most of the day, and rained considerably in the evening. Sowed and plowed in oats.

*March 19 Th.* Windy and quite cold. Hauled rails, made fence, cut & split rails, etc. etc.

*March 20 F.* Clear, calm and pleasant.

*March 21 Sa.* Pleasant, but a little windy. Sowed and plowed in oats, etc.

*March 22 S.* Quite a nice day. Don't feel well so I stay at home. There is to be singing and class meeting at Aylesbury, and I was requested by a friend to attend the singing, but I am rather ill; and I dont know that I should go were I never so well.

*March 23 M.* Finished sowing oats. I & M went to Pa's and staid all night.

*March 24 T.* Quite a warm day. Came home and comenced breaking my corn land. Went to Town in the evening, etc.

*March 25 T.* Quite cool and windy. Plowed.

*March 26 Th.* There was a considerable frost this morning, and the day has been cool. Plowed.

*March 27 & 28* Nothing of importance.

*March 29 S.* I & M. went to Mt. Moriah to meeting. The Rev. Z. Rush[2] preached quite a good sermon. His text was the 14 verse of the II chapter of Titus. On our way home we came by Pa's so it was near night when we landed at our cabin.

*March 30 M.* Hauled from Pa's another load of hay.

*March 31 T.* Plowed, March goes out cool. February was pleasant. March has been quite cool; so cool and dry that the oats sowed in Feb. are just beginning to appear. This is a cold, late spring, and I don't think I ever saw hay, shucks, straw, etc. as scarce before. In fact cow feed is not to be had.

*April 1 W.* Cool, cloudy and rained some.

*April 2 Th.* Clark staid with us last night for the first time since we moved.

*April 3 F.* Went to Wm Pardue's and got half bushel of sweet potatoes (mostly yams) for 37 1/2 cents. Sweet potatoes are high when they sell at from 75 cents to $1.25 per bushel as they do this season. Every planter should raise his own potatoes, and "chew his own tobacco," *if any.*

2. The politics of the Reverend Zebedee F. Rush would become of great interest in a few years, for Rush was a Unionist who became a leader of the underground organization the Heroes of America. Also called the "Red Strings," this organization consisted of supporters of the Union and had up to ten thousand members in North Carolina. See William T. Auman and David D. Scarboro, "The Heroes of America in Civil War North Carolina," *North Carolina Historical Review* 58 (October 1981): 327–63.

*April 4 Sa.* A little cloudy. April comes in quite cool and dry. When will it rain?

*April 5 S.* Somewhat warmer. Cloudy. Rained some last night.

*April 6 M.* Rainy and quite cold.

*April 7 T.* Cold as winter. Dont know that I ever saw such a freezing time in April before. Fruit killed. Manured and plowed up my sweet potato patch.

*April 8 W.* Weather moderated. But little frost this morning. Cut and split some rails, dug two hog troughs, etc.

*April 9 Th.* Quite warm. Cut and split a few more rails, dug a soap trough, fed the pigs, etc. Commenced raining just before night, and now, (about 1/2 after 8 p.m.)

"The rain comes down with—(*not*) all its might,
Darkening the face of the sullen night."

*April 10 F.* Cleared off pleasant. Went to "the stump digging" on the new road leading from H[ampton]ville to Y[adkin]ville. Digging up large stumps is not very agreeable work. I did not get home till near 8 o'clock at night.

*April 11 Sa.* Looks a little like spring.

*April 12 S.* Easter, but no eggs! When I was a child I usually gathered in a good supply of eggs about Easter, dyed them various colors, and then celebrated that blessed day by eating them. What egg feasts we had in days of yore! We were happy then, but we did not know it. We lost much of the present in anticipating the future, and promising ourselves much pleasure when we were grown to be men and women. Our acquaintance then with this vexatious world was very limited. We are now grown up, our family divided and our lots are being cast in differant places.[3] Our childish sports and our egg feasts have long since been done away with, but our promised future happiness has never yet been realized. We now find that

3. By modern standards Strong's family was still close together and quite close-knit.

"Dark and thorny is the desert
Thro' which pilgrims make their way,"

but amid the thorns and darkness we remember the mottoe, "Hope on—hope ever," and still look to the future for better days. Trust they'll come.

One of Mollies pet lambs died to day. Poor little sheepie.

*April 13 M.* Went over to Pa's, and hauled (on the buggy) my bacon over home. We have but a few small pieces, yet if we use economy enough in the meat eating line they will be sufficient for us.

*April 14 T.* Cool, cloudy and rained some. Jodie has symptoms of the croup. Went over after Pa, but just before I started I gave Jodie a tea spoon full of the tincture of Labelia, which is an excelent remedy for croup, etc. A tea spoon full was an enormous dose for a child of his age and size. When I returned with Pa Jodie was considerably better. No wonder!

*April 15 W.* Took Pa home. Wind blew very hard.

*April 16 Th.* Went to Reinhardt's after wheat. Got only 2 bushels.

*April 17* Went to Nicholson's mill, but did not get my grinding done. Clouded up in the evening and rained considerably.

*April 18 Sa.* Cloudy, and rained a part of the day. This has been a very cold week for April. Old men say they never saw the like before. The trees look almost as ded as winter. When will spring come?

*April 19 S.* Cool, cloudy and rainy. Rained a flood last night but, thanks be to the Ruler of heaven and earth, it was not such a shower as was poured on the earth in Noah's time. Then it rained forty days and forty nights. What a shower that was!

*April 20 M.* Cold and windy. Went to N's mill after my meal and flour in the forenoon, and helped Mr. Beall roal logs in the evening.

*April 21 T.* Still cool and windy. Hauled wood off Mr. B's new-ground.

*April 22 W.* Windy and cold. Mollie and I went over to Pa's.

*April 23 Th.* Not quite so cool. I and M. went to Nicholas Ball's burying.

*April 24 F.* Quite a nice warm day. Hunted my hogs, but did not find them. I've not seen "hair or hide" of them since last Sunday.

*April 25 Sa.* Laid off corn land. Begins to look like April.

*April 26 S.* Warm and smokey. Caroline and her husband came up yesterday, and I & M went over to Pa's to see them.

*April 27 M.* Very windy. Commenced planting corn. This is a remarkably late spring. We've had but few warm days yet.

*April 28 T.* Tolerbly warm.

*April 29 W.* Cool morning. My hogs (5 of them) came home. They had been gone 9 or 10 days. Jodie is 10 months old to-day. He can stand alone, and can walk after a chair. He has no teeth yet.

*April 30 Th.* April goes out cool and windy. April departs and leaves the woods leafless, and looking almost as ded as winter. The fruit trees and poplars seem to say that "Spring is coming," and I also heard an old *Whipperwill* the other night. "Hope on hope ever."

*May 1 F.* What a glorious month is May! Well, we are glad each year to see her come, and especially so this season as her warm sun and gentle showers are much needed. She comes in with a cool south east rain. I am only about half done planting corn. The pan cakes are done and Mollie says "come to dinner."

*May 2 Sa.* Rained yesterday, last night and this morning abundance. Warmest day we've had this year.

*May 3 S.* Rained a fine shower this morning. Caleb & Eliza paid us a visit.

*May 4 M.* Rained a flood yesterday evening and last night, and is still raining this morning.

*May 5 T, 6 W.* I & Alex. Johnson went to Isaac Long's after a load of wheat for Mr. Bell. We got 10 bushels only, at $1.10 per bushel. This is a high figure for wheat, but if the crops should fail this season its not a beginning.

*May 7 Th.* Sick to-day. Was taken last nigh[t] with a chill. Had high fever most of the day. Pa came over and gave me a wet sheet pack, and a rub off with cold water and dry towels; then something to make the sweat run. And it did run too, and I soon began to feel better. I lay in the wet sheet about two hours "still as a mouse," but did not sweat any till I was taken out, washed off and put to bed.

*May 8 F.* Feel better.

*May 9 Sa.* Still on the mend. Dont care about eating. Crave vinegar, and use a good deal of it.

*May 10 S.* Not so well. Have fever again.

*May 11 M.* Feel better.

*May 12 T.* Some better I think.

*May 13 W.* Began to stir out.

*May 14 Th.* Went to work tho' I felt quite weak, and not exactly well. Think it has been cloudy all this week, and we've had a little rain.

*May 15 F., 16 Sa.* Nothing of importance on Friday. Saturday I planted our cotton, beans, ground peas, etc.

*May 17 S.* Cloudy and cool.

*May 18 M.* A cool, rainy day.

*May 19* Still cool and cloudy.

*May 20 W.* Cool. Replanted corn.

*May 21 Th.* Still cool. Commenced plowing my corn.

*May 22 F.* Tolerbly warm. Plowed.

*May 23 Sa.* Quite a warm day. Went to Nicholson's mill, and had five bushels of wheat ground. Think we have flour enough now to last us till August.

*May 24 S.* Clear and quite warm. We went over to Pa's. Martha came to see us yesterday for the first time since we moved over on this side of the "big branch," as Vic. calls the creek, and this morning we all *took* and went over to pa's *we did,* & *we* did, & *we did.*

*May 25 M.* Warm. Plowed our potatoes, replanted corn, etc. etc. etc.

*May 26 T.* Sheared our sheep. Some of our neighbors have lost their sheep this spring by shearing them before the clod [cold] "May rain" which did not come as soon this May as usual.

*May 27 W.* Finished planting corn. Rained a shower this evening. It also hailed some while raining. Went to Town and bought a grass sythe and swath. Price of the blade $1.25. Swath 62 cents.

*May 28* Plowed and hoed corn.

*May 29 F.* Jodie's birth day. Jodie is eleven months old to-day. He can walk, but has not any teeth.

*May 30 Sa.* Replanted corn, etc.

*May 31 S.* Last day of May. We started to Mt. Moriah but the creek not being foardable we turned and came to Flat Rock. Rev. T. Howel preached. His text was the 18 & 19 verses of the II chapter of Ephesians. "For through him" etc. This is a damp day. This evening we had a considerable shower of rain and hail. "The weather can not be foretold." At any rate not by me.

Yesterday was fair, and I thought looked favorable for dry weather, but last night it "clouded up" and rained, and rained.

*June 1 M.* Warm and showery.

*June 2 T.* This is growing weather. We (Mr. B., Shade, and myself,) hauled rails and made the pasture fence; (part of it)

*June 3 W.* I & Mr. B. finished our fence.

*June 4 Th.* Commenced plowing my corn the second time.

*June 5 F.* Plowed the corn.

*June 6 Sa.* Hoed corn.

*June 7 S.* Went to Sunday school at Flatrock, and to Grant's to meeting. Heard a Rev. gentleman read a tolerbly good discourse.

*June 8 M.* Warmest day we've had this season. Hoed corn.

*June 9 T.* Another warm day. Rained a nice shower in the evening. Plowed some. etc.

*June 10 W.* Warm. Plowed. Rained heavily in the afternoon.

*June 11 Th.* Cooler than yesterday, and quite windy. Plowed.

*June 12 F.* Nothing of interest more than usual.

*June 13* Clear and hot.

*June 14 S.* Went to the singing at Aylesbury.

*June 15 M.* Very warm. Rained a few drop[s] in the evening. Plowed.

*June 16 T.* Mr. B's bees swarmed. This is the first swarm I've seen this year.

*June 17 W., 18. Th.* Nothing of interest on Wednesday. Thursday evening at 4 o'clock the Rev. Z. Rush preached at Aylesbury. We went to hear him. R. G. Green tried to get back into the church. I objected. He was expelled for lieing, and I think he should make some acknowledgements before joining the church again; this tho' he is unwilling to do.

*June 19 F., 20 Sa., and 21 S.* To day we went over to the stand and heard the Rev. Haith, a preacher in the M. P. Church, preach. His text was, "All are yours;" etc. I Cor., part of the 22d and all the 23d verse of the III chapter. Congregation small. Sermon good. The door of the church was opened to receive members, but none went in.

*June 22 M.* Plowed.

*June 23 T.* Cool. I & Pa went to Wm Pardue's meadow to make hay.

*June 24 W.* Plowed in the forenoon, and hauled my hay home in the afternoon. My bees swarmed, for the first time this year to-day.

*June 25 Th.* Plowed. *26 F.* A hot day.

*June 27 Sa.* Worked the road.

*June 28 S.* Staid at home and read my books, papers, etc.

*June 29 M.* Jodie's birthday.

*June 30 T.* Went over and bound wheat for Pa. Wheat is as well filled as I ever saw it.

*July 1 W.* Plowed, etc.

*July 2 T.* Cool for July. Went over and bound wheat for Pa. Clark cuts & I bind.

*July 3 F.* Cool. Worked the stage road from E. Willburn's to Hamptonville, Is. Cazy overseer.

*July 4 Sa.* I & Clark cut and put up some of my wheat. This is a great day in the U. S.[4]

*July 5 S.* Staid at home as usual and read my books, papers, etc.

*July 6* I & Clark cut & bound wheat.

*July 7 T.* Had quite a fine rain last night, and more this evening.

*July 8 W.* Commenced laying by my corn with the plow.

*July 9 Th.* A. C. Johnson cut wheat for me to-day. We were favored this evening with another fine shower of rain.

*July 10 F.* Went to H[ampton]ville in the morning, and plowed p.m.

*July 11 Sa.* Warm & cloudy. Finished cutting wheat. I sowed between 4 & 5 bushels of wheat, and have 135 doz. [bundles of cut wheat.]

*July 12 S.* A pleasant morning. Dont think I ever saw as much cool weather in July as we've had the past two weeks. Farmers are blessed this year with great crops of wheat, and also with pleasant weather to save it in,

Abundant crops and pleasant weather,
Are so very cheering to the farmer,
May they continue on forever,
While in the world remains a tiller.

*July 13 M.* I & Caleb finished hauling in our (mine & Pa's) wheat.

4. Thomasson's patriotic pride in the Fourth of July contrasts with his silence on sectional issues, such as the Dred Scott decision or controversy over Kansas.

*July 14 T.* This would be a very hot day but for the cool breeze that blows from the east. Plowed.

*July 15 W.* Went to Doweltown to borrow some money of Dr. H. Willson, but did not get any, as he said he had not "a particle." Dr. Franklin, I think it was, said "If you wish to learn the value of money try to borrow some." I've tried a few times in my life to hire money, and I've found it hard to get. Hope I shall find it necessary but seldom to make the tryal.

*July 16 Th.* Visited the [common-school] Com[mittee] in Dis. No 30 in order to get the school, but I met with but poor encouragement from two of them, Bell & Johnson, Esqs. This evening we had quite a fine rain.

*July 17 F.* Plowed hard. My oats are turning some.

*July 18 Sa.* Went to Aylesbury to hear the Rev. Z. Rush preach. He reached a good short sermon. His text was the 1, 2, 3, 4, 5, & 6 verses of the 15 chapter of Rom.

*July 19 S.* Went to Flatrock and heard a youn[g]ster by the name of Carter hold forth. Carter said his text was recorded *by the Acts of the Apostles.* Ha, ha, ha, And who was this "Acts of the Apostles"? Mr. C. made a mistake in the outset, for the "book of Acts," from which he read out his text, Rev. Justin Edwards says was written by Luke, and addressed to Theophilus. The text was "But none of these things shall move me," etc. Acts XX, 24. Such a mixt up affair I hardly ever heard from the pulpit. The word sir was used a great many times. Now this word, to my ear, sounds badly in a sermon. Carter went it for about an hour and a half.

*July 20 M.* A hot day. My bees swarmed yesterday, and while I was trying to hive them they took a notion to go back to the old gum again, and back they went. This morning, "bright and early," I rob[b]ed them of about a gallon and a half of their "hoarded sweets." Between 12 & 1 o'clock this evening they swarmed out again, and settled in a pile of brush near by in two bunches. I was at a[ ]loss, at first, to know how I should proceed in order to hive them. I finally concluded the only plan would be to spread a cloth near by on the ground, set the gum on it, and then get the dung-shovel and shovel them out of the brush pile and pour them down at the mouth of the gum. This I did, and succeeded finaly in hiving them. Yesterday one stung me and several others tried to but failed. To-day I did not get a single sting. Some say that bees will not swarm if you rob them. I do not believe any such doctrine.

*July 21 T.* We went over to Pa's, and I spent most of the day in search of a run-away hog. I found one in Mr. Parson's plantation, and I think it was mine. I could not get a view of him so as to be certain. We had, this afternoon, a splendid rain. Corn looks well, and is growing finely. Oats are ripening and some of them, I see, are being cut.

*July 22 W.* Plowed. Ground in nice order.

*July 23 Th.* Tax gathering at Jim Green's. I did not go as I dont go to Greens when I can stay at home or go any where else. I'm told they had no less than two fights, and that one poor wretch, Bill Foster, was taken off to jail. Finished plowing my corn. This is late in the season to plow corn, but as it is later than usual it has to be plowed later. Corn looks well, and is growing finely.

*July 24 F.* Went over to Pa's and I & Caleb went down the creek to Mr. Parson's farm to look again for my hog. We found him, and such a race as we had! After running up and down the creek thro' bushes, briers, etc. till his steam began to get up he splunged the water, went out at the lower gap and struck off for home. We crossed the creek on a sowerwood pole and put off on his trail. We caught him just before he got home, took him on and lodged the runaway in jail. All I want him to do now is to eat and grow fat. Hoed corn in the evening.

*July 25 Sa.* Hoed out my ground peas. Rained a shower, "it did."

*July 26 S.* A warm day with a rainy evening. I like to spend the Sabbath in reading when it is not convenient to attend church; so I've spent this and hope to spend many others in "the good old way."

I notice the Bee Birds about, and I guess they are after my bees. I think they only catch the drones, as I do not recollect seeing any of them (the B. B's ) near my bees at any time save the middle and latter part of the day—the time the drone, or male bee, always flies out. Then perhaps the birds are doing the workers a favor by helping them to get rid of the drones. Poor drones! I pitty you. I should not like to be a drone either in society or in a bee-hive.

From July the 26 to Aug. 3d. things passed on about as usual, without the occurance of any thing very notable that I now, about the middle of Aug., have any recollection of. However, brother Wiley paid us a visit in the time, but it was of but "few days," and not entirely free from "trouble." I do not mean that said visit troubled us, but him.

*Aug. 3 M.* Commenced school in Dis. No. 30.

*Aug. 4 T.* Had about 16 scholars I think.

*Aug. 5 W.* Staid last night at Pa's. The big rain yesterday raised the creek so that it was past crossing.

*Aug. 6 Th.* Election day. Went to Town and voted for A. M. Scales, Jr. of Rockingham. I clerked at the C[ounty] C[ourt?] box.[5] Esqs. Holcomb & Carter were the judges. John Johnson & Geo. D. Holcomb were the judges at the Congress box; R. S. Green clerk. Scales received 82 votes, and Puryear 285.[6] I'm told there was sixteen gallons of liquor on the ground, but that, in so large a crowd, was only enough to make drinkers feel rich and powerful. It was soon dispatched, and the Town was, when I left after the poles were closed, unusually calm.

*Aug. 7 F.* Taught school.

*Aug. 8 Sa.* Helped Pa and Clark stack our hay at Wm Perdue's meadow, etc.

*Aug. 9 S.* Staid at home & read.

*Aug. 10 M., 11 T.* Warm weather. School small.

*Aug. 12 W., 13 Th., 14 F., 15 Sa, 16 S.* We went to Shilo to meeting, but Jodie was not well and we had to leave before the services were closed for the day. The hat passed round and the sum of *one dollar & fifteen cents* was thrown in for the preacher.

*Aug. 17 M.* We have had very hot weather for a few days.

5. In North Carolina the governor and legislature appointed prominent local persons to the position of justice of the peace. These justices (often called squires) held office for life, and a portion of them sat each quarter as the county court, or Court of Pleas and Quarter Sessions. The county sheriff, however, was an elected official, and Thomasson may have meant here that he clerked at the election box where voters cast their ballots for sheriff.

6. In this election Alfred M. Scales finally defeated his rival Richard C. Puryear.

*Aug. 18 T.* Rained a nice shower at the school house, but not much here.

*Aug. 19 W.* Thrashed our wheat. Reece & Hanes' machine.

*Aug. 20 Th., 21 F.* Taught school only 4 days this week.

*Aug. 22 Sa.* Cleaned my wheat. I sowed, I think, about 4 3/4 bushels of wheat, and raised 31. This is a small turnout, but the wheat is very good. Owing to its being sowed late, and not much pains being taken in preparing the soil, it stood (a great deal of it) very thin on the ground.

*Aug. 23 S.* Killed another stand of bees last night. Got about 2 gallons of beautiful honey. Bees are little things, but they are worth taking care of. "Despise not the day of small things." Staid at home & read.

*Aug. 24 M.* Had 22 scholars. The weather has been cool for a few days.
Cost of my wheat crop.

| | | |
|---|---|---|
| Sowed 4 3/4 bushels | $4.75 | |
| Plowing in | 3.50 | |
| Cutting one day . . . . . . . . . . . . . . . . | 1.00 | 9.25 |

Raised 31 bushels.
Paid rent 9 1/2 bushels
Paid for thrashing 1 3/4

Then 9 1/2 + 1 3/4 = 11 1/4, 31 − 11 1/4 = 19 3/4
The number of bushels I still have which is worth $19.75. Then 19.75 − 9.25 = 10.50 worth of wheat left after expenses are all paid, except my own labor.

I have not written any in my Diary since about August 24. This is Sept. 19. Since that time but few things of importance have taken place within the limits of my affairs. On Saturday the 12 inst. the Examining Committee met at Yadkinville. I & Clark went down. I was examined, and received No. 3 in every thing but Geography and No. 4 in that. J. H. Kinyon & R. F. Armfield are the Ex. Com.

On S. the 13 we went to Senter to meeting and heard the Rev. Wm Barringer preach.

Dismissed school on the 17 till fodder time is over, or till the 7 day of

Oct. anyway. I also commenced pulling fodder on the 17. Corn is late this year, but very good.

*Sept. 20* Cloudy, and rained very hard in the evening. We went to Flat [Rock] to meeting. Heard the Rev. Z. Adams preach his farewell sermon to his Flatrock congregation. The sermon was short, but appeared to affect both preacher and people very much. The idea of old friends being separated to meet no more this side of the bar of God fills their hearts with sorrow.

*Sept. 21 M.* Cool and cloudy; Pulled fodder.

*Sept. 22 T.* A rainy day. Went to H[ampton]ville, got the news and bought a new book, the title of which is *Fields Scrap Book,* price $1.75.

*Sept. 23 W.* Clear and pleasant. Pulled fodder.

*Sept. 24 Th.* The weather and work of yesterday was repeated to-day.

*Sept. 25 F.* Finished pulling fodder.

*Sept. 26 Sa.* Tied fodder, cut tops,[7] etc.

*Sept. 27 S.* A very pleasant day. We went to Nelson Messick's stand, on the road just above his house, and heard the Rev. Laughlin, a preacher in the M. P. Church, preach quite a good, plain sermon. His text was "Wherefore seeing we also are compassed about with so great a cloud of witnesses let us lay aside every weight," etc. Heb. XII, 1 & part of the 2 verse. He first spoke of the "witnesses." And in "the multitudes who had lived and died in faith," he named Abel, Enoch, Abraham, Elijah, Daniel and the three Hebrew children. He then spoke of the "race;" then of the "weights;" and wound up with a good exortation. Think he spoke 1 1/2 or 2 hours.

*Sept. 28 M.* Cut tops.

*Sept. 29 T.* Went to Town and bought a few articles to the amount of 3.00. Cut tops, etc, etc, etc, etc, etc.

*Sept. 30 T.* Clear and cool. Cut tops.

7. He is cutting the tops of the corn plants to make more fodder.

*Sept. 31*[8] *W.* We staid at Pa's last night. Frost! this morning, but not enough to "bite" much. Finished cuting and tieing tops. And this is the last day of Sept.

*October 1 Th.* As clear a day, I think, as I ever saw, and pleasant. There was some frost this morning. Went over and helped Pa mow.

*Oct. 2 F.* Cloudy and rained some.

*Oct. 3 Sa., 4 S.* Cool and cloudy. Staid at home.

*Oct. 5* Took Pa's plows to the shop, plowed some, etc.

*Oct. 6* Went to mill; took 8 bushels of wheat—5 for myself and three for Jes. Casey.

*Oct. 7 W.* Plowed a few rounds, to[ok] Casey's flour to him, etc.

*Oct. 8 Th.* Sold Wm Pardue, Jr. my buggy, or swaped it to him rather for a filly a year & a half old, giving him $20 for the difference in value.

*Oct. 9 F.* Pardue came down this morning, and we exchanged property.

*Oct. 10 S.* Hauled my hay from Pardue's meadow. Had Jennie shod before with a pair of new shoes.

*Oct. 11 S.* Staid at home all day. I love to stay at home, and since I've sold the buggy I've a good excuse.

*Oct. 12 M.* Taught school to-day, but dismissed at night by order of the committee till the first Monday in November. Cloudy and rained a little.

*Oct. 13 T.* Borrowed old Med and Pa's big plow, and this evening I've been plowing my wheat land in grand style. Still cloudy.

*Oct. 14 W.* Got up quarter after four this morning. I wish to "thrive in the world" so I generaly "rise early in the morning."

*Oct. 15 Th.* Rained last night, so the ground was full[,] wet to plow but however I finished with the "big plow" about twelve o'clock, and took it home in the evening.

8. Since September has only thirty days, Thomasson must either have become confused when he tried to catch up for the days he failed to keep his diary or made two entries for September 30 (labeling one Sept. 31).

*Oct. 16 F.* Split a few rails, etc. Was invited to a still house raising (James Godfrey's) but did not go. I can not aid in the liquor traffic in any shape, form or fashion.

*Oct. 17 Sa.* Bought Mollie a spinning whell for $2. James Denny made it.

*Oct. 18 S.* Staid at home and read as usual.

*Oct. 19 M.* Commenced gathering corn.

*Oct. 20 T.* Cool and windy. Gathered and hauled in corn.

*Oct. 21 W.* There is this morning a killing frost. Commenced sowing wheat.

*Oct. 22 Th.* Another frost. Sowed wheat.

*Oct. 23 F.* General muster day at Yadkinville. I hired Mr. Bell (gave him 25 cents) to muster in my place, so I staid at home and sowed wheat.[9] Rained in the afternoon.

*Oct. 24 Sa.* Cloudy, and rained some. Shucked corn, went to Town, etc.

*Oct. 25 S.* Cold and windy. Clark came over and said Aunt was very bad.

*Oct. 26 M.* Still cool and windy. Gathered corn. Started over to Pa's and met Caleb coming after us. Aunt so bad, dont think she'll last long. We went over and found poor old Aunt Mary[10] strugling in the icy arms of death. She knew me, though, and asked if I had had my dinner.

*Oct. 27 T.* Aunt breathed her last this morning about one o'clock. She talked on till a short time before her death, but not very distinctly. She appeared to die easily. Death had lost his sting, and the grave its victory, for Jesus no doubt was there softening her dying bed. And bright angels no doubt were there ready to bear her blood bought spirit to "Abra-

9. North Carolina's militia law may be found in the *Revised Code of North Carolina, Enacted by the General Assembly at the Session of 1854* (Boston: Little, Brown and Company, 1855), 396–426. It does not specifically allow men to send a substitute to musters, but section 76 allowed captains to receive substitutes for militiamen who were drafted when the central government requisitioned some of the state's militia. The procedures of many local militias were often relaxed and informal, and local commanders probably permitted men to send substitutes to muster.

10. This Aunt Mary must have been a sister of Strong Thomasson's mother, for his father's sister Mary had died in 1846.

*Dec. 7 M., 8 T., 9 W., 10 Th., 11 F.* Taught school all the week. Had two heavy rains this week. Wheat looks finely. The weather has been warm this month so far.

*Dec. 12 Sa.* A cold day. Went to the Chairman's after money. Received $32. Came by Wm Pardue's and paid him $5. Pa finished raising his cow stable.

*Dec. 13 S.* A clear, pleasant day.

This is for the week ending Saturday the 19. Taught school. Another washing rain. We have a shower every week.

*Dec. 19 Sa.* Quite a pleasant day. Hauled over my top fodder, etc. I & M. went to her pa's.

*Dec. 20 S.* Cloudy and cool. We came home this morning and Mr. & Mrs. B. came with us. Lill lost her calf. Think it came a few days too soon. I saved its hide.

*Dec. 21 M.* A cold rainy day. School small.

*Dec. 22* Cleared off warm.

*Dec. 23 W.* Quite frosty this morning. Dismissed school till after Christmas.

*Dec. 24 Th.* Cool and cloudy, Think its fixing to snow. Helped Pa saw some timber for gate polings, and finished covering my cow stable in the forenoon; and in the evening I cut wood, and Caleb hauled it. I cut about as fast as he hauled with both horses.

*Dec. 25 F.* Christmas day. Got up this morning 1/4 after three, and found it snowing fast.

*Dec. 26 Sa.* Cold and windy. The snow fell yesterday morning as fast and in as large flakes as I ever saw it. Cleared off about 12 o'clock. If the snow had not melted so fast we should have had a deep snow. Made Jodie a pair of shoes yesterday, and half soled Mollie's to-day. It was too cold to-day for the snow to melt fast.

*Dec. 27 S.* A very cold morning. Staid at home all day.

*Dec. 28 M.* Somewhat cloudy, a[nd] several degrees warmer than yesterday. According to the old saying to-day rules March. Had ten scholars to-day.

*Dec. 29 T.* Warm, cloudy all day. Rained this morning. If to-day rules April we shall doubtless have many a shower in that month to the great delight of farmers generaly, and especialy to those who believe in the old adage,

"A wet April and a dry May
To fill the barn with wheat and hay."

Brother Wiley came up to-day.

*Dec. 30 W.* Raining this morning also. This has been one of the rainiest of days. Should it rain thro' May as steadily as it has rained to-day the people living along the Ohio will have no occasion, as the news paper says they now have, to sprinkle the bed of the river with water to keep the boats from kicking up a dust.

*Dec. 31 Th.* Last day of Dec., and of the year '57. This page closes another Diary, and this day another year of my life.

"Thus far the Lord hath led me on,
Thus far his power prolongs my days."

# Diary 1858

*January 1 F.* A clear, pleasant day. Hauled over one stack of my fodder. Brother Wiley went back to Smith Grove, the place he calls home.

*Jan. 2 Sa.* Clear, or nearly so, and pleasant. I & Caleb cut and hauled wood in the forenoon. In the evening I went to Madison & Son's store, bought a few articles, (see "Cost of living" for this year) paid them off in full and took a clear receipt.

*Jan. 3 S.* Still clear, and quite pleasant. Staid at home all day, and read.

*Jan. 4 M.* Cloudy and tolerbly cool. Had 12 scholars to-day.

*Jan. 5 T.* Got up this morning before 4 o'clock; find it cloudy and raining. This another very rainy day.

*Jan. 6 W.* Old Christmas.[1]
Clear and warm. If to-day rules Dec. we shall have a pleasant month. So may it be. I'm reading Pope's *Essay on Man,* and like it well.

*Jan. 7 Th.* Clear and cool.

*Jan. 8 F.* Somewhat cloudy, and cold enough, I think, to snow. John Poindexter, of Forsyth, staid with us last night. He told us a great many things about our old friends and neighbors. He had two horses, (worked one and rode the other) a little wagon, about 5 boxes of tobacco and his son. Was going, he said, to Alexander County. Hauling tobacco with one horse over the muddy roads these hard times is, I guess, a poor business.

1. Thomasson refers to the day on which Christmas would have been observed under the Julian, rather than the Gregorian, calendar.

*Jan. 9 Sa.* Cloudy and quite cool. I & Pa built a shelter for our sheep. I went to the shoe shop (John G. Johnson's) and halfsoled my shoes.

*Jan. 10 S.* A cloudy, drizzly day, but not so cold as yesterday and day before. Rose had her fourth *bull calf* to-day. Its a fine one too. I commenced reading the Bible about the first of this month, and expect to read it regularly through this year, and every year as long as I shall live.

*Jan. 11 M.* Warm, and very rainy.

*Jan. 12 T.* Warm, and nearly clear.

*Jan. 13 W.* Warm and cloudy. Pa killed his hogs. School small, small.

*Jan. 14 Th.* Still cloudy and cool.

*Jan. 15 F.* A rainy day. Bought, of a book peddler, *Clark's commentaries on the New Testament;* price $3.25. Dr. Adam Clark was a wise, and no doubt a good man. He has written a great deal in explanation of the Bible. What he says about a passage of scripture is, I believe, generally admitted to be its true meaning.

*Jan. 16 Sa.* Cleared off about the middle of the day quite pleasant. I & Caleb hauled a big load of leaves and put them in our cow and horse stables to make manure. In the evening I fixed a stable for my filly.

*Jan. 17 S.* A nice day. Staid at home, and read my books.

*Jan. 18 M.* Another nice day.

*Jan. 19 T.* School closed. Taught nearly four months this term, and received $59.

*Jan. 20 W.* Went to mill (B's) with corn. Went to Town and closed my acct. with the firm of J. & A. C. Cowles.

*Jan. 21 Th.* I & Mollie staid at her pa's last night. I & Pa put up a gate, in the evening, at the mouth of the cowlane. Tuesday, Wed. & Th. have been remarkably fine days. Dont think I ever saw finer weather in January.

*Jan. 22 F.* Another beautiful day. I & Pa measured off my land, etc.

*Jan. 23 Sa.* Cold and cloudy.

*Jan. 24 S.* A cold drizzly day. The Rev. Samuel Calaway staid with us to-day. I bought the *Pilgrim's Progress* of him yesterday. Price 40 cents.

*Jan. 25 M.* Cut saw-logs, etc.

*Jan. 26 T.* Nice weather.

*Jan. 27 W.* A beautiful day. Hauled saw-logs in the forenoon, and choped in the p. m.

*Jan. 28 Th.* Cold, cloudy day, and snowed like *forty* for a few minutes. Hauled logs & plank.

*Jan. 29 F.* Cold and windy. Went to mill. Mary and Williams[2] came up.

*Jan. 30 Sa.* Cool. Hauled wood, etc.

*Jan. 31 S.* Cloudy and cold. Mr. W. went home and left Mary up to stay a week. Finished my task of Bible reading for Jan., (Gen. & Exodus) and must now commence for February.

The Bible is the book of books, and should be read through by every person, after they have learned to read well, at least once a year. Our Savior said "Search the Scriptures;" the preachers say "Search the Scriptures;" and all good men (reading men I mean), of every denomination say "Search the Scriptures."

*Feb 1 M.* One of the coldest, and rainiest days we've had this winter. Worked in the shop, making us a large table of pine wood.

*Feb. 2 T.* Cold and windy, and somewhat cloudy. Shrubed off our house place, cut wood for the fire, etc.

*Feb. 3 W.* Tolerbly cool and a little cloudy. Cut firewood in the forenoon, and in the evening I & Alex.[3] hauled sawlogs and plank.

*Feb. 4 Th.* Snowed last night, and to-day too. Worked in the shop.

*Feb. 5 F.* Snowed some to-day too.

2. Thomasson refers to his sister Mary and her husband, Ezekiel Williams. They married in Davie County, according to the genealogical records of Jean Harris Thomasson, and may have been living there at the time. By the 1860 census they were living near Basil's father and mother.

3. Alexander Johnson, the husband of Thomasson's sister Martha.

*Feb. 6 Sa.* Clear. A pleasant day overhead. Sister Mary started home on horseback, met Mr. Williams a few miles below Hamptonville coming after her. Pa who was with her came back with the horses, and she went on home.

*Feb. 7 S.* Cool and windy. Staid at home and read the Bible.

*Feb. 8 M.* I cut wood, and Pa & Caleb hauled nine loads.

*Feb. 9 T.* Windy, a little cloudy, but not very cold. I & Mollie went to the Eagle Mills, and bought a few articles. See "Cost of living."

*Feb. 10 W.* Went to the pigeon roost last night; saw a great many pigeons, but only killed three. Went to mill, to Mr. Bell's and to Town.

*Feb. 11 Th.* Cold, and clouded up in the evening. I & M. staid at her pa's last night.

*Feb. 12 F.* One of the snowiest of days.

*Feb. 13 Sa.* Snowed all day yesterday, and is hailing this morning at a rapid rate. The snow is about 10 inches deep, or perhaps deeper.

*Feb. 14 S.* Cleared off this morning quite pleasant over head. This is one of the "big snows," and will, I guess, long be remembered by those whose wood piles are small.

Read the life of John Bunyan. Let all who read these lines, read his life, and his *Pilgrim's Progress* also. J. B. was a good, and hence a great man; and though he was imprisoned twelve years for new opinions sake done a vast amount of good it is said by writing, and by preaching to those who were disposed to visit him at the prison.

*Feb. 15 M.* Nearly clear, and quite pleasant over head. The snow melted fast. Went to Town and bought a few articles. See "Cost of Living."

*Feb. 15* Among other things I bought 6 of the best steel pens I've seen in many a day. I am now writing with one of them. I also saw the *Enterprise,* a small news paper, published in Jonesville, N. C. The price of the *Enterprise* is $1 a year.[4] Each subscriber may get the worth of his money in some shape or other, but I am rather afraid to subscribe myself. Think

4. This newspaper has survived, though the North Carolina Collection at the University of North Carolina at Chapel Hill has only two issues that appeared in June 1858.

I would like to edit a paper myself if I had the "skill and ability." Dont know if I shan't send out a "Prospectus" one of these days, and see if I can't get "all the world and the rest of mankind" to subscribe. I once made up a M.S. but never sent it to press.

*Feb. 16 T.* Clear and pleasant over head. Snow melts fast, but there is lots of it yet. I killed, to-day, the last hog I had. The Jews, in the time of Moses, were forbidden to eat swines flesh, and I don't know if I sha't deny myself that luxury before a great while. Dont you think the probability is, that I shall? The last hog dead, and no likelihood, that I know of, of my getting any more. There are a great many young hogs in the country, but the people who own them won't sell at any reasonable price. So I guess I shall be minus a pig for many a day to come. Well, one fine thing is, if I can live at all, I can live without the grunters.

*Feb. 17 W.* Worked in the shop, making a writing desk of pine. A tolerbly warm day, and a little cloudy.

*Feb. 18 Th.* Cloudy, and colder than yesterday. Worked on my desk. The snow melts, I believe, night and day, but the ground is still covered, save a few places where the snow has melted or been moved, as on the south sides of houses, barns, stacks, etc, etc. Guess its bed time.

*Feb. 19 F.* Cloudy, cold and drizzly. Finished my desk, all to putting on the lock and painting it. I have a pretty considerable notion of turning cabinet maker. I fear tho' that trade would not pay as well as farming. Dr. Ben. Franklin said, "Keep thy shop and thy shop will keep thee." The bible says, "He that tilleth his land shall have plenty of bread." Now I believe I had rather risk the Bible and the farm; tho' the man who keeps his shop *may* live well, but the man that tills the soil will be certain to have bread to eat.

*Feb. 20 Sa.* Clear, save a few wind clouds that occasionly pass over from the North West, Cold. Went to Benbow's mill, etc. etc. etc. If you wish to be healthy, wealthy and wise, go to bed early at night, and be sure to rise early in the morning; and then "make hay while the sun shines.

"Plow deep while sluggards sleep,
And have corn to sell and keep,"

Dr. Franklin. Sone one else has said,

"Improve each shining hour,
And gather wisdom all the day
From every leaf and flower."

As we can't gather honey from the "lilies of the field" we should gather wisdom by closely observing them, for we are informed by the Bible that even Solomon, tho' a great king, was not "arrayed like one of these."

*Feb. 21 S.* A pleasant day. Staid at home and read my books the Bible, *Scrapbook,* etc.

In the Bible I read the "repetition of sundry laws," etc. Lev. XIX. In the *Scrap Book* I read several nice pieces. Among others a beautiful piece of poetry written by the Rev. S. Grayham, during a passage from N Y. to Richmond, in Nov. 1826—three years before I was born. Poems are not affected by age like men. They are handed down from generation to generation, while men soon grow old and pass away,

"Onward we swiftly glide, and leave
Our home and friends behind."

Yesterday and last night the air was quite cold, this evening and to night it is quite warm. The snow is not all gone yet, tho' it is going fast.

*Feb. 22 M.* Rained, hailed and snowed, from the North East. This has been, so far, quite a cold February. The snow falls fast, fast.

*Feb. 23 T.* A little cloudy and very cold. Cut firewood in the evening, and Pa & Caleb hauled.

*Feb. 24 W.* Cloudy and very cold till evening when the weather moderated considerably. Mr. James Armstrong sent a negro man with an order and $11 for my cow, Rose, and promised to pay the ballance ($2) at court, next week. I received the order, the money, the promise, and let the man take the cow. Hope it will all end well as I sold her very cheap.

*Feb. 25 Th.* A pleasant day. Nearly clear. Went to Town, and had Jinny shod all around. Went over to Winecauf's in the evening, and bought of him two small stacks of hay at $1.25 each.

*Feb. 26 F.* Clear, and a few degrees colder than yesterday. Hauled home my hay. Received Nos. 1 & 2 of the first vol. of the *N. C. Planter.* Have not examined it much, but think the *Planter* is a tolerbly good paper, and well worth the subscription price—$1 a year in advance. Long live the *Planter.*[5]

*Feb. 27* February draweth to a close. It came in "like a lion," and seems to be going out "like a lamb." This day was several degrees warmer than yesterday. Hauled one load of hay for Pa, helped Alex. haul a load of straw from his stack at Pa's barn to the shuck pen near his stable. Now that I have bought I have hay plenty but it is so rough my horses will not eat it. I shall have to swap it to Pa, or buy more.

*Feb. 28 S.* Last day of February. Cloudy. Wind blows from the S. W. at the rate of, I cant tell how many, miles an hour. I have no barometer with which to ascertain the weight, or variations of the atmosphere.

I am many pages behind, this month, in my Bible reading. "Life is strangely fraught." So many things have we to do on the morrow, which we are not aware of to-day that what we have marked off to do, or a portion of it at least, is neglected. But no matter what else is to be done, we should read the Bible daily, and avoid "the counsel of the ungodly," "the way of sinners," and "the seat of the scornful," for "blessed is the man" whose "delight is in the law of the Lord; and in his law doth he meditate day and night." Think I will read my Bible task for next month.

*Mar. 1 M.* Cloudy and very cold, strong North wind with an occasional sprinkle of snow. Went to T. A. N[icholson]'s mill, and had a cold time of it too. J. D. Johnson paid me $1.37 1/2 for a book I let him have some time ago.

*Mar. 2 T.* Coldest day, I think, we've had this year. We cut and hauled 3 loads of wood.

*Mar. 3 W.* Windy and very cold—too cold to work out of doors, so I made fires and sat by them. I was not idle, tho', for I made a bridle for old Med; read an interesting piece or two in the *Sentinel;* wrote a few lines, etc. This is court week, and I expected to have gone to Yadkinville yesterday, but it was too cold. To-day is to[o] cold also, as it is, I

5. The *North Carolina Planter* was a monthly magazine "Devoted to Agriculture, Horticulture, and the Mechanic Arts" published in Raleigh by A. M. Gorman. It was thirty-two pages long and was full of detailed advice, articles, and diagrams about all aspects of farming.

believe, co[o]ler than yesterday. March is a spring month, but we have, at present, winter weather.

*Mar. 4* Clear, and considerably warmer than yesterday. Went to court. Paid my tax, $1.37, for the year 1856. Received the balance of my school money, $27, for teaching in Dis. No. 30.

*Mar. 5 F.* Clear and cold. I and Caleb cut and hauled Wood. Caleb received a No. (vol. IX, No. 26) of the *Age*—the first one for his subscription year. It contains a very good essay, on the subject of education, by Invalid. He, or she—I know not whether Invalid is male or female—say "Our early impressions are our latest and our strongest ones," and I agree with him exactly, for impressions that were made on my mind when I was only five years old have lost none of their strength yet, tho' I am now in my 29 year. O! that all impressions, and especialy those made on youthful minds, were good ones.

"The pebble in the streamlet scant,
Has turned the course of many a river;
The dew-drop on the baby plant,
Has warped the giant oak *forever.*"

*Mar. 6 Sa.* Cloudy and cold. Helped Pa and Alex. fence the plant bed, split wood and rails, etc.

*Mar. 7 S.* The snow fell yesterday evening to the depth of about an inch. Great rabbit snow, but this is Sunday.

Well, "whatever is is right;" that is, whatever the Lord does is right. If it snows on Saturday night, it is all right, and we should observe the Sabbath just as strictly as tho' there was no snow on the ground, and no rabbit tracks in it.

*Mar. 8 M.* The snow fell yesterday evening and to day to the depth of 5 or 6 inches. I & M. went to her pa's yesterday evening, and came home this evening. It is snowing again to-night. Who ever saw so much snow in March? Dont think I ever did.

*Mar. 9 T.* Clear, cold and windy. Snow melted but little.

*Mar. 10 W.* Clear and pleasant over head but very sloppy underfoot as the snow melted fast.

*Mar. 11 Th.* Quite a warm day. Packed away my bacon in ashes, went over to the saw mill and packed up my plank, etc.

*Mar. 12 F.* Clear and warm. Cut and split firewood.

*Mar. 13 Sa.* Clear and warmer, I believe, than yesterday. Muster day, but I did not go as I fell quite unwell—have a bad cold. We—I and Caleb—hauled 3 loads of wood in the forenoon, and rested in the evening. Clark received a few specimen copies of the *Waverly Magazine* published in Boston, Mass. by Moses A. Dow, Editor & Proprietor, at $2 a year. The *Waverly* is a large weekly, being about 10 by 15 inches, 16 pages, and is a real "store house" of interesting tales, etc. The paper is good, print fine and *tolerbly* clear. Dont see how Dow can affoard to publish it at $2 a year, nor I dont see how any one who has any thing else to do can affoard to read it. Who could stand such a weekly, literary gorge for a whole year? Mr. Dow, "That cant be did in these parts." Your *Magazine* is to[o] large, and published to[o] far from home—the south.

*Mar. 14 S.* A little cloudy, and a smokey, and a little windy. I & Mollie went on horse back to Mt. Moriah and heard Rev. Isaac W. Avent preach. His text was the 3, 4, & 5 verses of the first chapter of the first Epistle of Peter:—"Blessed be the God and father of our Lord Jesus Christ," etc. The sermon was good, of course, but I don't think it contained any new ideas. His theme was the reserrection of Christ—a subect in which the whole human family is, or should be much interested.

*Mar. 15 M.* Thrashed a few oats.

*Mar. 16 T.* Warm and windy. Went to Eagle Mills to see about getting a school. Was told to wait till Saturday week and then go down.

*Mar. 17 W.* Thrashed oats in the morning. Wind blew very hard in the evening.

*Mar. 18 Th.* A little cloudy and quite warm. Spent the day trying to trade my filly to Tom Godfrey, but did not succeed. Poor business.

*Mar. 19 F.* Planted about a peck of potatoes, on the lower side of the path, below the spring. Rained in the evening.

*Mar. 20 Sa.* A drizzly, cool day. Worked in the shop in the forenoon, making me an old fashioned straw cutter, with box and knife; and in the afternoon I went to Town. Paid off Pa's blacksmith bill—$2, mailed a letter to sister Mary, etc.

*Mar. 21 S.* Cloudy in the morning, but about twelve, or a little before, the sun shown out quite warm. Wind blows moderately.
Read a good piece in the *Waverly Magazine,* headed "Annie Bently[.]" Annie was a poor girl and went to be "parlor maid" for a rich lady. While she was yet with her mother she was somewhat discontented with her lot, and, like many others, failed to appreciate the many blessings a kind providence has bestowed even on the poor of this world, and fancied that to be rich was to be happy. Alas! Poor Annie's notions of matters and things soon changed, for when she saw that the rich have many trials and difficulties to contend with that the poor know not of, she came to the conclusion that "Better is a dry morsel, and quiteness therewith, than an house full of sacrifices with strife."

*Mar. 22 M.* Cloudy, and rained late in the afternoon. Wm Weatherman split rails (456 I think he said) for Pa. I commenced sowing oats.

*Mar. 23 T.* Still cloudy. I & Pa hauled sawlogs to, and plank from the mill.

*Mar. 24 W.* Cleared off about 12 o'clock to-day. Helped Alex. get the timber for his shed this evening.

*Mar. 25 Th.* Clear and warm. Helped Alex. frame his shed in the morning, and went after my corn (ten bushels I bought of Jesse Tulburt) in the evening. See "Cost of living for 1858."

*Mar. 26 F.* Sowed oats.

*Mar. 27 Sa.* Tolerbly clear, and quite warm. Plowed in oats in the forenoon, and in the afternoon went out to ______ in ro[a]dhands. The peach trees are blooming.

*Mar. 28 S.* Clear and a little cooler than yesterday. Staid at home.

*Mar. 29 M.* Went to Town in the morning, got Jenny shod before, received a letter from Rev. B. Cleg, Principal of Olin High School, in Iredell Co. N. C. In the afternoon I sowed and plowed in oats.

*Mar. 30 T.* Nearly frost enough this morning, I think to kill the peaches. We have had nice weather for a few days, and the peach trees are out in full bloom. Hope the fruit will not get killed this year. Sowed oats.

*Mar. 31 W.* Last day of March and I am but little over half done sowing oats. I put my oats in this year with a "Twister," a kind of turning plow. Cool, cloudy and rained some this afternoon. I sowed and plowed in one bushel of oats, and thrashed out 2 or 3 bushels to-day. This has been a pleasant week so far, March came in with cold weather—*very* cold—but wound up pleasantly.

*April 1 Th.* A very rainy day, and tolerbly cool. I had ______ in the road hands for to-day, but it was to[o] rainy to work. Went out in the evening and notified them to work the ro[a]d on M. the 5 inst.

*April 2 F.* Another cool, rainy day. Old Mr. Reinhardt, an aged Dutchman,[6] says for it to rain on "Good Friday" is a sure sign of a dry summer.

*April 3 Sa.* Hauled over my last stack of fodder. Cleared off to-day quite warm.

*April 4 S.* Not exactly clear, but quite warm. Hope we shall have an abundance of fruit this year.

*April 5 M.* Worked the road for the first time this year. We got thru about 12 o 'clock. In the afternoon I sowed and plowed in oats. Warm weather.

*April 6 T.* A blustery day, and not exactly clear. I & M. staid at her pa's last night, and to-day I went on to court. I paid Jessey Tulburt $5 for 10 bushels I bought of him a few days ago, and I also paid off an old acct. made in Oct of 1853, for some books bought of Hauser & Willson. The account was $3.70, and tho' I called 2 or 3 times to settle it, from the fact that it was over looked, it stood open till to-day.

*April 7 W.* Cloudy and cool. Finished sowing oats. Clark plowed them in. I went to Leonard Messick's and helped dig a grave in the forenoon

6. Mr. Reinhardt almost surely was of German descent, perhaps one of the Pennsylvania "Dutch" (from "Deutch") who migrated in large numbers to North Carolina over the Great Wagon Road.

for one of Louis Messick's children that died yesterday evening; and in the afternoon I helped Pa put a new gate at the head of the lane.

*April 8 T.* Still cloudy and cool. Plowed for Pa with the big plow.

*April 9 F.* Windy and quite warm. Rained last night a fine shower. Went to Shilo to meeting, and heard the Elder, Rev. N. F. Reed, preach a choice sermon. His text was, "Simon son of Jonas, lovest thou me?" John XXI, 17. Reed is quite a small man, but a great preacher. Don't know that I ever heard a sermon that I liked better. Plowed this evening after meeting.

*April 10 Sa.* Warm. Planted our our garden beens. Received the April No. of the *Planter.*

*April 11 S.* Cloudy and a little windy. Ma, Eliza & Caleb are gone on horseback to Mt Moriah to meeting. Pa went yesterday and did not come back. I intended to have gone yesterday and to day too, but as it is I stay at home.

*April 12 M.* Rained last night, and to-day a great deal. Mended my shoes.

*April 13 T.* Cloudy and quite cool. I and Caleb cut & hauled some wood and pine, and I went in the afternoon and helped "Uncle Wattie"—Mollies grand pa—roll a few logs.

*April 14 W.* Cloudy and cool. Worked in the shop on the fan mill.

*April 15 Th.* Worked in the shop. Pa and Caleb went to mill.

*April 16 F.* Plowed for Pa with the "big plow." Rained a nice shower just after 12 o'clock.

*April 17 Sa.* Cloudy and rained in the afternoon. Worked in the shop, etc.

*April 18 S.* A dark drizzly day. Read, in the *Chain of Sacred Wonders,* the history of the destruction of "Ai," by Joshua and the hosts of Israel. I also read of the five kingdoms that were destroyed in one long day. The day was longer than usual because Joshua commanded the sun to stand still and it obeyed him. Reader, do you wish to know anything more about the doings of Joshua? If so, "Read and then you will know,"

Jenny brought her first colt—a horse mule—last night. It is or would be, about 2 ft. 10 in. high if its fore legs, which bend forward at the knee joints considerably, were straight. It is now a dark bay.

*April 19 M.* Cloudy and rained a light shower about 12 o'clock. Worked in the shop. I have two wheat fans "on hand."

*April 20 T.* Cloudy, windy and rained some to-day too. Worked in the shop.

*April 21 W.* Clear and warm. Took some bacon (35 lbs.) to Eagle Mills and bartered it for molasses, getting 12 1/2 cents per lb. and giving 50 cents a gal. for molasses.

*April 22 Th.* Feel very unwell, tho' I worked some in the shop.

*April 23 F.* Cool and rained a shower this afternoon. Lay by to-day.

*April 24 Sa.* Cool and windy. There was some frost this morning but I don't think there was enough to do much damage. Hope not. Lay by to-day also. Think I feel some better tho' than I did yesterday.

*April 25 S.* Cool, somewhat cloudy and quite windy. Staid at home.

*April 26 M.* Rained and snowed. Dont know that I ever saw as cold a day this late in April. The snow fell as fast as tho' it had been the dead of winter, but the ground not being as cold it (the snow) melted about as fast as it fell. Some of our garden beens got frost bit Saturday morning. Dont think the fruit is killed yet, but fear it will be.

*April 27 T.* Cool, cool.

*April 28 W.* A big frost. Peaches killed.

*April 29 Th.* A little warmer. Pa commenced planting corn. I worked in the shop on my fan mills.

*April 30 F.* A warm day. Went over to Wm Pardue's neighborhood to see about getting a school, but did not succeed. How will that man escape Divine vengence who refuses to send his children to school, to take a family newspaper, or give any thing for the support of the gospel, and then spend 30, 40 & even 50 dollars a year for coffee and tobacco? Does such a man deny "the flesh?" Is he being faithful over a few things? and laying up for himself treasure in heaven? Verily, he is wasting his Lord's goods, He is faring to[o] sumptuously every day, and it may be that he will be called on, at no very distant day, to remember that he had his good things in this world.

*May 1 Sa.* Sheared our sheep.

*May 2 S.* A warm day. Went to Thomas Pardue's to meeting. Pa preached a temperance sermon. His text was, "For if ye do these things ye shall never fall." II Peter, I, 10. It rained tremendious this evening. The fields are washed in gullies, the fences, where they cross the branches, are washed down. The milk is spoiled, the butter damaged, and *the wash board is clean gone.* It also hailed considerbly.

*May 3 M.* Cool and drizzly. Helped Pa haul rails and lay them on the pasture fence, haul a load of wood, etc.

*May 4 T.* Cool and rainy. Stabled the sheep. It rained, and rained, and rained.

*May 5 W.* Cleared off warm. Worked in the shop.

*May 6 Th.* Cool and cloudy. Went to Town and bought 2 lbs nails, one bottle of ink, one box of tacks and received 2 No's of the *N. C. Planter*—April and May. Worked in the shop this afternoon. Nearly cold enough, I think, to frost. The May No. of the *Planter* is out in a new and beautiful dress. It is well worth $1 a year, and should be read by farmers generaly. Every person in N. C. that cultivate, or *tries* to cultivate the soil—even a garden—should subscribe for the *Planter.* Subscribe, every body!

*May 7 F.* Clear and cool. Went to Col. M. Willburn's and paid him $3, the interest on my note up to the 3d inst. Came back by Clarks and he handed over $2 that Mr. James Armstrong had left with him for me.

*May 8 Sa.* Nearly clear, and tolerbly warm. Went to Shilo to meeting, and heard the Rev. Isaac W. Avent preach a very good sermon. His text was, "And as Moses lifted up the serpent in the wilderness, even so must the Son of man be lifted up; that whosoever believeth in him should not perish but have eternal life." John III, 14 15.

*May 9 S.* Staid at home all day.

*May 10 M.* Rained late yesterday evening, and is still raining this morning. Did not rain long. I went over to E. A. Windsor's in the evening to trade him my filly, Pres, for one of his mules, but we did not trade.

*May 11 T.* Rained again last night. Helped pull cockle out of Alex.'s wheat in the forenoon, and in the afternoon I hauled some plank from the sawmill, worked in the shop, etc.

*May 12 W.* Went to Nicholson's mill.

*May 13 Th.* Worked in the shop.

*May 14* Worked in the shop in the forenoon, and in the afternoon I helped Pa plant his sweetpotatoes. We have had cool weather this week—almost cold enough, I think, for frost. It is, perhaps, best for the wheat; but the corn and oats are not doing much.

*May 15 Sa.* Hauled a load of plank and a load of wood.

*May 16 S.* I & Mollie went to her pa's yesterday evening, & came home this morning.

*May 17 M.* Worked in the shop.

*May 18 T.* Swaped my filly, pres, to E. A. Windsor for a young mule, giving $4.50 to boot. Rained a shower this evening.

*May 19 W.* Cool and rainy.

*May 20 Th.* Clear and cool for May. Plowed and hoed our potatoes.

*May 21 F.* Clear and cool. Sprouted oats.[7] Think this is fine weather on wheat, some of which is now in bloom.

*May 22 Sa.* I & Mollie went up to Clark's yesterday evening, staid all night and came home this morning bringing a load of corn from Henry Tulburts. (I bought the corn of Jessy Tulburt.) This was a cool morning for May. I saw a little frost on the hay in Clark's stack yard.

*May 23 S.* Cloudy, windy and cool.

7. The editor has not been able to discover exactly what Thomasson meant by this phrase. Since he planted his oats at the end of March, perhaps he was weeding and tending to the young plants.

*May 24 M.* Warm and a little hazy. Finished sprouting my oats.

*May 25 T.* Helped haul the rails and fence in the cow lot, etc.

*May 26 W.* The pine worms came this spring in greater abundance I think than they did last spring. Some of the pines are entirely striped of their leaves, and there are but few that have not been attacted. The worms, when grown, are about one inch long, and about 1/16 of an inch in diameter; of a greenish color, and look clear and oily. I dont think they pester anything but the pines. They come in early, and leave about the middle or last of May. The chickens eat them with great relish.

*May 27 Th.* Went to Town and got Jenny shod before.

*May 28 F.* A little cloudy. Went to Eagle Mills and took our wool to get it carded. Bought a bunch & 1/2 of cotton yarn, and a pair of shoes for myself. See "Cost of living" for May, 1858. Uncle Jo. & Aunt Polly Hauser came up to see us this evening.

*May 29 Sa.* Cloudy, cool & drizzly. We, Uncle Jo. Aunt Polly, Eliza, Caleb & myself, all went to the mountains to get May cherries; and we got them too.

*May 30 S.* Still cloudy and cool. Several of us went to Mt. Moriah to meeting. There was two sermons preached. The Rev. B. Clegg preached first. His text was "Wilt thou be made whole?" John V, 6. Rev. W. L. Van Eaton preached in the afternoon. His text was, "Wisdom is better than weapons of war; but one sinner destroyeth much good." Eccl. IX, 18.

*May 31 M.* Still cloudy and damp. Uncle Jo. & Aunt Polly left for home this morning. It cleared off to-day about noon, and the evening was very warm. I helped Pa roll some logs off his tobacco patch. If men will and women will chew tobacco, I think they should raise it for themselves, but I think they could spend their time and strength in a better way. It takes a great deal of close attention and hard labor to raise tobacco, and after it is made it proves a curse instead of a blessing to the human family. It has no doubt been the ruin of many a poor soul.

*June 1 T.* Warm, and a little cloudy.

*June 2 W.* Quite warm.

*June 3 Th.* Warm and cloudy. Rained a little shower in the afternoon. I went to "Trumpet Branch Foundry," and got a set of fan mill irons, for which I paid $1.25.

I caught a tick behind Jodie's ear this morning, and to-night his ear and the side of his head is swelled very much. I never saw such a swelling from a tick bite before. I put sweet oil, salt & tobacco juice on it this morning. There is now, a poltice made of the roots of wild indego, on it. Hope the swelling is stoped.

*June 4 F.* Went to Casey's shop in the evening and got some irons for my fan mill. Received the June No. of the *N. C. P.* It came in a nice blue dress, and well filled.

A nice shower of rain fell this evening. Jodie's ear is considerably better.

*June 5 Sa.* Rained a fine shower last night. Pa set out his tobacco plants. I worked on my fan-mill.

*June 6 S.* Cloudy and rained a little. Staid at home and read, among other things, two of Wesley's sermons. One on family religion, from the text, "As for me and my house we will serve the Lord," and the other on redeeming time, from the text "Redeeming the time." Wesley, I think, generaly preached short sermons, but they very plain and very good, What he said was to the point

*June 7 M.* Cloudy in the forepart of the day, and quite warm. Worked on my wheat fan again to-day. It is rather small, but think it will blow out the chaff tolerbly well.

*June 8 T.* Quite warm this morning. Why did'nt you write more Mollie?[8]

*June 9* Went to Eagle mills and [got] our wool rolls.

*June 10 Th.* Went to D. Johnson's saw mill and got a load of lumber to make wheat fan frames.

*June 11 F.* Alex. Caleb and I went to the cherry orchard and got *lots* of cherries. There are fore kinds—Read harts, Blackhearts, honey & May cherries—all of which are very good.

8. Thomasson is teaching his wife to write.

*June 12 Sa.* Rained last night, and this morning too. Went to Shilo to hear the Rev. Avent preach, and was not disappointed. His text was, "Blessed is the man that endureth temtation;" etc. James I, 12. Avent is somewhat tedious; but he is, I think, a very good preacher. He came home with us for dinner, and staid all night. Had another nice shower this evening.

*June 13 S.* Warm, and not quite clear. All gone to M't Moriah to meeting but Mollie, myself and Jodie. There are more flies, it seems to me, this summer than usual. They torment me no little by craulling over and biting my hands and face.

Were there any flies when the Lord look[ed] down from heaven upon all that He had made and pronounced it good? If so they surely have degenerated very much since that day. Of course they are not now what they were in the garden of Eden, if they, too, suffered by the fall of man.

*June 14 M.* Worked in the shop. Clark came down and cut wheat—about forty doz. I think. Feel quite unwell in the afternoon.

*June 15 T.* Sick. Cold chills run over me, and I have some fever.

*June 16 W.* Feel better to-day. Just before going to bed last night I bathed my feet in warm water, drank some composition, and after I got in bed Mollie placed a warm brick to my feet and soon the sweat began to pour out at a great many if not at every pore. Sweated out the fever.

*June 17 Th.* Dont feel so well to-day.

*June 18 F.* Not well yet.

*June 19 Sa.* Went to the sawmill and got a load of plank. In the afternoon I & Mollie went to her pa's & went to Town.

*June 20 S.* I & M came home in the forenoon, and in the evening we went to Shilo to prayer meeting.

*June 21 M.* A warm day. Went down to Rev. T. A. Nicholson's and sold him my mule for $52.50.

*June 22 T., 23 W., 24 Th., 25 F., 26 Sa., 27 S.* Bound wheat every day this week except Monday. We have had a little shower of rain every day, I believe, this week. We had a charming rain yesterday evening. The weather was very warm last week. Some of our wheat is badly injured by the rust. This, Sunday 27, is a clear, still, warm morning.

*June 28 M., 29 T., 30 W.* Very warm weather. Jodie was two years old the 29. He is not as large as some children at his age, but he is a pert little fellow; cant be still many minutes when he is awake. On the 30 we worked the road. Had only six hands but we worked thro'.

*July 1 Th.* Went to Town in the forenoon, and in the afternoon I helped Frank Madison pull flax. Had quite a nice shower.

*July 2 F.* Warm weather.

*July 3 Sa.* The hotest day, I think, that we have had this season. Went to the exhibition, near "South Troy," on Hunting Creek, in Iredell. It was mostly an exhibition of nonsense, very wicked and very disgusting. I had other business else I should not have been there at all. I left before the close, resolved to attend no more such shows. Had quite a fine rain in the afternoon.

*July 4 S.* Pa preached a temperance sermon at Jo. Johnson's. The congregation was small. I & M. went to her pa's after meeting. We had a fine rain this evening. I & M. stoped in at Jo. Johnson's as we came on home to get out of the rain. It rained till near dark, and we had to come home in the dark.

*July 5 M.* Planted some potatoes, cut and hauled a load of wood, etc.

*July 6 T.* Made a gallon of blackberry wine. I measured one gallon of berries, poured them into a large crock and mashed them with my hands. I then added half a gallon of boiling water, and let it stand 24 hours. I then strained the juice, and added two pounds of brown shugar.

*July 7 W.* Hauled over my wheat, and packed it away in the stable loft. Received the July No. of the *North Carolina Planter.*

*July 8 Th.* A cool rainy day. The 6 & 7, yesterday and day before were also cool days.

*July 9 F.* A warm day. I worked for Pa to-day—plowed in the forenoon, and helped him put up hay in the evening. This was the preacher's day at Aylesbury & Shilo, but he did not go, I'm told, to either place.

*July 10 Sa.* I & Alex. cut, bound and put up *30 shocks* of my oats to-day. Oats generaly are but little account this year, and mine in particular are hardly worth saving. Oats have the rust this year, a thing that old men say they never heard of before.

Caleb went to mill, and there heard that James Patterson, an aged minister in the Methodist connection, was buried yesterday at Snow Creek.

*July 11 S.* Cloudy in the morning and rained a nice shower.

*July 12 M.* A rainy day. Worked in the shop.

*July 13 T.* Warm, and not exactly clear yet. Finished cutting our oats.

*July 14 W.* Clear and not very warm. Last night there fell a heavy rain.

*July 15 Th., 16 F., 17 Sa.* On W. the 14 we (Pa, Alex., Caleb & myself) commenced mowing Wm Pardue's meadow, and finished it on the 15 about 12 o'clock.

On the 16 we stacked it, and I & Alex. mowed in Pa's meadow about 1/2 the day.

*July 18 S.* Quite a warm day. Went to Shilo to prayer meeting.

*July 19* Went to Samuel Haze's and to James Cashe's, and made an engagement to teach the School in District No. 2, Iredell Co. N. C.[9] Rained a little early this morning.

*July 20 T.* Warm and showery.

*July 21 W., 22 Th., 23 F., 24 Sa., 25 S., 26 M., 27 T., 28 W., 29 Th., 30. F.* To-day, Friday 30, I & Mollie went over to A. H. Thomasson.[10] We had a long hot ride over some rough road. Thomasson lives in, I think, about the roughest part of Iredell Co. He paid $800 for 90 acres of hilly land, part of which is thickly covered with stones. We found them all hearty, and in good spirits.

*July 31 Sa.* Feel quite unwell. I went over to Statesville to be examined. The committee met in the afternoon and examined about 10 or 12 of

9. This employment will lead Strong to move his family a few miles away to Iredell County.

10. Jean Harris Thomasson's genealogical research does not make clear exactly who this individual was, but he may have been a cousin of Basil Armstrong Thomasson's father.

us two or three hours. I think all got certifficates. I got No 1 all round. One chap got a certif., a No 1 he said, that could not spell wagon! He said there were two ges in the word,—thus waggon. I spelt it wagon. He contended strongly for the gg, and refered me to Webster!

*August 1 S.* Clear and hot. We came on our way home through Olin, and by S. Hicks' where we staid till after dinner. We got to Pa's about sundown. Found brothers W[iley] W., W[illiam] C[lark] & Sister Mary & Mr. Williams, W. C.s wife all there. *Well we did.*

*Aug. 2 M.* Came down and commenced school. Had 23 scholars. Went over to Tommy Holcombs and staid all night.

*Aug. 3 F.* Had 26 Scholars to-day. Staid at John Barnard's to-night.

*Aug. 4 W.* Had 30 chaps to day. Dismissed till Monday, and went home to move down.

*Aug. 5 Th.* Hot and dry. Dug our potatoes. Went over and borrowed Mr. Bell's wagon bed to move in.

*Aug. 6 F.* Hauled down a load of our "dry goods."

*Aug. 7 Sa.* Caleb hauled us down and took the horses and wagon back.

*Aug. 8 S.* Clear, hot and dusty. We have had no rain for some time. The corn is failing very fast.

*Saturday 14* I've been quite unwell all the week, but did not stop school. Had from 20 to 30 Scholars—enough for so small a house.

*Aug. 15 S.* Rained a light shower or two last week, but not enough to do the corn much good. Staid at home all day.

*Aug. 20 Friday night.* I think my School has averaged bout 30 scholars every day this week. Think I am getting on tolerbly well with my school, on good terms. Some of the scholars study very well, while others are exceedingly careless. I like to teach school if I could only have things to my notion. Have a good [school]house well furnished, and books of the right kind, and enough of them. This has been a warm, dry week. The corn is drying up. Don't think there will be much upland corn made in this country.

*Aug. 21 Sa.* Went up to Pa's. Started early and got there just at seven o'clock. Clouded up from the north this evening, but no rain. I think it rained the fourth Sat. or Sunday in Aug. for three or four years past, but dont think it will rain to-day, (Sunday)[.] Pa's cows, two of them, and a yearling have the "black tongue." He washes their mouths with pepper, and salt, and—I dont know what else. The cows seem to be very sore and refuse to eat. Their noses get dry and hard, tongues crack open, etc. Pa came on his way to Clarksbury, as I came on home, and carried my apples, flour, etc. for me.

*Aug. 22 S.* Clear and hot, though there is a pleasant wind blowing up the creek.

*Aug. 23 M.* Clear and a little cool. There were forty scholars at School to-day.

*Aug. 24 T.* Cool and dry, dry, dry.

*Aug. 25 W.* Clear, and a few degrees warmer, I think, than yesterday was.

*Aug. 26 Th.* Had a row in the school house this evening. F. M. Barnard had accused Bob Parks & Pete Faulcomb of taking his watermellons, and when I rang the bell for book, after dinner, Parks came in the house and gave Barnard a tremendious cursing. I sent off for the committee, and told Parks to hush or leave the house. Four of them, Barnard, Faulcomb, Press Parks & Bob, then left the house & settled the matter between themselves. The Com. came over, but did not expell Bob as he agreed to do better in the future.

The Parkses are trying to break down the School. I see that plainly. Bob Parks, this day, put a blot upon his already bad character that will not be removed soon.

*Aug. 27 F.* Worked the road. I & M went up to Pa's last night. Soon after we commenced the road it commenced raining, and rained a nice shower. We finished the road about sun set.

*Aug. 28 Sa.* Rained another nice shower early this morning. In the afternoon Caleb brought us home in the wagon.

*Aug. 29 S.* Clear and pleasant. Think the rain is over with for a while.

Mollie has gone off to Mr. Coxes, and left me to mind Jody and the beans. Jody is out and gone, and the beans—let me see if they are not dry. Not quite, but as "an ounce of preventive is better than a pound of cure" I'll throw in another dipper of water.

"Single or double,
This life is full of trouble."

And now the fire has burned down, and the beans—I don't know whether they are done or not.

*Aug. 30 M.* Clear and pleasant. School not so full. Fodder time is at hand, but the fodder is quite indifferant—hardly worth gathering. I dont know that the fodder, any year, will pay for the time required to save it in, and the damage done to the corn by gathering the fodder. Corn would no doubt be heavier if the fodder was left on the stalk, for soon after it is taken off the stalk, corn and all, dries up, and frequently the corn is swiveled [shriveled].

*Aug. 31 T.* Pleasant, but not exactly clear. Alex. worked Jenny down to Alfred Johnson's yesterday, and about 8 o'clock last night Alfred came over after me to go and see her die—said she had the colic, or the bots one. I left Alfred here with Mollie and went over, and when I got there Jenny was lying stretched out in the yard like she might be at least half dead, though they said she was a good deal better than she had been. I got her up and "dutched" her, and in a few minutes she went to eating grass, and seemed to be about well. This is the first hard spell she has ever had, to my knowledge, and I hope it will be the last one, for there never was a better nag, of her size, to work, I don't think.

*Sept. 1 W.* On Tuesday night the 13 of July there fell a good rain. Since that time I dont think we have had enough, to wet the ground an inch deep, at any one time. Hence the vegetables have failed. It is raining some to-night, and I think the prospect good for plenty of rain soon. It is to[o] late to do the corn much good, but the beans, sweet-potatoes and cabbages would grow some yet, and the fodder would not dry up so fast.

*Sept. 2 Th.* Clear and pleasant. Dismissed school till the 16 inst. I dont like this stopping school to gather fodder. The fodder ought to be saved to be sure, but then there are small children enough, in almost every District, that can't go so well in the dead of winter, to make a full school during fodder time. The weather is pleasant now, and now is the time for them to go to School, but because the "big boys and galls" have to stop to save the fodder, the school must stop too. It seems that the people think the Common Schools are intended for the special benefit of the large scholars, and that the little ones have but a very small claim, if any at all, on them.

Now the Bible says, "With all thy getting get understanding." Then I contend that the schools ought not to be stoped for fodder, but that while the fodder is being saved by the large scholars, they (the schools) should be kept in opporation for the special benefit of those that are too small to do execution in the fodder field.

*Sept. 3 F.* Quite a warm day. Went over to Eagle Mills, in the evening, and bought me a new slate, price .25. Rained a light shower.

*Sept. 4 Sa.* A few degrees cooler than yesterday. Rained another little shower. Andrew Baggarley[11] came to Mr. Coxe's this evening, and, with a little instrument, showed us something of the power of velosity and circle.

*Sept. 5* Walked over to Mt. Moriah to meeting. Rev. T. A. Nicholson preached Newton Barron's wife's funeral. His text was, "Who is on the Lord's side?" This question was put to the Israelites by Moses when he came down from the mount, and found that they had been worshiping an idol. Moses had been on the mount so long that the people, it seems, had dispaired of his return, so they ordered Aaron to make them Gods to go before them. And now what was Aaron to do. Who should he obey. God or the people? He doubtless knew that the Lord had said, "Ye shall serve the Lord your God," but perhaps the people, as they were "set on mischief," threatened Aaron with death if he did not comply with their wishes. Being thus situated he determined to obey the people, so he told them to "Break off the ear-rings which were in the ears of their wives, their sons and their daughters," and take them to him. Here was a very favorable opportunity to rid the people of their jewelry and Aaron imbrased it. Now did Aaron commit a sin by making that golden calf, or not? This is a question which I dont believe I am able to answer, even

11. This is the same man who was the promoter of Eagle Mills and South Troy. However, his Eagle City never developed to match his dreams.

to my own satisfaction. I am of the opinion however that it was not, because we have no account, that I recollect, of his being directly and personaly punished for it; and because the people had as well worship the gold in one shape, perhaps, as in another; and by making the calf of their ear-rings he caused them to get rid of their gold.[12]

*Sept. 6 M.* Caleb came after us and we went up to Pa's.

*Sept. 7 T.* Sick. Kept my bed a good part of the day.

*Sept. 8 W.* Feel some better. Went over to Mr. Bells in the afternoon.

*Sept. 9 Th.* We came back to Pa's.

*Sept. 10 F.* Mollie came down home after her Sunday clothes, and I cut tops for Pa.

*Sept. 11 Sa.* Cloudy, and rained a light shower. Went to Town and got Jenny shod all round.

*Sept. 12 S.* A two-horse load of us came to Mt. Moriah to meeting. Heard our preacher Isaac W. Avent preach two very good sermons.

*Sept. 13 M.* Went back to meeting, and heard the Elder N. F. Reed preach a choice sermon.

*Sept. 14 T.* Went to the shoe shop, etc.

*Sept. 15 W.* Came by meeting, and then on home. The meeting broke to-day.

*Sept. 16 Th.* Recommenced school with, I think, 12 scholars. Cool & windy.

*Sept. 17 F.* Quite cool.

*Sept. 18 Sa.* Some warmer, I think.

*Sept. 19 S.* Clear, and cool in the morning. I have not seen any frost yet, but have heard of it. The latter part of the summer, and the fall so far has been very dry, and a killing frost this early in the season would be a death blow. It is a wonder that the earth is not cursed on account

12. Evidently Thomasson disapproves of the wearing of jewelry.

of the wickedness of the people, and it may be that "the end of all flesh is come before the Lord," and that these dry and dusty times are the beginning of sorrows.

Rev N. F. Reed's text last Monday was, "Turn ye, turn ye from your evil ways; for why will ye die, O house of Israel?" Ezekiel XXXIII, and part of the 11 verse. He asserted that lost souls were the authors of their own destruction, and told the sinners of that congregation that if they were lost, and that they would be there was no doubt unless they repented, they would be the authors of their own damnation. He then proved to the satisfaction, I think, of all present that his assertions were true. He said that there were but three influences that acted on man; viz. The influence of God, of the devil, and of his fellow. He then proved that God did not damn the sinner; and to do this he quoted the first part of the verse containing his text, "As I live, saith the Lord God, I have no pleasure in the death of the wicked." He next proved that the devil did not damn the sinners, but if they were not damned they would not be under any obligations to the devil for their escape. He next proved that man could not, and did not damn his fellow, and to do this he quoted the 28 verse of the X chapter of Matthew.

"Fear not them which kill the body, but are not able to kill the soul." Man is not able to damn his fellow, but it is hard to tell how much of the damnation of the drunkard should be set to the accou[n]t of the grog seller. O ye fathers and mothers who neglect to train up your children in the way they should go, if they are lost it will be hard to tell how much of their damnation should be set to your acct. Man is a free agent. He can choose life or death. If he chooses life God will help him to work out his soul's salvation. If he chooses death the devil will help him to work out his damnation. No man would say he chooses death. But actions speak louder than words. By neglecting the "great salvation," by refusing to be saved on the terms of the Gospel, they choose death, and go down the sides of the pit. "Turn ye, turn ye from your evil ways; for why will ye die, O sinner?

In the evening, about sun set, we went down the creek, and found a small quanity of choice muscadines, a few rods below the foard. The vine comes up near the edge of the water, and runs up an old mapole among the branches of which it bears its fruit.

*Sept. 20 M.* Clear and warm.

*Sept. 21 T.* A little hazy, and quite warm. We went again this evening and got more muscadines. They are very deliscious abot this time.

*Sept. 22 W.* Warm and smokey. We went again this evening to the muscadine tree, but did not get much fruit, as it was not there to get. "Sound sleep," says Dr. N., "is the sign as well as condition of health." Then I must be quite healthy to-night, for I can scarcely read a page or write a line without going off into a sound nap. Wish I was not such a dull, sleepy-headed animal.

*Sept. 23 Th.* Clear and quite cool. Think there will be frost in the morning. I think it generaly rains every year about the time the sun crosses the Equator, but it seems we are not to have any this year. If it dont rain soon there will be squally times among these hills in less than twelve months. If the righteous are not forsaken it may be that his seed will have to beg bread. Well, the "Great Teacher" knows what is best for us and if He has seen fit, in his wisdom and goodness, to give us a lesson in the shape of a famine, we must take it. This is not the abiding home of man, and the sooner his probation ends the better, if he is prepared for a higher and a nobler state of existance. I am rather of the opinion that this earth never was intended to be man's eternal home, and that he would have be[e]n removed from it in some way or other, even if Adam had not sinned. This earth is, and ever has been, according to Geology, changing. Hence it is not a fit place for animals whose natures are not adapted to it. Hundreds of years ago there were races of animals which are now extinct. When the productions of the earth, the climate, etc. were adapted to their natures they lived, but when these changed they died, and now they can nowhere be found.[13] So it will be, I suppose, in a few hundred years to come with many races that now exist. As man was the last animal created, I guess he will be one of the last found on the earth.

*Sept. 24 F.* It was not as cool this morning as I expected it would be. I did not see any frost. It has been a little cloudy to-day, and I saw a circle around the sun, (as the phrase is, but the circle was, of course, along way this side of the sun.) which is said to be a sure sign of rain. There are but two places which will eventually contain "Adam's race," and yet

13. Like other well-read people of his time, Thomasson was aware of some of the facts that Darwin soon interpreted in his theory of evolution.

it takes hard strugling to get to either place. Life is a war-fare. Every indiv[id]ual, as soon as they cross the line of accountability, are soldiers in one or two great armies.

If you make up your mind to "dwell in the tents of wickedness" that you may "enjoy the pleasures of sin for a season," the devil will enlist you and send you forth in battle array. Then O, what a strugle you will have. What an up-hill business it will be for you to fight against the author of your being.

How your conscience will sting you, and how awful you will fe[e]l when, in the dead hours of night, you think of death, the cold grave, and the eternity beyond. As you travel on in the way of sin and death, you will be met, at every turn and corner of your life, with the offer of salvation. Jesus who has blead and died that you might live will offer you salvation, freely, "without money, and without price." Will you be able to refuse it? It will be a tender time, and unless you "harden your heart, and stiffen your neck" you will be overcome—you will be constrained to "ground your puny arms of rebellion, and fight against your God no more." When you view Jesus in all his loveliness, and think what he has done for you, and hear him say "Behold I stand at the door and knock," will not the door of your heart fly wide open to receive Him? I tell you it will unless you strain every nerve to keep it shut. And when you hear His ministers reasoning of "righteousness, temperance, and judgement to come," O how you will be made to tremble, and when the question is put "believeth thou the prophets?" you will perhaps be compelled to acknowledge that you are almost persuaded to be a Christian, that you are almost overcome; but with a mighty effort you rouse up and say, as did Felix of old, "Go thy way for this time."

*Sept. 25 Sa.* Warm and cloudy, and in the evening it commenced raining sure enough. Perhaps a circle round the sun is a sure sign of rain sure enough.

*Sept. 26 S.* A cool, rainy day. We went to Harmony Hill to the campmeeting, expecting to come home at night, and staid two days and a half! It happened thus:—Uncle Sac Albea let us have his horse and buggy, and when we would begin to talk about starting home, he would rather insist on our staying, and so we did not get home till Tuesday evening. This was a tolerbly good campmeeting, and had it not have been so smokey I think it would have been one of *'em*. Now, as to the smoke, I had just about as soon be burning off new ground. There were 35 or 40 tents, and, I recon, as many camp fires. I think people ought to do their cooking at home, and eat "*cold vittles*" during campmeeting. It cleared off Monday morning without raining much at last.

*Sept. 29 W.* Clear and pleasant. School quite small.

*Sept. 30 Th.* Clear and quite warm. Ma came to see us to-day for the first time since we moved here. Mollie bought a brown worstered dress pattern for herself, for 25 cents a yard. 9 yds 25 a yd. $2.25. Jane & Cash came to see us to-night. Fashions are ever changing. Hoops have of late taken the place of petticoats, I look upon Hoop Skirts as being a great modern invention—a great improvement on petticoats. Women now can assume the shape of a hay stack without putting on twelve or fifteen under dresses;—enough to load an ______ "animal of burden," much less a little, slender woman. Verily the "good time is coming." Women are unloading themselves, and we now have some reason to hope that they will yet be "help meets" for man, as they were origionaly intended.

*October 1 F.* Warm, and cloudy part of the day, and rained a few drops in the evening. Pa started over to Milton Campbell's after $27 of my school money. Mr. Williams came up this evening after his watch, and staid all night with us.

*Oct. 2 Sa.* Clear and warm. Mr. Williams left early for home. I went up to Nicholsons's mill, had half a bushel of wheat ground, and carried the flour (about 20 lbs.) home on my shoulder. Bought 10 lbs of tallow at 12 1/2 cents a pound.

*Oct. 3 S.* Clear, except a little smoke or fog, and quite warm. Staid at home all day and read & wrote. Jodie is quite unwell. He wants a *yink* (a drink) as he calls it, about every five or ten minutes. Hope he'll be well soon.

O that it would rain and cool the air, settle the dust and soften the ground so that people could plow.

"O may it all my powers engage," not to "lay up treasure upon earth where moth and rust doth corrupt, and where thieves break through and steal." But "To do my Master's will!" and thereby "lay up treasures in heaven," that my heart may be there also. O God grant that I may so live this life that when it comes to a close I may leave the world without regret, and go to that blessed world where the weary are forever at rest.

*Oct. 4 M.* Clear and very warm for October. Pa came by, on his way home with $28 of my wages,—one dollar more than I ordered as Campbell could not make the change. I saw the comet this evening about dark for the first time. It has been seen, I learn, these several days.

*Oct. 5 T.* A little cloudy, and warm as August. Bought David's *Elementary Algebra* yesterday for 75 cents, and commenced studying it last night. Took another lesson to-night. I think I've made a pretty good start.

*Oct. 6 W.* Rained two or three light showers. The prospect seemed to me to be good for an abundance of rain, but it did not last long. When was there ever such a time. A few of the farmers are sowing wheat, but unless it rains soon they had better let it stay in the granary. The comet seems to be traveling south. It was plainly to be seen to-night, (Oct. 7) as the sky was quite clear. It has an enormous tail, bending to the north.

We went to Mr. W. Colvert's to night meeting. Rev. Wm Carter preached a choice sermon, about an hour and a half long, from, "I tell you, Nay: but, except ye repent, ye shall all likewise perish." Luke XIII, 3 & 5 too. He told us that it is customary, and ever has been in all ages of the world, for mankind to look upon the afflictions and misfortunes of others as being judgements for some wickedness or other. He instanced Job, the "Galileans whose blood Pilate mingled with their sacrifices," and "those eighteen, upon whom the tower in Siloam fell," and told us that it was wrong to look upon the unfortunates and afflicted as sinners above all others.

After he dismessed this part of the subject he told us what repentance is, and then urged the necessity of it. Repentance, he said, is said by some to be, "going out of one state of mind into another," exactly opposite. And it is said by some to be "self knowledge." I think these deffinitions are both very good ones.

*Oct. 7 Th.* Warm, cloudy part of the day, rained a little, and then cleared off. What I wrote about the comet, on [Oct. 6], was intended for to-night.

*Oct. 8 F.* Clear, windy and cool. The comet appears to be higher in the elements, and further south every night. Where it came from and where it's going is more than I can tell.

*Oct. 9 Sa.* Quite a considerable frost this morning. The day clear and pleasant. Bought a bureau of Johnathan Cash, price $9.50. I gave him my note five months after date. So now we have a "chest of drawers."

*Oct. 10 S.* Another frost this morning, and now, about one o'clock in the evening, it is clouding up appearantly for rain.

*Oct. 11 M.* The rainiest day we've had in a great while, and tolerbly cool. Sold Jenny to John G. Johnson for $70, due 12 months after date.

*Oct. 12 T.* A cool drizzly day.

*Oct. 13 W.* Cleared of[f] nice and warm. Saw the comet again to-night. When I first saw it it was rather north of west, now it is south west.

*Oct. 14 Th.* Heard the cannon over at Statesville, a distance of about 20 miles. The cars rolled up to that place, I'm told, to-day, and a large crowd gathered in to see the sight, hear the music, help eat the dinner, ride the Rail Road, etc.[14]

*Oct. 15 F.* School small!

*Oct. 16 Sa.* Went up to Pa's early this morning. Got old Med and the wagon and took all my wheat (6 bushels) to Nicholson's Mill and had it ground up. I got home with the flour a little before sunset. I unloaded, took the mare and wagon home and staid all night at Pa's. Mollie and Jodie staid at home by themselves!!

*Oct. 17 S.* We have had clear, warm weather now for a few days. We went to Mt. Moriah to meeting. Rev. T. A. Nicholson preached, but I

14. The citizens of Statesville were celebrating the arrival of the western extension of the North Carolina Railroad, which brought to their doors the excitement of rail travel and its stimulus to trade and commerce. Towns often celebrated the completion of the railroad; a few years earlier Salisbury staged a barbecue featuring twenty-four hogs, sixteen sheep, six cows, ten opossums, and fourteen hundred pounds of bread. For more detail and an analysis of the extent to which improved transportation encouraged farmers to grow market crops, see Paul D. Escott, "Yeoman Independence and the Market: Social Status and Economic Development in Antebellum North Carolina," *North Carolina Historical Review* 66, no. 3 (July 1989): 275–300.

did not get to hear the sermon, as Jodie was unwell, and I had to be out with him.[15]

*Oct. 18 M.* Warm and not exactly clear all day. The winged ants came out this evening, which is said to be a sign of rain. Hope it will come soon.

*Oct. 19 T.* Warm and a little cloudy. Bought us two new chairs for .50 each. I think I am getting on very well with my school. I've not heard any complaint yet.

*Oct. 20 W.* Cloudy part of the day, and rained a few drops. Still quite warm. Clark gave us a call to-day; the first since we moved to Troy. I and Clark settled our accts, and I sold him my hand saw for $1.75 cts.

*Oct. 21 Th.* Rained a light shower last night. Had 20 scholars to-day.

*Oct. 22 F.* Pleasant weather. School small.

*Oct. 23 S.* I went up to John G. Johnson's early this morning and we finished our trade. So Jenny is sold. I hated to part with so good a nag, but I do not have much use for a horse, and times are heard [hard]. I went over to Pa's and took dinner and then struck out for home.

*Oct. 24 S.* I went to Mount Moriah to meeting. Mollie being unwell she & Jodie staid at home.

There has been a considerable revival going on at Mt. M. for some weeks. There is meeting there every Saturday night, Sunday, and Sunday night.

*Oct. 25 M.* Cloudy and cool.

*Oct. 26 T., 27 W.* Still cloudy and cool.

*Oct. 28 Th.* Cool and rainy. I've not seen as rainy a day lately. Think people can plow after this.

*Oct. 29 F.* Rained a great deal last night. Guess the ground is wetter now than it has been since about the middle of June. The creek is tolerbly flush, but fordable. Jodie is two years and four months old to-day. He is

15. Evidently Thomasson did not leave all the duties of child care to his wife.

tolerbly well grown,—has black eyes, light curly hair, pearly teeth and rosey cheeks. He improves very fast in talking, and is a great papa boy.

*Oct. 30 Sa.* Warm and foggy. Bought a waistcoat (black) for $1, and the trimings—

*Oct. 31 S.* Clear and warm. The Rev. Mr. Holler, a Baptist minister, baptised 2 women, in the creek just above the ford, this morning.

*Nov. 1 M.* Warm and a little cloudy. Nov. comes in like a lamb; then according to an old saying it will go out like a lion. School small! small! What do the people mean? Every man "that layeth up treasure for himself, and is not rich toward God," *is a fool,* according to Scripture. Then what a world of fools this is!

*Nov. 2 T.* A rainy day.

*Nov. 3 W.* Clear and pleasant. A great time for the young wheat. Mollie is going ahead with her Grammar;[16] she can fraise such sentences as, "John ran through the house into the garden," very well.

*Nov. 4 Th.* Tolerbly cool this morning.

*Nov. 5 F.* Cool, cloudy, and rained some in the forenoon. Borrowed *Watson's History of the U. S.*, of Mr. Cox. Think it is a very good History.

*Nov. 6 Sa.* Pleasant weather. I & Mr. Jn. Cox started to Mr. Reason Cashe's to grind our axes. I expected to cross the creek at the head of the pond, but the boat being on the other side we could not cross without wading, so we came back home through t[he] woods and killed three squirrels. Shadrack[17] and his lady came to see us this evening for the first time since they were married.

*Nov. 7 S.* A cold frosty morning. Our visitors staid till late this afternoon, and then left for home.

*Nov. 8 M.* Quite frosty.

16. Evidently Strong has talked Mollie into regular lessons.

17. Shadrack Beall, Mollie's brother. The 1860 census recorded that he was twenty years old, a house carpenter, and living near his parents.

*Nov. 9 T.* A tolerbly cool day. Clark & his wife, mother-in-law and step-daughter, all came to "our house" this evening.

*Nov. 10 W.* This [is the] coldest morning we've had this fall, I think. Bought 3 bushels of dried apples of Clark. Two bushels pealed at $1. per bu. and one bushel unpealed for .50 cents. Mr. Samuel Hays brought us some beef to-day. Good neighbors are good to have.

*Nov. 11 Th.* Nothing new.

*Nov. 12 F.* Somewhat cloudy and quite cool. Caleb came down after us late this evening and we went home with him. Sister Martha is quite ill. She has something [like] the consumption. She coughs a great deal. She is staying at Pa's.

*Nov. 13 Sa.* Very rainy in the fore-part of the day. Cleared off in the afternoon, and the wind blew strong and cold.

I went to Town and bought a time piece for $5. I also received the Oct. & Nov. Nos. of the *Planter,* and two little books *The Little Ones Ladder* and *Wouldst Know Thyself,* from the publisher, George F. Cooledge, 329 Pearl Street New York. Think they are valuable books.

*Nov. 14* A pleasant day. We came on our way home by Mt. Moriah and heard Rev Isaac W. Avent preach his last sermon for this year at that place. He gave us a very good talk on the subject of the "Lord's Supper," after which he administered the "Sacrament." I communed for the first time in a good while as I do not believe that alcohol should be used in the Sacrament. The wine to-day was homemade—the pure juice of the grape I guess.

*Nov. 15 M.* Cloudy, and almost cold enough I think to snow. Had 24 or 25 scholars to-day. I generaly have about 20.

*Nov. 16 T.* Clear and cold. It snowed a small sprinkle last night. The wind blew very cold to-day.

*Nov. 17 W.* A very cold, frosty morning. A mountain boomer drove his team into the pond just above the ridge at Eagle Mills! He was in the wagon, and when he saw that his team was getting into "deep water" he made for land and left the horses to do their own driving. They swam across with the running gear, but the bed floated off, and lodged on the dam below. This teamster must have been drunk, or a fool, or perhaps both. The water in the pond is said to be 10 or 15 feet deep; and he foard it with a loaded wagon! It is well for him that the "fool killer" was not in Town.

*Nov. 18 Th.* Mostly clear, and tolerbly pleasant after the morning. Mollies ma & pa came to see us last evening, and staid till after dinner to-day.

*Nov. 19* Clear and cold.

*Nov. 20 Sa.* Cold. In the afternoon it snowed, hailed and rained. A bad time to be without wood. Made us a bed of straw and 16 lbs. of cotton.

*Nov. 21 S.* Cloudy but not so cold as yesterday. There was a considerable sleet last night, but it is melting off fast this morning.

*Nov. 22 M.* A very rainy day, and tolerbly cool. Had 16 scholars. Read to-night the history of the colonization of Pennsylvania. William Penn was no doubt a good man. He treated the red men as brethren, and thus secured their favor; while some of the other early settlers treated them as wild beasts of the forest, and suffered dearly for it. "Honesty is," and ever has been, "the best policy." If man would have friends he must show himself friendly. William Penn by a kind, honest and friendly disposition made friends of the savage Indians.

*Nov. 23 T.* Clear and cool. Wm White & Sally Nicholson I'm told were married to-day.

*Nov. 24 W.* Windy, cold and not exactly clear. Cold, dark clouds passing over from the northwest. Guess we'll have cold weather now for a few days.

*Nov. 25 Th.* School small—only 15.

*Nov. 26 F.* Clear and not very cold. Caleb brought a load of beef (2 1/2 quarters) to market this evening, but could sell no more than one quarter, (75 lbs) and that I bought. 75 × 3 = 2.25. Guess we've got beef enough to last us "a few days," if not longer.

*Nov. 27 Sa.* Cloudy and tolerbly cool. Got wood for Sunday; went to Eagle Mills twice, as usual on Saturday; put away the cabbages, (a few that Caleb brought us) etc, etc.

*Nov. 28 S.* Cool, cloudy, and rains a light shower every hour or so. We stay at home on Sunday and read our good books—the Bible—*Chain of Sacred Wonders*—*Prince of the house of David*—etc, etc.

*Nov. 29 M.* Cloudy in the forenoon, but clear and pleasant in the evening. Dont feel very well.

*Nov. 30 T.* Cloudy in the morning, nearly clear in the afternoon, and cooler than yesterday, Feel better to-day. I took four of "Wright's pills" last night. I dont believe in taking much medicene, but a few pills now and then may ward off a spell of fever, or headache, or some other ache no less disagreeable. "A stitch in time saves nine."

*Dec. 1 W.* Cold. School small. A little cloudy—fixing to snow perhaps. Came by the store this evening and bought me an "every day" hat for $1.25; one & a half yds. calico (14 cts a yd. = 21) to make Jody an apron.

*Dec. 2 Th.* As rainy a day as we commonly see, and tolerbly cool. Had 20 scholars.

*Dec. 3 F.* Cleared off warm—too warm, I think, to stay so many days.

*Dec. 4 Sa.* Cool and cloudy. Went over to Eagle Mills, and bought a side of upper leather for $2.[,] ten lbs. padding .03 = 30, and a whetstone (an oil stone) weighing one pound & five ounces for .20.

| | |
|---|---|
| Leather | 2.00 |
| padding | .30 |
| whetstone | .20 |
| | $2.50 |
| | .40 |
| | $2.90 |

Mr. S. Hays hauled us two loads of wood (two horse loads) for .20 a load.

*Dec. 5 S.* Still cloudy and tolerbly cool. Staid at home.

*Dec. 6 M.* A very rainy day, but not as cool, I think, as yesterday was. Had about my usual No. of scholars.

*Dec. 7 T.* Still cloudy and warm, but did not rain much, tho' it drizzled all day. Had 14 scholars. I get up these mornings half past three o'clock. I think to go to bed at 9 and get up at 4 would be about right. I retire sometimes as early as half past eight, but generally about nine o'clock. Make a good, and proper use of your time, reader, if you wish to be "healthy, wealthy and wise."

*Dec. 8 W.* Another very rainy day. Rain! Rain! Whoever saw so much rain? Guess Noah saw more than has fallen this week.

*Dec. 9 Th.* Clear and cold. School out. Taught four months, at eighteen dollars per month. $18 × 4 = $72.00. What I shall do next I hardly know. I must get another school if I can.

*Dec. 10 F.* The co[o]lest coldest day, I think, we've had this season. Went over to Eagle Mills, and bought Mollie a pair India Rubber over shoes; price $1.

*Dec. 11 Sa.* A very cold, frosty morning. Clark staid with us last night. Turned warmer in the afternoon & clouded up. What next, rain or snow?

*Dec. 12 S.* Warm, cloudy and rained a light sprinkle early this morning. Looks now (10 minutes past 9) like it might clear off warm. We went up to Mr. Bell's this evening and staid there till Thursday. It rain[ed] again the first of this week and the creeks got up. We (Mollie, her Ma and myself) went to Hamptonville on Wednesday 15th. Mollie bought her [material for] a gingham dress, 25 cts a yd.[,] a pair of scissors 25 cts. and I bought the *History of Greece,* 1.00. Amount 25 × 8 = 2.00 + 1.00 + .25 = 3.25.

We started over to Pa's this (Wed.) evening but the creek was not foardable.

On Thursday 16 we went over to Pa's and staid there and at Alex. till New Year. The week before Christmas was rainy. I made Mollie a pair of shoes, as the pair I had made at Johnson's shop was to[o] small. Christmas passed off rather dull, I spose on account of rain and mud. Christmas day tho' was a nice, pleasant day. The next day, S., it rained.

*Dec. 27 M.* Clear and warm. Set out in search of a school. Came by home and put on my Sunday, went over and got Hays & Cash to sign my Draft & Report as I had an idea of going to N. Campbell's before return. I staid at Alford Johnsons on Monday night, and on Tuesday morning, bright and tolerbly early, I set out for John Gaither's. I was told, by one that did not know, or wished to deceive, that No. 9 wanted a school, but when I got to Gaither's I found it was a mistake. I staid at Gaither's till after dinner, a good one it was too, and then I was off for Cranbary Albea's near Olin. I got there about feeding time, put up and staid all night. Next morning I went out to see Committee of the Olin District, but did not bargain with them. One of them, Mr. Alford Sharp, promised to write to me soon. Whether he will or will not I am not able to say.

This, Wed. 29th, was a cloudy, damp day. I went on over to Campbells, got my money ($44.) and came back to A. H. Thomasson's two miles this side, and staid all night. Thursday morning I set out for home. I was to have been at Pa's, where I left Mollie and Jodie, on Wednesday, but not being able to find a single man at home with whom I had business, I got belated. Think it rained all night Wed. night. It rained a good deal on Thursday. Dont think I ever saw a muddier time. I got home in the evening about 3 I think.

*Dec. 31* A very rainy day. I was sick. And this day closes the year 1858.

# Diary 1859

*January 1 Sa.* Rained a light shower in the morning, and then appeared in the north west, a beautiful rainbow. Brother Wiley brought us home in his buggy, went over to the Mills [Eagle Mills] & bought some yarn & cloth and went back to Pa's. We left Pa's about 2 o'clock this afternoon, so I guess it was after night when brother W. got back there.

*Jan. 2 S.* Feel quite unwell. Have a cold, and soreness in my right breast. Staid at home all day.

*Jan. 3 M.* Clear and cold. Snowed some last night I think. Went over to Eagle Mills and closed my store acct. with a note. For amount see "Store Bill" or Note. Bought goods to the amount of .75 cents.

*Jan. 4 T.* Clear, windy and cold. Feel better to-day tho' my breast is quite sore yet. Went to Eagle Mills to-day too and bought 2 lb. white sugar, 15 cts per lb., and 1 lb. of butter .10 ct. 15 × 2 = 30 + 10 = .40 cts to-day. Mollie is sick, and I have supper to get, the dishes to wash, etc. It is well enough for a man to know how to cook, wash dishes, etc. Had I have been ignorant of these things, I would have been likely to have gone to bed to-night minus my supper. "Knowledge is power." Another sprinkle of snow fell last night.

*Jan. 5 W.* Clear and cold—very frosty but the morning seems to bespeak a pleasant day. Molie is better but not well. My breast is very sore.

*Jan. 6 Th.* Cold and cloudy. Think it will snow this time surely. Went out this morning to see how many scholars I co[u]ld get for a school in this District. Wm J. Colvert signed two—all I got. Mr. Samuel Hays hauled us another load of wood this evening.

*Jan. 7 F.* Warm and rainy. My cold is not much, if any better, and the soreness in my right breast hold[s] on.

*Jan. 8 Sa.* Clear, windy and very cold. Sent on for the *Age.*

*Jan. 9 S.* Cold. Looked early this morning like it might snow soon, but the sun is shining now (half past eleven). The wind does not blow as hard as it did yesterday, but I guess its very cold out of doors. Jodie had the croop last night, but I think he is better now as we have been doctoring of him the best we knew and could. Like most children of his age, he does not like to take medicine.

*Jan. 10 M.* Clear and very cold. I went up after Pa and some medicine for Jodie, Pa came and gave him a few doses of "Jayne's Expectorant" and it seemed to help him wonderfully. Alex. & Caleb came down and got a load of cotton seed for the cows.

*Jan. 11 T.* A little cloudy and very cold. Jodie is very ho[a]rse yet. Pa staid with us last night. I went to Eagle Mills and bought a few articles for which see "Cost of living 1859."

*Jan. 12 W.* Nearly clear, and several degrees warmer than yesterday morning was. Jodie, I fear, is not much if any better.

*Jan. 13 Th.* Warm and cloudy. Jodie don't seem much better. Mr. Hays hauled us two loads of wood at 25 cents a load.

*Jan. 14 F.* Warm and rainy. Rained a great deal last night. Jodie seems better this morning. I took him up and washed him in tepid water. He kicked and tried to scream, but he is so ho[a]rse he could not do much at it, and I washed. He ate a few mouth fulls of breakfast, and is now, 1/4 past 9, driving nails in a stick of wood. He delights in a hammer and nails. Hope he'll soon be as well as ever.

*Jan. 15 Sa.* Cleared off last night quite warm for January, but the wind blows pretty strong this evening, and I think every blast is a little cooler than the preceeding one. Paid Mr. Jo. Cox $1.25 for ten lbs. of tallow I bought of him some time in the summer. The tallow was here in store house, and belonged to John Gray, but Mr. C. said he had leave to sell it, so I bought it of him as above stated. Bought at Eagle Mills one bushel of corn for which 1 paid fifty five cents.

Last Saturday we sent on for the *Age,* and to-day received the paper, No. 20, Vol. X. The *Age* is a good journal, the best for the price, extant. Mollie's Aunt Mary Chapel came to see us this evening for the first time. She lives in Ashe County and does not often visit over this way. She is a very talkative, and I rec[k]on, industrious woman. She had a pretty little blue-eyed girl with her.

*Jan. 16 S.* Clear and pleasant. Our visitors (Mrs. Bell, Mrs. Chapel and Catherine) have just left, Jodie is a sleep, and I have a chance to write a line or two, but as I am at a loss to know what to write about I'll lay down my pen and get my book.

*Jan. 17 M.* Cloudy, but not very cold. Went up to Nicholson's Mill and had my wheat, 5 bushels, ground. I brought a small sack of the flour home on my shoulder.

*Jan. 18 T.* Clear and tolerbly pleasant after the morning. Made another effort to get a school in this Dis. but failed. There is no chance for a school here unless one would teach for nothing and board himself. In the afternoon I went over to Alford Johnson's and traded him the *Prince of the House of David, Confessions of an Attorney,* and vol. one of the *N. C. Planter,* for 8 lbs. tobacco.
The books cost me about $2.17. Think I can sell the tobacco for $3.30. I and Mollie both read last night by moon light. Dont think I ever saw lighter nights than the two last. The moon shone very bright indeed.

*Jan. 19 W.* Clear and very cold in the forepart of the day.

*Jan. 20 Th.* Warm and cloudy.

Went up to see the Dobbins land. Dont blame Mr. D. for wanting to sell it, for it is hard to have to pay tax on such a mixture of flint rocks and thin soil. It wont pay.

*Jan. 21 F.* Warm and rainy. Rained very hard in the afternoon.

*Jan. 22 Sa.* Clear and cool. Received another No. of the *Age,* and I was anxiously looking for a letter or two, but none came. I wrote to brother W. C. two weeks ago—time enough, I think, for me to have received an answer to-day, but none came. The wherefores I know not.

*Jan. 23 S.* Clear and *very* cold—the coldest day, I think, that we have had this year. There is but very little, if any thawing on south sides where the rays of the sun have a fair sweep. They seem to lose nearly, or quite all their heat, before they reach this ice bound globe of ours, and fall coldly and almost powerless. Think the earth will remain bound with fetters of ice to-day; and to-night they will be more firmly welded. "Old Winter" has been in our neighborhood for some time, and a few weeks ago he gave our sabulous mother, the earth, a *pretty tight squeeze.* That, however, didn't satisfy him, so last evening (Saturday evening) he paid the old lady another visit—one she won't forget soon—for he threw his icy arms around her and—. About this time I went to bed, but when I got up this morning the old coon was still holding on. Guess the old fellow is in possession of no small shear of the "bone and sinew of a man," or something equally as tenacious, for his *hugs* sometimes last a month or two.

*Jan. 24 M.* Clear and cold, tho' a few degrees warmer, I think, than yesterday. Borrowed Mr. Z. Albee's gray horse, Bill, and went to Nicholson's Mill after a sack of our flour. I went u[p] last Monday and had the wheat ground.

*Jan. 25 T.* Cold and cloudy. Looks like it might snow, and we are about out of wood.

*Jan. 26 W.* Rained some last evening, and is still cloudy to-day. The weather is very changeable. The old man Dinglar lay in Albea's still house Monday night, and about day Tuesday morning he came out to Mr. Wallace's and went to bed. In the afternoon he had a fit of some kind and run Mr. W's folks all off, (Mr. Wallace not being at home) threw some of the bed clothes out in the yard, broke all the glass in the shed window, etc. etc. He then started for Troy, without his shoes, coat or hat, aiming to put up at Mr. Coxe's. There were four men here from Alexander Co. with cotton, and when they heard of the extra's Dinglar was cutting up at Wallace's[,] two of them started up there and they met him about half way minus his shoes, coat and hat, with a stone in his hand. He told the men, so they said, that his clothes were back there at that (Mr. W's) house, and that he swore he would kill that bitch with the stone. The men told him to throw down the stone, which he did, and to go back with them and get his things. The men then brought him on to Troy, but, Cox not being at home, Mrs. Cox would not take him in knowing of his capers at Wallace's, so we took him in Gray's store house and seated him before the fire. Soon he began to call for water, and I took a cup and brought him about a quart, nearly all of which he drank at two draughts. He soon after became speechless, and died about 9 that (Tuesday) night. "As the bruit dieth, so died he"—only worse. He was one of king Alcohol's subjects, who had served him long and well, and while approaching his end no efforts were made, that I know of[,] to prolong his life. All seemed willing for him to depart, as it was thought he never would be any better.

His body rests in a drunkard's grave, but his soul—where is it?" From his life no hope *can* we borrow.

*Jan. 27 T.* Warm and very rainy.

*Jan. 28 F.* Still warm and rainy. Rained a great deal last night. The creek is up, but not past foarding yet.

*Jan. 29 Sa.* A nice, pleasant day. I'm not well.

*Jan. 30 S.* Pleasant to-day too. Staid at home.

*Jan. 31 M.* Last day of January. Still unwell. Helped Mollie quilt. January goes out calmly and pleasantly.

Don't know that I ever saw a warmer winter than this has been so far. We have had a few cold days, but no snow, and hail, and sleet of consequence. We have had a great deal of rain, and mud, and foggy, unwholesome weather. but for all this we have been greatly blessed with health.

I've set in this year to get out of debt, and if we live and have health and strength enough to "push along, and keep moving," I think I'll come it. I do not intend to buy any thing on time this year, unless its a bit of land, if I can help it. I must stop going in [debt] if I would get out.

"Better go to bed supperless," said Franklin, I think, "than get up in debt." Its bad to get up in debt, but I guess it would be well enough, generaly, to go to bed supperless, or nearly so. We should eat to live, and not live to eat. O that I could govern myself in this matter of eating.

*February 1 F.* Tolerbly cool, and a little cloudy. Not well yet, tho' I'm better.

*Feb. 2 W.* Cool and very rainy. Can make out to stir, and that is about all.

*Feb. 3 Th.* Cloudy and cool. Went over to the store to-day and bought some pills, peas, butter, etc. See "Cost of Living."

*Feb. 4 F.* Cloudy and cool. Sam Hays hauled us 4 loads of wood.

*Feb. 5 Sa.* Clear part of the day, and cold. Received 2 Nos. of the *Age,* and a catalogue of the publications of H. Dayton, New York.

*Feb. 6 S.* The clouds look like all things were about ready for a snow storm. Will it come? We've not had any snow, except a light sprinkle or two, this winter, and I guess the majority of our people will be glad if the winter passes of[f] without snow, for corn is very scarce generally. But notwithstanding the scarcity of breadstuff many are boiling up 4 and 5 bushels of corn a day, converting the "staff of life" into a stuff the vilest of the vile, and dealing it out to their neighbors. Such men are a curse to the land they lie in. Instead of feeding the hungry, and clothing the naked, some of them will take the last peck of corn from a family of starving children, and give their drunken father in exchange a quart of nasty liquor. Remember, O man! that for all this God will bring thee into judgement.

*Feb. 7 M.* A clear, nice day, and only tolerbly cool after the morning, which was very cold.

We are all nearly 'on the lift' with bad colds. Dont know that I ever heard so much complaint about colds, "in all my born days," as I've heard this winter. Dont think there can be a man, woman or child, in all the land that has not had a *spell.* Paid Sam Hays $2 for 8 loads of wood. When we burn it up we must leave this Town, if we dont get work here. Gray charges us 50 cents a month house rent, and it take 4 loads of wood, $1 worth, a month; so our house rent & wood amounts to 1.50 a month. Cant stand that and nothing to do. No Sir.

*Feb. 8 F.* Cold and rainy.

Mollie keeps her bed most of the time to-day; so I had to kill, clean and cook the chicken for dinner. I had dinner to get out and out, so it was not ready till about 2 o'clock! Well, "the fact is," it was about 12, or later, before I set in, so I think I did—hard to beat—to get up a dinner of stewed chicken, corn bread, peas, butter, fruit, etc. etc. all in two hours. Who among you regular, do-nothing-else cooks, can beat that?

*Feb. 9 W.* Cool and very rainy. Mollie not able to cook yet. I got the peas done to-day about, or a little after one o'clock! I'd make a great cook if I had enough of two things:—viz. practice and something to cook. Mr. John Gray, an old, grayheaded man, says he thinks that a certain amount of rain falls every year, and if it falls in the winter the summer will be dry, and if it falls in the summer the winter will be dry. So I guess he looks for another dry summer.

*Feb. 10 Th.* Nearly clear and pleasant. Went over to Alfred Johnson's in the afternoon and got 4 lbs, 11 plugs of tobacco. I got Alfred to subscribe for the *N. C. Planter* this year, and sent on the dollar for him, and he paid me in tobacco at 25 cents a pound. If I can sell the 11 plugs at 15 cents a plug I'll make about 60 cents. Caleb came down to-night to get money to buy medicine for sister Martha. I fear that hir disease has got to[o] firm a hold for medicine to do hir much good, but it may be that she will get well yet if she can get medicine enough of the right sort. And what *sort* would that be? "Jayne's Expectorant" is the kind she uses.

*Feb. 11 F.* Cloudy and cool.

*Feb. 12 Sa.* Rainy in the forenoon. Mr. Dobbins came up and we went and look at a seventy one acre tract of land which he has about 3 miles west of this place. I like the place tolerbly well, and a[s] Mr. D. offers it cheap, $2 1/2 per acre, I think now that I shall trade for it next week. Most of the land is worth $5 per acre and I think that in a few years it will sell for that, if not more.

*Feb. 13 S.* Clear. The sun shines warm, but the wind blows cool. Mollie is quite unwell yet. I had the "Sunday bread," four nice, large biscuits to bake last night. I did a nice job of it too,—hard to beat.

*Feb. 14 M.* Went [&] borrowed a horse and buggy of Mr. Z. Albea and went up to Pa's. Sister Martha is very low. Dont think she can get well.

*Feb. 15 T.* I & Pa came over to look at the land I am on a trade for. Then on to Albea's to take the horse and buggy home. We staid there till after dinner while it rained a shower. After dinner, and after the rain we started for home on foot. We came by Troy, Eagle Mills, and then went on to Lee Shoemakers, as Pa wants a wagon of Lees make. We got home about dark.

*Feb. 16 W.* Went over to Mr. Bell's. Cloudy and damp.

*Feb. 17 Th.* Cool, cloudy and damp. In the evening I went to Milus Dobbins' to finish our land trade.

*Feb. 18 F.* We traded but did not bind the bargain. Dobbins said he would meet me at Eagle Mills Sat. or Tuesday and make me a deed. I landed back at Pa's about 12 o'clock.

*Feb. 19 Sa.* Cool, cloudy and *very* damp. Came home and went over to Eagle Mills but Dobbins did not come.

*Feb. 20 S.* Staid at home last night by my self. Warm and cloudy. Went back up to Pa's. Cleared off in the afternoon quite cool.

*Feb. 21 M.* Clear, cold and windy. Caleb brought us home this evening in the wagon, and loaded back with cotton seed. He took about 15 or 16 bushels.

*Feb. 22 T.* Mr. Dobbins did not come to-day either. This is a nice day.

*Feb. 23 W.* Cloudy and sprinkled rain a little. Mr. Dobbins came up to-day, and we finished our land trade. I have the notes tho' yet to get Mr. Bell to sign them as additional security. The deed we left with A. F. Gaither of Eagle Mills. When I get Mr. Bell's name on the notes I am to leave them with Gaither, for Dobbins, and get my deed.

*Feb. 24 Th.* A nice, pleasant day. Went up to our intended home to see if those old women, Ann Sudivan & Till Cass, had left. Old Mr. Wm Coffin went over with me. I do not like to go among such stock so I got Mr. C. to go along as a kind of "bodyguard." The ladies, if ladies they be, are in our house yet, tho' they were all from home to-day but two of the children, a little boy and a little girl. I told them to tell their mother to look out for other lodgeings. I then went over to Pa's. Sister Martha is very low. I got home about, or a little after, sunset.

*Feb. 25 F.* Cloudy and very rainy in the forenoon. Rain, more or less every week.

*Feb. 26 Sa.* Cleared off to-day. I went up and got Ily Campbell to go and help me daub our house. The women are still there. Received the *Age* and two Nos. of the *Advocate*,[1] both choice papers and only cost $2.50.

*Feb. 27 S.* Cool and cloudy. Read my papers, went out and sat awhile with old Mr. Gray who has just moved to this Town. Mr. G. is a fine old man in his way, but I fear he does much harm with his "good Brandy" which he keeps on the sideboard all the time, to keep up his own spirits, and treat his friends with. He has frequently invited me to drink, but I have as often refused. I don't drink spirits.

*Feb. 28 M.* Went up to Pa's to see sister Martha. Found her very low. She says she cant live. She requested me to pray for her, and said it would not be long till we would meet where there would be no need of prayer. "O may we meet in heaven."

*March 1 T.* Clear and warm. Moved to our new home.

*Mar. 2 W.* Went over to Pa's and found Martha just alive—unable to speak above a low whisper. I asked her how she was, and she said, "I am here yet." I got old Med and the wagon and came back after Mollie & Jodie. We went over and staid all night.

*Mar. 3 Th.* Sister Martha is no more. She breathed her last about five o'clock this morning. Martha lived a Christian and died happy. Just before she left the world she talked about 'Jesus.' She said, "He is close by." O that my last end may be happy too.

This was a very cold, rainy day. I came by home to see if things were all right there. Did not find anything amiss. Went on to Eagle Mills, got the burying clothes, then on by Tommy Nicholson's, up to Pa's. Dont think we have had a colder rain this winter.

*Mar. 4 F.* Clear and windy, but not very cold. Sister M. was buried about 2 o'clock this afternoon. We mourned, but not as those who have

1. This reference probably is to the *North Carolina Christian Advocate*, a Methodist newspaper that no longer survives. The August 8, 1855, issue of the Raleigh *Spirit of the Age*, however, carried a prospectus for the *Advocate*. According to the prospectus, the *Advocate* was to be a weekly family paper emphasizing the doctrines of the Methodist Church and the interests of North Carolinians. It would be published by the Methodist Episcopal Church, South, would cost $1.50 per year, and was to appear in January 1856.

no hope, for we believe that Martha is now in heaven, and we hope to meet her there in that good world where

"Sickness, sorrow, pain and death
Are felt and feared no more."

*Mar. 5 Sa.* Clear and warm. Got wood, etc, for Sunday in the forenoon, and in the evening I went to the Eagle Mills after our papers, the *Age* and *Advocate* two choice journals. I received also six specimen Nos. of the *Weekly Day Book*, a democratic paper published in New York by "Van Evrie, Horton & Olney."

*Mar. 6 S.* Staid at home and read. Warm with some appearance of rain.

*Mar. 7 M.* A very rainy day. The heaviest shower fell about night that we've had in many a day.

*Mar. 8 T.* Commenced fixing up my fences. I've a lot of that work to do. Windy, but tolerbly warm.

*Mar. 9 W.* Clear and warm. Finced.

*Mar. 10 Th.* Warm with signs for rain again. Sowed my clover seeds, some cabbage seeds, finced some, etc. Eliza & Caleb staid with us last night. I think that the farm suits me better than school teaching, and if I were able to pay my debts I dont think I'de teach any more soon.

*Mar. 11 F.* Warm, cloudy and quite rainy in the afternoon. Made fence. Ily Cambell finished the thousand rails he has been getting for me. He boards himself, and I gave him $5 a thousand to cut & split.

*Mar. 12 Sa.* Clear and warm. Made fence, etc. The buds are swelling smartly. If the weather continues warm I think the peach trees will be in full bloom by next Saturday. Did not go to the [post] office to-day, so I have no late news.

*Mar. 13 S.* A nice day. Ma & Caleb came to see us, staid till after supper, then we went home with them and staid till Monday evening.

*Mar. 14 M.* Another *very* rainy day. Came home late this afternoon. Caleb came with us bringing the wagon and team to haul rails for me.

*Mar. 15 T.* Clear & windy. Hauled rails.

*April 4 M.* Cool and windy. Went in the afternoon to E. Johnson's log rolling.

*April 5 T., & 6 W.* Cold weather. Peaches killed. Think the apples, most of them, are alive yet. Went over to Pa's. They are planting corn this week. Think the ground is rather cold yet.

*April 7 Th.* Clear and warm. Went to meeting, but the preacher, Rev. C. N. Anderson, did not come.

*April 8 F.* Warm and windy.

*April 9 Sa.* A little cool. Wind from the north east. Left home early (before breakfast) and went over to Pa's. Found them, (all but Pa who was not at home—gone to John Nicholson's after sweet potatoes for seed) all on foot. Their heiffer, Rose, died last night. A considerable loss. From Pa's I went on to Hamptonville. I stoped at Mr. Bells as I went on and got him to sign my land notes. Sold my mule colt, Jim, to A. C. Cowles for $45. Was that enough. He was folded the 18th of April 1858, and has had but a poor chance for a living. Paid my blacksmith bill $2.05; paid my tax $1.56. Came back to Mr. Bell's, ate dinner, and then struck out for home via Eagle Mills. Received 4 papers two *Ages* & 2 *Advocates*. Got home a little before sun set.

*April 10 S.* Staid at home. John G. Johnson, Jes. Tetroe & his wife came to see us this evening. I do not approve of Sunday visiting just merely to chat and pass off the time. Much better stay at home and read the Bible, with other good books and papers, and endeavor to grow in grace and in the knowledge of our Lord and Master.

*April 11 M.* A warm day with some appearance of rain. Went to Nicholson's Mill in the forenoon and brought home on my shoulder, 23 pounds of flour. "The borrowerer is a slave to the lender," says the good book, so it is better I think to carry flour from the mill on one's 'bone cart' than it is to borrow, or do without.

In the after[noon] I cut sprouts where I hope one of these days before a great while to plant corn. Corn is now very scarce—hard to get at $1 per bushel. Flour is selling at three cents. Hard times, about now.

*April 12* Warm. A heavy rain passed a little south of us this evening.

Sprouted, planted our garden beens, some onion buttons, etc. Mollie took off a hen this morning with 11 chickens, the first she has had hatched this year.

*April 13* Warm weather. Went to J. S. Hane's near Hamptonville, to see if he would not take up the note I hold against him but he did not do it. Said he would meet me at Nicholson's Mill tomorrow morning and let me have some wheat. I came home by Pa's. They are planting corn.

*April 14 Th.* Went to mill and waited, and waited for Hanes till I began to think he was not coming, but at last he came. And now what about the wheat? How much can I have? The amount of the note is $10.73, and I agreed to take wheat for it at $1 per bushel. Well, I got two bushels only, this time. I'm to have more, said Hanes, if he can spare it.

*April 15 F.* Grub[b]ed, sprouted, etc.

*April 16 Sa.* Went to Eagle Mills in the afternoon and bought one half gal. jug; one half gal. molasses, and three fish, mackeral. I gave 10 cts for the jug, 25 for the 'lasses' & 25 for the fish.

*April 17 S.* Cloudy and cool. Rained last night and early this morning. Snowed a little too. Mr. & Mrs. Bell are here to-day. This is their first visit since we moved.

*April 18 M.* Cold and windy. Went over to Pa's after Bet to plow. Weather moderated a little in the evening, and we commenced planting corn. I lay off & cover, and Mollie drops the corn.

*April 19 T.* Planted corn. Mollie now has 25 young chicks. Lill had a calf yesterday evening. This is her second one, a nice little heifer, with white face, white back and red sides.

*April 20 W.* Warm and cloudy. Went to mill, finished planting corn, except what I shall plant with the hoe, etc. Found a hen's nest with one egg in it!

*April 21 T.* Warm, and a little cloudy. Took Bet, Pa's young mare, home. Pa and Caleb went after their new wagon, and I came home and went to work at my other gate.

*April 22* "Good Friday." Rained a light sprinkle. Then according to old

Mr. Reinhardt's saying, the coming summer is to be a dry one. But I hope, as it only rained a little to-day, next summer will only be a little dry. O that this may be a good crop year. Finished my gate, etc.

*April 23 Sa.* Windy and cold. I see men out to-day wraped in their over coats, blankets, etc, as tho' it was the dead of winter, We have some fears lest the fruit should be killed and the wheat injured to-night. Went to Eagle Mills and to Troy. Had not been to the latter place since we moved.

Paid Mr Gray $6.75 the amount I owed him for house rent, bacon, etc. Got our papers, and a few little articals at E[agle] M[ills]. See "Cost of Living." Came home and covered up our beans, beats, cabbage etc to keep off the expected frost.

*April 24* Easter. Clear and warm. Sister Eliza & Edney Pardue came over to see us to-day.

*April 25 M.* Warm and a little cloudy. Mad[e] a hand wagon, or cart, to haul manure on the cotton patch, garden, etc. We set out a few cabbage plants, watering them as the ground is quite dry and hard.

*April 26 T.* Windy and somewhat cloudy. Hauled out manure on my hand wagon, which is, I think, better than a wheelborrow.

*April 27 W.* Cloudy and cool. Pa staid with us last night, and this morning we 'bundled up' and went home with him.

Alexander, Caleb and myself sheared our sheep this afternoon. I have 4, Pa 7, and Alex. 5, old ones. In all 16, I do not know the number of lambs. Had a fine rain.

*April 28 Th.* Cool and cloudy. Plowed up our garden, and set out about 200 cabbage plants.

*April 29 F.* Still cloudy and cool. Jesse Tulburt came over to-day, and I paid him off in silver, and took up my due bill. See bundle of "old notes, receipts, etc."

*April 30 Sa.* A cool, rainy day. Guess this is the "black berry rain." Went over to J. G. Johnson's and had my shoes half soled.

*May 1* Clear and cool. A brisk wind from the North East. "So saul died for his transgression which he committed against the Lord, and also for asking counsel of one that had a familiar spirit, to inquire of it." I Chron. X, 13. Then it must be wrong to inquire, of those who profess to be witches & wizards, concerning the future. "If any of you lack wisdom, let him ask of God," "in faith, nothing wavering," "and it shall be given him." James I. 5 & 6. Staid at home and read in the Bible.

*May 2 M.* Went over to Pa's and got Luse, a little two-year old gray mule, to plow.

*May 3 T.* Plowed, breaking out middles.

*May 4 W.* D[itt]o. Do.

*May 5 Th.* Went to Mt Mariah and heard the Rev. C. N. Anderson preach for the first time. His text was, "Brethren, if any of you do err from the truth, and one convert him;" etc. James V, 19 & 20. The sermon was short, but tolerbly good. Went over to Pa's and brought home my bees, one stand, on my shoulder, and Pa's, who came with me and staid all night.

*May 6 F.* Went up to bro Clark's, John Willburn's, Col. Willburn's, Harrison Felts', and by Pa's on my way home, where I left Luce, and came home on foot.

Clark has an heir, a girl, born Tuesday morning 3d inst. so "they say." Paid Col. W. the interest on my note.

*May 7 Sa.* We've had a clear, dry week. Nights and mornings cool. Days warm. Wheat looks promising. Oats backward—too dry for them. There is, these mornings, a great chance of honey dew; so I guess the bees are doing finely. I have but one stand.

*May 8 S.* Clear and beautiful. We, as usual, staid at home.

*May 9 M.* Warm weather and dry. Replanted corn, etc.

*May 10 T.* Went to mill in the forenoon. Had a great rain this afternoon. Hope the oats will grow now.

*May 11* Cloudy and warm. Borrowed Jenny of John G. Johnson, the man to whom I sold her, and went after my flour. I went to mill yesterday on foot and had the wheat, two bushels, ground. We, Elison Johnson, Elwood Coffin, and myself, tore down, moved and rebuilt the stable.

*May 12 Th.* Clear and very warm. Planted the broom corn, sugar cane, pop corn, etc.

*May 13 F.* Clear and warm. Worked the road in the forenoon, and plowed our cotton patch in the afternoon.

Mollie headed this page.

*May 14 Sa.* Warm. Rained a light shower in the afternoon. Planted our cotton seed, commenced plowing my corn the second time, etc. Think we have some four or five hundred cabbage plants set out, and the most of them are growing finely. Our beans too, and in fact our garden stuff generaly, look well. Hope we'll have plenty of vegetables this year.

*May 15 S.* Nice, growing weather. Caleb came over to see us and staid till late in the afternoon.

*May 16 M.* Plowed our corn, replanted, etc.

*May 17 T.* A rainy day. Fine for oats, but I fear the rust will take the wheat, and if it does, what will the people do for bread?

*May 18 W.* Warm and cloudy. Plowed some in the afternoon.

*May 19 Th.* Another rainy day. Went over to Alexander Benbow's and got 92 sweet potatoe slips and set them out in our garden. Pa was over to-day, and says that he has planted his tobacco crop this rainy weather.

*May 20 F.* Showery. It rains & rains. Commenced sprouting the meadow. I was knocking the sprouts off a stump, when a yellow jacket got after me and tried to sting, but he failed. Glad he did.

*May 21 Sa.* Went to Hamptonville, to Mr. Bell's and to Pa's. Bought at Town, one jack plane . . . . . . . . . . 75

one chisel, 3/4, 8 1/3 a 1/4 . . . . . . . . . . . .25

one spring ballance . . . . . . . . . . . . . . .25

1.25

Ate dinner at Pa's, ground my plane bit & chisel and was off for home.

*May 22 S.* Clear and pleasant. At home.

*May 23 M.* Plowed in the forenoon & took Luse home in the afternoon.

*May 24 T.* Hoed corn, sprouted in the meadow etc.

*May 25 W.* Put a gum, made of heart pine plank, two feet long & eight inches wide, in our spring; also a box between the two springs in which to keep milk, fruit, or anything that we wish to keep cool. Sprouted in the meadow in the afternoon.

*May 26 Th.* Hoed corn in the forenoon, and sprouted in the meadow in the p. m.

*May 27 F.* Sprouted in the meadow. This has been a week of very fine weather. Cool mornings and great quanities of honeydew.
Robed our bees the 25th and got some very good honey. I have not heard of any bee swarms yet this year. The worm is doing great damage to the wheat crop. Dont think I ever heard of the "wheat worm" till last year. Think they are confined this year to the fields that were in wheat last year. Then it wont do to sow wheat on wheat on wheat stubble land.

*May 28 Sa.* We went to Mr Bell's. Mollie had not been there since Christmas.

*May 29 S.* I went to Flatrock to Sunday School, and took a[ ]part with them in reading, singing, etc. Came home about night.

*May 30 M.* Sprouted in the swamp, etc.

*May 31 T.* Pa sent after me. I went over and found him very ill. Came back in the afternoon and took Mollie and Jo. over and staid all night.

*June 1 W.* Left Mollie & Jo. at Pa's, and cut out after Dr. Ellis for Pa. Found the Dr. at home but he would not come. Said it was "perfectly impossible." It is, I think, at least 17 miles to Dr. E., and when I found that I could not get him I struck for Smith Grove, 14 miles farther, after Dr. Cash, and Wiley & Mary. Dr. Cash would not come either. Wiley came up that night.

*June 2 Th.* Early this morning I, Mary & Mr. Williams started to Pa's. We got there about the middle of the afternoon. Had a fine rain at Pa's this afternoon. It rained but very little here.

*June 3 F.* Staid at Pa's. Hope he is better and that he will soon be well.

*June 4 Sa.* Came home and plowed our cotton.

*June 5 S.* Cool! Cool! Heard some talk of frost, but I did not see any. I expect there was frost in places.

*June 6 M.* Clear and cool. Plowed, hoed cotton, etc. Went over in the afternoon to see Pa. Think he is some better, or hope so at least.

*June 7 T.* Warm & dry. Cut rail timber. Zac Jefferson & Noel Cass split three hundred rails for me to-day. I engaged Zac. yesterday to get some rails for me, but I got one hand more than I bargained for and only got 300 rails made at that. I do not like to board so many hands, these hard times, and get so little work done, so I dismissed them after supper.

*June 8* Warm, with signs for rain. Went with Jesse Teatro to Eagle Mills and got our can of molasses. I also bought six of the best pens I've seen in many a day, for which I paid five cents. Went over to Pa's in the evening, and found him worse than he had been. Had a tolerbly good rain in the early part of the night.

*June 9* Warm. Pa some better. Left Mollie and Jo. at Pa's. Came by Alex. Benbow's and got 200 or more potatoe slips. Think I have enough now to finish setting out my patch. Had a fine mess of potatoes yesterday—the first we've had this season. Heard something said about frost bitten corn, so I guess there was frost Sunday morning sure enough. Frost in June is a rare occurrence in this climate. Set out the potatoe slips and watered them.

*June 10 F.* Staid at Pa's last night. Think Pa is on the mend slowly. Weather warm and dry.

*June 11 Sa.* Cool and dry. Went to the "Trumpet Branch Foundry" to see about getting a thrashing machine. Jo. Weasner, the superintendent, said we might come after the machine Friday week.

*June 12 S.* Went over to Pa's, *we did,* and staid till Monday morning. Wiley and his lady were up.

*June 13 M.* Came home and went to plowing.

*June 14 T.* Plowed.

*June 15 W.* Staid at Pa's again last night. Pa seems to be a good deal better. He is now using "Wistar's Balsom of Wild Cherry."

*June 16 Th.* Plowed, hoed corn, etc.

*June 17 F.* Cloudy and warm, but not much rain yet. Think we would soon have plenty of snaps, potatoes, etc. if it would rain soon.

*June 18 Sa.* Warm and dry. We have some flattering prospects for rain, but it dont come. The oats are drying up, and the garden stuff would be the better of a shower occasionaly. Wheat is very good. It, as a general thing, does not stand very thick on the ground, but it is well filled. We had, to-day, our first, and I expect our last, mess of green peas for this year as we have but very few planted and Mollie is bound to "save seed." Well, that is all right. Save a bountiful chance for seed, and plant a fine patch early next spring—in February. Our bees swarmed about one o'clock to-day. They settled on a peach tree, and I hived them nicely.

*June 19 S.* Cloudy and a little cool. Took Mollie and Jo. to John G. Johnson's and left them there while I went over to see Pa. Think Pa is still on the mend. We got back home about night.

*June 20* Warm and dry. Hoed corn.

*June 21 T.* Finished hoeing corn for this time. Our garden stuff is drying up. O that it would rain a shower. Think oats will be cut short in this part of the country at least. Three o'clock in the afternoon. Had my bees to hive over again. They took exception to their new home somehow, and to-day about eleven o'clock they "swarmed" out and settled among the branches of the 'cedar.' I thought they were doing well. I, however, observed that they "lay out" yesterday and this morning, and that, I dont think, young swarms often do so soon. They had made two or three pieces of beautiful comb, each of which was as large, I think, as a man's hand. I would like very much to know the cause of their discontent. The box or gum was new, tight and roomy. I put them this time in the salt gum, as I had no bee gum ready. Mine are the best tempered bees I ever saw. It is a rare thing for them to show signs of ill temper by offering to sting. And now, half past three o'clock, it is raining gently—about right to wet the ground without washing it. The rain comes just in time to save our vegetables. A few more hot dry days would have fired them. Mollie is baking "berry pie" for supper so I don't care how soon that meal is ready.

*June 22 W.* Cloudy. Went over to Pa's, got Lucy, the mule, and plowed our cotton, broomcorn, pop corn, sugar cane, and roastenear patch.

*June 23 Th.* A little cloudy and cool. Wind from the south-east. Went to Hamptonville, to M. Bell's and to Pa's. Bought a few articles at town. See "Cost of Living."

*June 24 F.* Warm and a little cloudy. We were blessed last night with a fine shower of rain. Guess our cabbages, beans, potatoes, etc, etc. will grow now.

*June 25 Sa.* Mostly clear, and quite warm. Cut some rail timber in the forenoon, went over to Pa's in the afternoon. Pa still seems to be mending slowly.

*June 26 S.* Clear and swealtry. Shadrack & his lady came to see us. This is their first visit since we moved to this place.

*June 27 M.* Clear, still & warm, (1/2 past 10 P.M.)[.] Went to Nicholson's mill and bought 25 lbs. flour, 3 cents per lb. Carried it home on my shoulder.

*June 28 T.* Allen Coffen commenced cuttng here to-day. Had a light shower of rain this afternoon. Corn and garden stuff looks well and is growing very fast.

*June 29 W.* Caleb came over and got a small load (12 doz) of oats. Jefferson and Cass split another hundred rails for me. I paid Jefferson off and discharged them.

*June 30 Th.* Had a fine rain this afternoon. Coffen cut more oats. The oats have the rust some.

*July 1 F.* We, Mr. Bell, Alexander, Clark & myself, went to "Trumpet Branch Foundry" to get a thrashing machine, but as Weasner would not insure the machine further than to mend it as the casting broke, provided he had mettle & coal at the time, we did not get one. Weasner's casting is not fine and smooth, and is said not to be good. Rain this afternoon too.

*July 2 Sa.* A very warm day. Hauled some rails, went over to Pa's etc.

*July 3 S.* Staid at home all day.

*July 4 M.* Cool. Plowed & hoed cotton.

*July 5 T.* Cool. Plowed & hoed cotton, took the wool to the carding machine—Eagle Mills.

*July 6 W.* Cool and cloudy. Hoed cotton, went over to Pa's, etc.

*July 7 Th.* Still cool. Cut and split some rails, hoed cotton, corn, etc.

*July 8 F.* Cool, cool. Wind from the north east. This has been a cool drying week for July.

Hoed corn in the forenoon, and in the afternoon I made a hay rake. The head is ashe, the teeth braces white oak, and the handle pine. Think this is the second time I've tried my hand at rake making. The first one was a rough specimen, but this one I call hard to beat, tho' I did most of the work with my ax, auger & pocketknife. Clouded up this afternoon and sprinkled rain a little. Wish it would rain

"By the way-side let it fall
That the rose may spring up by the cottage gate,
And the vine on the garden wall."

*July 9 Sa.* Cloudy. Went over to Pa's, got a bushel of wheat and took it to Grant's Mill, but did not get it ground as the miller, Mr. Crouch, said if we could do without it I had better not come after it till Monday, tho' he might get it in late this afternoon. I left it. Came on by Pa's and borrowed eight pounds of flour to last us till Monday, got the jug (half gallon) of vinegar and came on home.

*July 10 S.* Still cloudy and rained a light sprinkle this morning. Think rain would do a great deal of good now. Our garden beans are failing fast. Had our first mess of cornfield beans for dinner to-day.

*July 11 M.* Went over to Pa's. Clark came down and we concluded to start to Salisbury in the morning after a thrashing machine. I toped the first plants of Pa's tobacco while Caleb brought my flour from the mill.

*July 12 T.* We, Mr. Bell, Alexander, Clark & myself, started after a thrashing machine. Clear and very warm. We rolled on till a while after night and then took up camp five or six miles below Mocksville, at the Rev. Mr. Bessent's.

*July 13 W.* We left camp early this morning and got to Salisbury about nine o'clock. Saw the rail road,[2] (for the first time) the gass works, depots, machine shops, and in fact the town itself. Staid there till about 2 p. m. and then rolled off empty. Boyden had but 2 machines on hand, and they were both sold, so he said.

*July 14 Th.* We camped last night between Mocksville & Smith Grove as we thought of going by the latter place to see if we co[u]ld get a machine there, but we were told this morning that Green Clifford, the machine maker at Smith Grove, had made but one thrasher this year, so we turned our course homeward. We got to Pa's a little before night. I've now spent 6 days trying to get a machine but we fail at last. Dont think I'll ever spend another week in the same way.

*July 15 F.* An exceedingly warm day. Had a nice little shower this afternoon, and another one after night. They were not very extensive tho'.

*July 16 Sa.* Warm, warm. Went to Eagle Mills got our rolls, three dirt crocks, Jody a waist band, etc. See "Cost of Living."

*July 17 S.* Clear and hot. We went over to Pa's and staid till evening. Pa is not so well as he has been.

*July 18 M.* The anniversary of our marriage. Mollie scalded and scoured.

*July 19 T.* Had a fine rain last night. Hauled up most of my oats with my hand cart and put them in hand stacks.

*July 20 W.* Went over to Pa's and got old Med, to haul some rails, to build a pasture fence, or a cross fence between my corn and pasture, but got disappointed in getting a wagon. I have a hard tug to get them rails hauled.

*July 21 Th.* Went over to John G. Johnson's, got his wagon, hauled my rails and put them up; so we have a pasture for our cow at last.

2. He refers to the North Carolina Railroad, which had been built from Goldsboro in the east through Raleigh, Hillsboro, and Salisbury to Charlotte in the western Piedmont.

*July 22 F.* Took the wagon home, and then went on to Pa's. Jodie went with me.

*July 23 Sa.* We went to Mt. Mariah to the quarterly meeting. Heard Rev. N. F. Reed preach. His text was "For none of us liveth to himself, and no man dieth to himself." Rom. XIV, 7.

*July 24 S.* Cooler. We have had very warm weather for a week or so, but this morning is more pleasant.

We went to meeting. Heard the Rev. N F. Reed preach a good missionary sermon from the same old text, "Go ye therefore, and teach all nations," etc. Matt. XXVIII, 19 & 20.

*July 25 M.* Rained a little last night. We staid at Pa's to-day. We went over there to see our kinfolks (Uncle Flem, and Aunt Theny, Aunt Jatha & Jimmy, cousin),[3] who came up on Sat. last. Cousin Jimmy brought us home in his buggy. Mollie bought two lbs. coffee, at Madison's as we came on, the first that has been bought in our family. I am opposed to the use of "Dram," Tobacco and coffee, as they are used. I believe they were made for some good purpose, but not to eat & drink.

*July 26 T.* Mostly clear and warm. Commenced making hay.

*July 27 W.* It is raining beautifully, 1/2 past 7. Went over to Pa's to get some wheat, but there was none thrashed. I came, on my way home, by John G. Johnson's and borrowed ten pounds of flour.

*July 28 Th.* Cured my hay and then put it up in small cocks. Pa & Ma came to see us today, for the first time since Pa was taken sick.

*July 29 F.* Caleb came over and helped me stack my hay in the forenoon, then I went home with him to help him redaub the old dry house, but we did not get it done. A cloud coming up seemed to threaten us with a shower, so Wiley took us off to rake hay.

*July 30 Sa.* Went to Eagle Mills, in the forenoon, after our papers—the *Age* & *Advocate.* We went to Mr. Bell's on a visit this evening. Noel met us with the mare & buggy and hauled us over.

*July 31 S.* Warm, dry weather. We came home the afternoon.

3. Flem and Jatha are siblings of Strong's father, Andrew Thomasson.

*August 1 M.* Mowed off our turnip patch, etc. etc.

*August 2* Went over to Pa's and helped thrash the wheat.

*August 3 W.* We cleaned up 13 bushels of wheat, with my much abused fan, and I took it to mill. Had five bushels ground for myself and the balance for Pa. The wheat is so cheaty this year it lost over a bushel (so said the miller) in cleaning the 13. Got home about dusk.

*August 4 Th.* Election day. Took the wagon and team, and Pa's flour, home, and then went over to Williamsburg, to the election. I was in Salisbury the day of the giving in at Williamsburg, so I gave in to Perry Tomlin to-day and he charged me 25 cents for his trouble. I thought hard of him for it as I did not hinder him, nor put him to any extra trouble, *Perry Tomlin, Eqr.* I'll not forget him.

*July 5 F.* Rained a fine shower yesterday evening, and this afternoon it rained a great deal. Paid off my sowing bill, .97 1/2 cts. at Johnsons.

*July 6 Sa.* Warm and cloudy. Went over to Pa's, to Jim Grant's, and to the sale of Jo. F. Johnsons property. There was nothing sold but his wheat. The highest of it went, I think, at 80 cents.

*July 7 S.* Pa, brother Wiley & his lady, came over to-day. For dinner we had cabbages boiled and fried; Mutton or 'sheep meat' as Jodie says; pickles, beets & cucumbers; milk, coffee, for Pa, bread, butter, mollasses, etc. We have the finest cabbages I have seen anywhere this year, and every one who sees them gives them the praise. Our beets also excell anything in that line that has come under my observation in a great while.

*August 8 M.* Commenced plowing for wheat. Plowed till dinner, and then tore up our turnip patch, in the afternoon.

*August 9 T.* Sowed and plowed in the turnip seed, plowed (turning oat stubble) for wheat, etc. etc.

Pa & Wiley came over again to-day and took dinner with us. We had green corn and beans for dinner. After we 'got thro' with dinner, Pa & Wiley went home, and we went with them as Pa wanted me to help get up the hay. After we got over I & Wiley hitched up and hauled a load, about half of which fell off in the meadow as we were crossing the branch. We reloaded and rolled off for the barn. We got there in due time with our hay, considering our bad luck. I & Alexander hauled the next load. We boomed it on, so that the hay did not fall off, but the wagon came near upsetting when we crossed the branch.

*August 10 W.* Warm weather. Finished getting up the hay, and Caleb brought us most home after supper, in Wiley's buggy.

*August 11 Th.* A big rain fell this afternoon.

*August 12 F.* Warm and showery.

*August 13 Sa.* Sprouted in the field south of the house, after I came from Johnsons, till the rain came. A heavy rain it was too. Rained most of the afternoon.

*August 14 S.* What a growing time! "He that tilleth his land shall have plenty of bread." If we do our part the good Lord will be sure to do his, so that we shall not lack any good thing. Seek first the kingdom of God, and His righteousness, then all these things (temporal and spiritual blessing) shall be added unto you. Had another heavy rain commencing about twelve o'clock. Guess the ground is wet enough to plow now.

Does it always rain on the thirteenth of August? By refferance to my Diaries for '54, '55, & '56, I think, I see that it rained on that day in each of these years. It also rained yesterday, the 13 of Aug. 1859. We are generaly glad, and should always be pleased to see it rain, for "He hath done," and ever will do, "all things well." "He maketh His sun to rise on the evil and on the good, and sendeth rain on the just and on the unjust."

*August 15 M.* Went over to our Common School house, early this morning to meet the Com. men, Mr. Bart Morgan & Mr. J. B. Johnson. Morgan was there, but Johnson did not come, so I did not agree to teach the school. I do not like Morgan's terms. He only offers fifteen dollars per month, and wants 24 days taught for a month at that! We cant trade unless he amends his terms.

I came back home and went over to Pa's, got Luse, and went to plowing.

*August 16, 17, 18* Clear, or nearly so, and warm. Plowed. The little mule works finely.

*August 19 F.* Plowed part of the day. Rained a light shower.

*August 20 Sa.* Clear and pleasant. Plowed a few rounds, cut sprouts, choped out the weeds and grass in our cabbage garden, trimed down the weeds in the yard, got wood and pine for Sunday, etc. Our cabbages are very fine having large white heads.

*August 21 S.* Staid at home and read my books—good books, I have no others—and my papers.

*August 22 M.* Went over to Pa's and got old Med to plow. Pa let Clark have Luce to ride to School a while, so now I have to plow "the gray mare."

*August 23 T.* Warm and cloudy. Plowed till it commenced raining in the afternoon.

*August 24 W.* Cleared off to-day about twelve. Rained a great deal yesterday afternoon and last night.

We went over to Pa's with our peaches and put them in the dry house.

*August 25 Th.* Clear, or nearly so and warm. I came home and plowed. Mollie & Jo. staid at Pa's to tend the peaches.

*August 26 F.* Staid at home last night by myself. Finished breaking up my wheat land this morning. Went over after Mollie & Jo. We all got home, safe and sound, about night.

*August 27 Sa.* Quite warm, and a little cloudy. Made a shoe bench, and half soled Mollies shoes. Finished toping our cotton, some of which is waist high.[4] It all looks well. This is a great place for gardening. If there was a good vegetable market nearby I could make money by raising "taters," cabbages, beets, onions, turnips, cucumbers, etc. etc. but as it

4. Thomasson is probably growing a small amount of cotton as a cash crop. Like many other small farmers in the Piedmont, he was always short of cash and interested in increasing his earnings. But a more important priority was holding on to one's farm and staying out of debt. For such reasons many nonslaveholders in the South put most of their time and energy into raising food crops and planted only a small amount of cotton. To put all one's effort into cotton was too risky, since the price of this market crop fluctuated. See Gavin Wright, *The Political Economy of the Cotton South* (New York: W. W. Norton, 1978), 62–74. In the 1850s North Carolina farmers who lived near the route of the North Carolina Railroad began to grow more market crops, but they also kept a "safety first" strategy in mind. For more information, see Escott, "Yeoman Independence," 275–300.

is we have the fun of eating our good vegetables to ourselves, so we have full bellies and empty purses. Well let us "eat, drink and be mary" in the right way, remembering that we have "to give an account for the deeds done in the body."

It is now (a little past ten p.m.) raining considerably. Guess it is bed time, so I'm off.

*August 28 S.* Cloudy. We had calculated on going to Mt. Mariah to meeting to-day, but it is so cloudy and damp we gave it out. We stay at home, as usual, and read our papers, (the *S[pirit] of the Age,* & the *N[orth] C[arolina] C[hristian] Advocate*) and books.
Sunday night. I have just been looking over some of my old *documents.* The fall of 1851 I taught school at Pleasant Valley, near Pfaff Town,[5] Forsyth County. I had a rude set of boys to deal with. I boarded at Martin Holder's. Mr. Holder was a short, rough dutchman,[6] not very good looking, but full of fun. He combed his hair once a week, viz: every Sunday morning. Then by the next combing day it was hanging about his head in small rolls, which looked somewhat like so many rat tails. Mrs. H. was a slim, dry looking little woman—good cook (dutch women generaly are that) and fond of good coffee and light bread. Mr. & Mrs. H. were, in their way, fine people. They treated me kindly. I shall always remember with pleasure the few weeks I boarded with them. During the summer of '52 I worked at John Crew's, putting up tobacco. The most of March of this year I spent at Sedge Garden, selling goods for Wagoner & Thomasson. This was a move in the wrong direction for me. I had better been at school. While at S. G. I received a letter from brother Wiley, who was then teaching school near Muddy Creek P. O., directed as follows:

"B. A. Thomasson,
Any where in the world but where he
ought to be."

This made me feel a little cheap, but however I staid my time out, and then left. I next bought Pa's crop of tobacco, striped it out, and had it "put up," or manufactured at John N. Crew's factory, working in the factory two months and, I think, about ten days myself.

5. Pfaff Town is now a part of Winston-Salem.

6. Mr. Holder was almost surely of German origin. Many people used this appellation, taking it from the word for German: Deutch.

*August 29 M.* A north east rain, and quite cool for this month. Made a pot cover for Mollie, dug some of our irish potatoes, etc.

*August 30 T.* Clear and warm. Cut grass to make hay. Caleb came over after old Med. He brought our dried peaches. After all our trouble I dont think we have half a bushel in that lot. Mollie has dried some, and is now drying more in the sun shine.

*August 31 W.* Warm, but not exactly clear. We have cool nights and very heavy dews.

Read from here down for the first day of September.
Put up two small stacks of hay. Mollie helped me put up one this morning. I threw up the hay and she packed it around the pole. This, she says, is the first *hay stack* she was ever on.
Caleb came over in the afternoon after some beets, and to ask me to the tobacco barn raising, so I got him to help me put up the other stack. I am done hay making, at home at least, for this year. This is the greatest year for yellow jackets I ever saw. The wood and fields are fairly alive with them. It was so warm last winter I guess they did not freeze out like common.

*Sept. 1 Th.* I made a mistake in writing for yesterday. We stacked the hay to-day instead of yesterday.

*Sept. 2* Clear and warm. Went over to Pa's to the barn raising. I and Warren W. Johnson carried up the corners, and bro. Wiley, Alexander, H. B. Godfrey put up the logs. We did not get done. We lacked hands. Louis Messick & Frank Madison were asked but they failed to come. I think people who want so much help of their neighbors ought to be willing to help in return. The Lord helps those who help themselves. Man helps those who help him.

*Sept. 3 Sa.* Clear and warm. Went to Wm Weatherman's to see if he would go and work some for Pa and to Wash. Mullace's to see if he would make up a crout tub. Weatherman said he would be on Monday morning, and Mullaces said he would make the tub soon as he could get timber and time. Some have commenced gathering fodder this week. I think the corn is full green yet to be striped of its leaves. Much injury has no doubt been done to corn by pulling fodder too early.

There appeared, last Thursday night, a very strange phenomenon in the elements. I got up in the later part of the night and on going to the door I discovered that it was light—thought it was day—and that the sky looked red. I did not know then whether the sky was realy red or whether it was my eyes as I had just got thro' a sound nap. But since I hear others talking about it. My neighbor, Mr. J. Teatroe, says it was light all night. This was certainly a strange night. Who can account for it? Was it the omen of something in the future? Who can tell? Guess it will long be remembered as the "light night." There was a very dark evening in March 1852. I was then selling goods for Waggoner & Thomasson, and when the dark cloud passed over, from west to east, the chickens went to roost, and I, thinking it was night, lit up a candle. By & by the darkness passed off and it was not night yet.

*Sept. 4 S.* Cloudy and rained a little in the morning, and a fine shower in the afternoon. Rev. T. A. Nicholson preached sister Martha's funeral at Pa's, not at the house, but in the grove between that and the road. The congregation was small, owing in part I guess to the meeting having been published so recently; only the Sunday before. The 14 & 15 verses of the VII chap. of Rev. was the text. Sermon good.

*Sept. 5 M.* A rainy day.

*Sept. 6 T.* Old Mr. Weatherman came over to me get boards to-day. We went over to Pa's and worked till dinner covering the tobacco barn. I went over and borrowed Jim S. Grant's saw, and we came home and sawed board timber in the afternoon. Cloudy and cool.

*Sept. 7 W.* We, I and Mr. Weatherman, sawed timber for boards, etc. Pa came over in the afternoon, but did not stay many minutes.

*Sept. 8 Th.* Mollie sick. Had to stay in the house most of the forenoon and wait on her. Rived a few boards.

*Sept. 9 F.* Clear and pleasant. Rived boards, on the branch, at the lower end of my meadow. A large black oak, about three feet in diameter, had been blown *down,* instead of *up,* some years ago—so long it had lain there that the bark and sap had pretty well rotted off, and the worms had commenced their work of cutting it to pieces. Such a sized tree as that would last a great while if the worms, flat redheaded borer, would let it alone. But soon after the "giant oak," or any other "giant" of the forst, has ceased to live the timber destroying insects commence their depredations. Water enters, and soon the hard wood is decomposed. "Dust thou art, and unto dust shalt thou return." All things earthy must return to the origional substance.

"This world is all a fleeting show,
For man's illusion given;
The smiles of joy, the tears of woe,
*Deceitful shine,* deceitful flow—
There's nothing true but heaven."

This is my birthday. For thirty years have I been in this world. Thirty years more, and where, O! where shall I be? Dispose of me, O Lord, as Thou seest fit, ad give me grace to *sincerely* say, "Thy will be done." Amen.

*Sept. 10 Sa.* Finished riving boards. I have 2[-]feet boards enough (about 1550) to cover a small house. Commenced pulling fodder.

*Sept. 11 S.* A very beautiful morning.

*Sept. 12 M.* Went over to Pa's to help get boards to cover the tobacco barn. I, Pa & Alexander went to Rob't Bell's woods and got the timber.

*Sept. 13 T.* Finished gathering fodder. Mollie helped me, and proved to be a good hand in the fodder field.

*Sept. 14 W.* Went over to Pa's and rived boards to cover the tobacco barn.

*Sept. 15 Th.* Cloudy and cool. Wind from the north east. We went over to Pa's this morning, and I and Alexander *most* covered the "*backer*" barn.

*Sept. 16 F.* A real "north easter." Came as near raining all day as common. Guess this is the equinoctial storm.

*Sept. 17 Sa.* Clear and quite warm. It rained a small flood last night, so the creeks are past foarding to-day. Guess the low ground corn is considerably damaged. The corn crop is not very good on the ridges this year, and if that on the creeks and rivers is much injured we shall have another hard year. This is a hard world to live in. Let us learn to take it as we find it, without a word of complaint, and improve it all we can, that when we leave it it may be better than it was when we came into it.

"O may it all my powers engage
To do my master's will."

Every person should try to make the world better, instead of worse, but how many are laboring daily to make demons, instead of christians of their fellow beings,—their brothers. Remember, O man, that thou art thy "brother's keeper." Then slay him not lest thou hear it said, "the voice of thy brother's blood crieth unto me from the ground," and thou be "cursed from the earth."

*Sept. 18 S.* Nearly clear, and quite warm. We went over to Pa's. Pa, Ma, and Caleb left home this morning for the Harmony Hill camp meeting. We got some choice apples, staid till after 3 p.m. and then came home. Home is the best place after all. We sometimes desire to go abroad, and take a peep at the world, but when we venture out we soon tire, seeing we can find "no place like home," or none at least that suits us as well as the "sweet secluded spot." This is well too. The man that loves not, and so neglects his home, if he has one, must be looking out for happiness in the wrong direction.

Home. There is music in the word. O that we may always have a good home;—a home on earth. "As long as life endures," and "—, when this flesh and heart shall fail, and mortal life shall cease," then, O then, may we have a home in heaven, and live "A life of joy and peace." Amen.

*Sept. 19 M.* Cloudy, and rained a few drops in the afternoon. Went to Wash. Mullace's to see if he had made us a crout tub. M. was not at home. His wife said he was "gone to the Factory," and that he had not made us any tub, as the timber he had procured for that purpose was wormeaten. What shall we do for a crout tub? Will our cabbages have to rot in the garden for want of a tub to pickle them in? I also went to Wm Mullace's & to a Mr. Reed's to see if they would hire me to teach their school. They were both gone to mill, so I did not see either of them. I also went to Jacob Fraley's tan yard and got a small bit (1 1/2 lbs.) of horse hide sole leather. Got back home before dinner. In the afternoon I went to J. S. Hanes' to see him on business. He was gone to mill too. This must be a great day for milling.

*Sept. 20 T.* Another rainy day. Went over to Pa's and drew the nails out of the boards which covered my blacksmith shop.

*Sept. 21 W.* Cut and tied up a few tops. Pa & Ma came by, on their way home from Harmony Hill camp meeting, and staid all night.

*Sept. 22 Th.* Cut & tied a few more tops, etc.

*Sept. 23 F.* Went over to Pa's and helped to cut and hang tobacco. Pa has a fine crop of the weed this year. We cut a fine chance to-day.

*Sept. 24 S.* Heard that Mullace had got other timber, and had made us a tub, so I went over after it this forenoon. I did not get it tho', as it was made of red oak & pine (the staves oak & the head pine) I feared it would not do for a pickling tub. A tub to hold a cabbage pickle *must be very* tight. I came home and we emptied our cucumber pickles and made that tub tub (one Leonard Messick made for me 2 or 3 years ago for a beer tub) full of the nicest crout I ever saw. Not a green leaf in the tub. We cut up one head which weighed ten & a half pounds. This is a great year for cabbages.

*Sept. 25 S.* A nice, pleasant day. Staid at home and read our papers, etc.

*Sept. 26 M.* Went over to John G. Johnsons and bought of his son Thomas, a black sow for $2.50. Went to Pa's in the afternoon and got Luce, the gray mule, to plow in my wheat.

*Sept. 27 T.* Went over to Johnson after the hog I bought yesterday, and bought another—a black and white spoted sow—for $2. Tom assisted me in getting them home. Commenced sowing wheat.

*Sept. 28 W.* Sowed wheat.

*Sept. 29 Th.* Do.

*Sept. 30* Warm and cloudy. Rained some last night, and also early this forenoon.

Sowed wheat. Washington Mullace came over and I sold him a white oat tree for staves to make tubs and kegs.

*Oct. 1 Sa.* Warm and cloudy. This has been a warm week for Sept. The news paper says there was a snow storm in New Hampshire on the 14th. Sowed some wheat, stacked some fodder, went to Eagle Mills after the papers, etc.

"Foot prints across the continent" or letters of G. F. Pierce, now being published in the *Advocate* are very interesting. I read No 7 to-night. (Sunday night the 2.) The Bishop in speaking of the prairie dog says: "The prairie dog is a marvel in natural history. These little animals live in communities. They burrow in the ground, and a pile of dirt marks the mouth of every habitation." He also says: "The rattlesnake, and a species of owl dwell with the dogs." and that "These all live together in peace and friendship." It seems to me that I have read, in some book or paper, of the snakes eating the dogs. This, however, they may not do only in cases of extream necessity. But I should think one or two occurances, of this kind, cause enough for the disolution of the partnership heretofore existing between *dogie* & *snakie,* and for the dogs at least to set up for themselves. Perhaps tho' they have no objections to the snakes occasionally dispatching one of their number in order to keep down population, as they are supposed to live on grass, and that, in some localities, is very scarce. I would like to see the dogs & their towns, but the snakes & owls—keep them out of "the show![")]

*Oct. 2 S.* Rained some again last night. Cleared off to-day a few degrees cooler I think than last week. Staid at home and read our books, papers, etc, "There's no place like home," especialy if one has enough good books & papers to read, but a home without books—ah me, what a void is there. No doubt there are many now in the road to ruin who never would have entered that track if the home of their youth had been supplied with good books, papers, etc. Home to them was too dull and monotonous, so they sought amusement elsewhere, and the world being hostile to all that is good, and they being prone to evil, they soon learn to hate righteousness, love sin, and are led on in the way in which they should not go.

Parents[,] furnish your homes with good books and papers, and give your sons and daughters a christian education if you never give them anything else.

*Oct. 3 M.* A very cold, heavy dew—almost frost. Sowed wheat. Got a bushel of white wheat, of Jesse Teatro, which weighed 67 lbs. Sisters Mary & Eliza, came over to see us to-day.

*Oct. 4 T.* Clear and pleasant. Went over to Pa's, and I and Caleb housed the tobacco, made a shutter for the barn door, etc. Pa's tobacco is very nice, but it hung out rather to[o] long. I very much fear it will get burned up, barn and all.[7]

The barn is worth $50, and it has 40 or 50 dollars worth of tobacco, about one third of his crop, in it now, and if barn, & tobacco should get burned up it will be a bad loss. *I do hope* it will not.

*Oct. 5 W.* Went over to Pa[']s and helped skin the beef. It weighed 198 lbs. I got 24 pounds at 2 1/2 cts.

*Oct. 6 Th.* Clear and warm. Sowed wheat.

*Oct. 7 F.* Some one, I guess, left the gate unfastened, and our hogs got out and went back to J. G. Johnsons. Sowed wheat.

*Oct. 8 Sa.* Brought our hogs back.

*Oct. 9 S.* A cool, rainy day. The Temple Hill campmeeting is now in session. Mr. Williams came over and staid till the afternoon.

*Oct. 10 M.* Went over to Pa's after our cow, (Lill) and I & Mr. W. went over to Jim Grant's to get some "fice"[8] pups. They are nice, but not old enough to take from their mother so we left them. The darkie (Becky) to whom they belong promised to save them for us at ten cents each. We ingaged two—one a piece.

*Oct. 11 T.* Went over to Pa's to help them cut tobacco. I cut till "dinner" by myself, as Pa's hands, Caleb & Edney Pardue, did not get home from the campmeeting till about that time. I rove tobacco sticks in the afternoon. At this meeting, they tell me, A. C. Johnson made a profession of religion.

*Oct. 12 W., 13 Th., 14 F., 15 Sa.* I do not now (Oct. 30) remember what was done on these days. Perhaps I sowed wheat part of the time.

*Oct. 16 S.* We, and Mr. Wm Coffins two daughters, went to "Niggar Jack," or Union Grove to the campmeeting. We saw a good many people,

7. At this time farmers dried or cured their tobacco by building a fire inside the tobacco barn, and Thomasson worried about the danger inherent in fire curing.

8. "Feist" pups are small, mongrel dogs. This expession is still used in piedmont North Carolina.

and heard the Rev. Canutt preach. His text was, "He that being often reproved" etc. Proverbs.

*Oct. 17 M.* A rainy day. Plowed some. Went to John Mullace's this morning and got a half bushel of his early wheat, for which I paid him 37 cents.

*Oct. 18 T.* Went back to the c[amp]m[eeting]. They are having quite a lively time.

*Oct. 17[9] W.* We went over to Pa's last night. To-day I helped them finish housing the tobacco, and I & Mollie gathered us some apples.

*Oct. 18 Th., 19 F.* Dug our sweet potatoes.

*Oct. 20 Sa.* Pack[ed] our potatoes away, finished sister Elizas shoes etc.

*Oct. 21 S.* A pleasant day. Went to Zion to the "big meeting." Heard the Rev. Swin preach from, "Prepare to meet thy God, O Israel!"

*Oct. 29 Saturday* This has been a nice week for business. We picked out our cotton,—not all of it; just went over the second time,—tore the old roof off the kitchen, got plates, rafters, etc, and put a new roof on the same. We have cold, frosty mornings.

*Oct. 30 S.* The frost this morning looked like a little snow, there was so much of it.

*Nov.* Diary

For the week ending [Saturday] Nov 5. This has been a *very* pleasant week indeed. "Indian Summer" is upon us with all its softness and beauty. The dark green of summer has given place to the yellow and red of autumn. What a pleasant, joyful, and yet solemn time! This is the gathering in of the harvest. This week we gathered our corn, finished sowing wheat, etc.

*Nov. 6 S.* Warm and smoky. Noel & Jo. & Caleb staid with us last night and till this afternoon. I halfsoled Caleb's shoes last night. I sowed my

9. It appears that at some point Thomasson omitted some entries and then, perhaps on Saturday the 29th, attempted to think back and make entries for days he had skipped.

1/2 bushel of early wheat on Th. 3d inst. I went to Mill with corn on F. 4th.

*Nov. 7 M.* Finished daubing our cabin.

For the week ending [Friday] Nov. 11.

This has been a warm week. We shucked our corn, made another tub (about three bushels) of choice sour crout, and I went over to Pa's and helped haul out manure one day & a half—Friday & Saturday till noon.

We have had no rain lately. The early sowed wheat, some say, is being killed by the fly. One man, Mr. Shadrack Myars, I'm told has to resow a portion of his crop. I fear mine is injured, but I hope not much if any.

*Nov. 12 S.* Windy and cold. We spent the day at home reading our papers & book. We have no time to visit our neighbors. We have to work for the body & soul too, and on Sunday we try to gain good and useful knowledge.

For the week ending [Saturday] Nov. 19.

Mollie has been sick all the week. Edney Pardue came after us on Mon. 13,[10] and we went over to Pa's; Mollie to weave her cloth, and me to work for Pa. That was a cold morning. I came home at night and staid by myself.

On Tuesday 14 I & Alex. Johnson went to James Ortons and got two hogs,—one for Pa & the other for Alex. J.[11]

On Wed. 15 I hauled out manure in the forenoon. We came home after dinner, or after 12 o'clock. On Thursday, 16[,] I took old Med home and got some medicine for Mollie. The weather has moderated very much. On Friday 17 it rained. We have had a long dry, warm spell. Some say the wheat is injured by the fly, but I hope not. Some people try to make things out as bad as they can but,

"This world is not so bad a world
As some would like to make it,

after all.

10. Actually the 14th.
11. Alexander Johnson, his brother-in-law.

*Saturday 19* was mostly clear, windy, and warm. I got wood, pine etc. for Sun., cooked our *vittles,* and performed the work of housemaid in general.

*Nov. Sunday 20* mostly clear, and quite warm. Mollie is not well yet, but some better I think.

*Nov. M. 21* Rained.

*Nov. T. 22* Cleared off warm. Put away our cabbages, etc.

*Nov. W. 23* Clear and warm. Cut some fire wood in the forenoon, and in the afternoon I went over to Pa's and brought home a load of my plank with old Med.

*Nov. Th. 24* Hauled two loads of the wood I cut yesterday, another load of plank, etc.

*Nov. F. 25* Cloudy & cool. Hauled rock to build our kitchen chimney, took old Med and the wagon home.

*Nov. 26 Sa., 27 S., 28 M.* Went to Pa's, to Grant's mill, with corn, and to Mr. Bell's with some yarn to be died [dyed] blue for Mollies winter dresses.

*Nov. 29. T.* Took Bet home (I had her to ride yesterday) and came back by old Joh[n]athan Barron's to get some pork, but he did not kill today. So Mr. Wm Coffin, who went down there with his mare and wagon to haul it home for me, had to come back empty.

*Nov. 30 W.* Clear and warm. Barron killed his hogs, and I bought one weighing 91 lbs. at eight cents per pound. Wm Coffin hauled it home for me. It is a fat little hog.

*Dec. 1 Th.* Clear and warm. Cut fire wood.

*Dec. 2 F.* Warm. Went to Pa's; got old Med and the wagon; came home and went to Nicholson's mill. The mill was just empty, so I got my grinding done right off.

*Dec. 3 Sa.* A very rainy day.

*Dec. 4 S.* Took old Med and the wagon home. This a rainy day too.

*Dec. 5 M.* Still cool and rainy. Met the Committee men (2 of them. Mr. Mullies & Mr. Reed) of Dis No. 4, at the School house, made a contract with them to teach their school for ($15.) fifteen dollars per month of twenty days.

*Dec. 6 T.* Commenced school in Dis. No. 4., Iredell Co. N. C. Had 5 scholars! The rain continues.

*Dec. 7 W.* This is the fifth day of the 'wet spell,' and the rainiest of them all, I believe. We have not seen the sun since last Friday evening when we went to mill, nor the moon and stars since Friday night.

*Dec. 8 Th.* Cleared off last night very cold.

*Dec. 9 F.* Clear and cold. School small—only 7 or 8 scholars.

*Dec. 10 Sa.* Clear and pleasant after the morning. Mr. Jesse Tetroe hauled wood for us half the day with his cart and stears. He hauled eight fine, big loads. His oxen are large and strong, so we piled on about as much as would "stand the storm." In the afternoon I went to Eagle Mills. Got 4 of our papers, half a gallon of clover seeds, 50 cts, some coat trimings, etc.

*Dec. 11 S.* A little cloudy, and cold. We staid at home. Clark came to see us for the first time since we left T[r]oy.

*Dec. 12 M.* Clear and cold, eight out at school to-day.

*Dec. 13 T.* Weather about as yesterday. P. Campbell is just now, this week, gathering in his corn. Guess they will see *ructions.* I fear I shall lose as much by teaching school as I shall gain. I ought to be doing up my fences, ditching and cleaning up the swa[m]p. But, I *must* have money.

*Dec. 14 W.* A nice day. School small.

*Dec. 15 Th.* Signs of "falling weather" visible. Not much cold.

*Dec. 16 F.* Cloudy, and cold.

*Dec. 17 Sa.* We had a real north east storm of wind, snow and rain, from about 9 o'clock last night, till 7 or 8 this morning. It snowed first, and then rained. The snow is nearly or quite all gone now, (1/4 till 9, p.m.)

I made brooms, mended my old boots, etc. to-day. Mollie now has sixteen nice, corn brooms,—enough I think to last her one year at least.

For the week ending Saturday 24.
School increased some this week, but small yet. Weather very cold. Dismessed school on Friday evening till Tuesday morning. I lectured my students on the importance of going to school, and learning all they can while young, and then left it to them to say how much holyday we should have; They said—*One day!* Think I must give them some apples in the way of a treat. On Sat. 24, I built an ash-house, in which to save the ashes, got wood for Sunday, etc.

*Sunday 25 Christmas.* A nice day, and not near as cold as yesterday. Caleb came after us so we went over to Pa's. Just as we were sitting down to dinner our old friend and neighbor A. H. Thomasson, Esq. drove up. We were truly glad to see him. He has sold out in Iredell, so he informed us, and has bought the Yadkinville hotel. We wish him great success in his new business. He is now in Town. May he fill his pockets with hard dollars instead of rocks.

*Monday 26* We staid all night at Pa's. I came home this morning to feed, etc. Mollie came home late in the afternoon. I mended one of my boots in the day, and the other one at night.

*Dec. 27 T.* Weather moderated very much. Yesterday and to-day were warm days. Clouded up this evening, and I believe it is raining now, 1/4 past 9 p.m. Had ten scholars to-day.

This year draweth to a close. The days of 1859 will soon be numbered. Then 1860. O! who can t[e]ll what will take place in these United States during the year 1860. The enemy of all that's good seems to be stretching every nerve for the destruction of the Union.

*Dec. 28 W.* A cool rainy day.

*Dec. 29 Th.* A North East rain and sleet. Have not seen the trees as heavy laden with ice in many a day.

*Dec. 30 F.* Cleared off warm, as usual after a sleet. School small. Wrote a land deed for E. A. Johnson to A. C. Johnson. The tract of land contains 204 & 58 poles according to L. V. Campbells survey of Dec. 8, 1859.

*Dec. 31 S.* Close of the month, and of the year '59. Commenced snowing this forenoon, and snowed very fast for a few hours. The ground being wet, and not frozen, a good deal of the snow melted, perhaps half, soon after it fell. I think the snow is four or five inches deep. Last winter we had no snow, but any amount of rain. Hope we will have less rain this winter, even it we have more snow.
X Caleb (brother) and Ezekiel Williams (brother-in-law) came over to-day. Mr W. took his dog, Frank, off. The snow melted some this afternoon. I think tho' it only melted next to the ground, as it has been cloudy all day.

Now I've made a misstake. The things about which I am now writing took place to-day, Jan. 1st '60, and not yesterday, Dec. 31, '59. So read from the cross mark down for New Year's day.

# Diary 1860

*Jan. 1 S.* Cloudy and cold. Staid at home. Caleb and Mr. Williams came over. Snow melted but slowly.

*Jan. 2 M.* Clear and very cold.

*Jan. 3 T.* Cold! cold! cold! The snow hangs on. Melts but very little where the sun has a fair sweep, and not any at all in the shade.

*Jan. 4 W.* Weather considerably moderated. Snow melted off rapidly. Had 16 scholars to-day. 12 boys & 4 girls.

*Jan. 5 Th.* I thought yesterday that we would perhaps have a few days of moderate weather, but to-day is about as cold as ever.

*Jan. 6 F.* Cold and a little cloudy. This is the fifth week of my school. I have not had to use "the green"[1] any yet. I read the "Law of School" most every day, and tell the scholars what I *will* do if they transgress that Law. Think my scholars are as attentive and try as hard to learn as any set I ever dealt with. If they had books enough they would learn a great deal this session. If I were able I would supply them with good school books at my own cost; but alas! alas! I am not able to supply myself, and "they" say "Charity begins at home." "Old Christmas" to-day.

*Jan. 7 Sa.* Another sleet. Rained lots to-day. There is more ice on the ground,—that being completely iced over—but not so much on the timber this sleet.

*Jan. 8 S.* Clear and warm. Elwood Coffin, Mr. Teatroe informed us this

1. A green switch.

forenoon, was married last week, to one Miss Dobson, of the Powder Spring neighborhood.

*Jan. 9 M.* Clear and pleasant. Pa and Alexander came by on their way to Eagle Mills. They brought us two bushels of chop meal, and took our cotton to the gin, at Troy.

*Jan. 14 Sa.* Cold and rainy. It rained moderately three or four days of this week. Taught school all the week. Went over to Pa's this afternoon, and brought home our meal, two bushels.

*Jan. 15 S.* Clear and pleasant. Mr. & Mrs. Williams and Caleb came over to see us to-day. They came up last Monday. "We like to have our friends visit us, but visiters should not make their visits to[o] long." Spelling Book.[2]

*Jan. 16 M.* Another nice, warm day. Had, I think, sixteen scholars.

*Jan. 17 T.* Rained some this forenoon, and then cleared off a few degrees cooler. Mollie is making her second woollin dress to-night. Green —yes, green suits her complexion perhaps as well as any color, and green it is. In this dress she will go—not to the fair, but to her dad's, if she can. Taught school four days this week.

*Jan. 20 F.* Went over to Pa's and got old Med & the wagon to haul straw. I bought a small stack of Elwood Coffin. Caleb helped me haul it. Went to Eagle Mills. Sold our cotton, 49 lbs. at 10 cts. per pound. No, I sold 39 lbs. & brought 10 pounds home. Got our ram skin from the tan yard, T. Holcomb's, I paid forty cents for taning it.

*Jan. 21* Clear and pleasant. We went to Mr. Bell's. Stoped at Madison's store, on our way, and bought a pair of cotton cards, for 50 cts; 6 yds. of sheeting for 11 cts. 50 + 11 × 6 = $1.16.

*Jan. 22 S.* Clear and pleasant. Came home in the afternoon.

*Jan. 23 M.* Nice and warm. Taught school.

*Jan. 24* Very nice, pleasant weather for January. Mr. Bell brought us

2. Presumably Thomasson means that he worked on a spelling book or helped Mollie study spelling.

the flour of 3 bushels and 3 pecks of wheat which J. S. Hanes let me have on a note I hold against him. $3.25 yet due on the note. Mr. Bell has it. Ketch me buying notes again will ye on *any* man!

For the week ending Sa. 28, or for three days of that week, viz; Wed. Th. & Fr.

Taught school five days this week. Weather a good deal colder than it was last week, tho' not very cold yet. We have been favored with some very pleasant weather right in midwinter. Some of the farmers have commenced turning the soil, and making preparations for the coming "seed time." Will they *all* be ready? Some no doubt will, while others will not be ready. "Seed time, and harvest" come and go every year, and yet how few are ready to plant in "seed time," or gather in "harvest" time.

"To everything there is a season." "A time to plant, and a time to pluck up that which is planted." If we fail to be ready to plant, and to gather in the proper seasons, O may we not fail to be ready when the "time to die" shall come. To this end, O God, help us to do the will of our Father who art in heaven. Amen.

Mollie cooked the last of our pumkins this week, & made *pumkin pie* for Sunday. We were looking for Mr. & Mrs. Bell, but they no come.

*Jan. 28 Sa.* Clear, cold and windy. Killed the "old sow." She was perhaps one year old—not much fat, weighed 87 lbs.

*Jan. 29 S.* Mollie and Jane[3] wanted to go to Zion to hear the Rev. S. Swaim hold forth, but I no go. So we staid at home most of the day. Caleb came over after his dog, Dash, which had been here most of the week.

We weighed ourselves Saturday. I pulled down 134,—16 pounds more than I weighed last winter, and nearly as much as I ever notched. Mollie weighed 116, Jane 138, and Jody, bless his black peepers, weighed 35 3/4.

*Jan. 30 M.* A pleasant day. A little smokey. Perhaps there is "fire in the mountains." Taught school.

*Jan. 31 T.* And last day of Jan. A cool, rainy day. Half soled and heeled one of Jane's shoes last night, and the other one to-night. I dont like to mend old shoes, but so it is, I have it to do. "A penny saved is a penny

3. Probably Mollie's sister, Jane Bell Benbow.

gained." I can save a goodly number of pennies in a year, by making and mending our own shoes at odd times. Then why not do it? We have to do many things which we do not like.

*February 1 W.* Cold weather.

*Feb. 2 Th., 3 F.* Taught school five days this week. Have about 20 scholars.

*Feb. 4 Sa.* Mostly clear and not so very cold after the morning. Went over to Pa's, and got him to cut me a sack coat.

Received four beautiful Primers from George F. Cooledge, N. Y. I like them well. Wish I had enough such to supply my school. I gave one to Victoria & Ellin.[4]

*Feb. 5 S.* A rough time. Ha[i]led, snowed and rained. Mrs. Bell, Catharon & Jo came to see us last evening. Mr. B. & Jo. left for home this afternoon, in the rain.

*Feb. 6 M.* Some warmer. Still cloudy. Had fifteen scholars.

For the week ending Saturday [February] 11.

Taught school all the week. Ma staid with us Thursday night.

*Feb. 12 S.* Clear and pleasant. We went to Mt. Mariah to meeting. Eliza staid with us last night and went with us to meeting. The preacher, Rev. T. A. Nicholson, was not there. Old Mr. Huff, of Hamptonville, gave us a short talk. In the afternoon Wm Coffin & wife came over and sat with us till about sun set. This the old man's first visit.

For the week ending Sat. [February] 18.

Taught school five days this week. Clark staid with us Wednesday night. Sold him about a bushel of sour crout. We have the nicest and best crout I ever saw. Clark broke his cart, and then broke the bit off my inch auger. I sold him two pigs—a black one and a red one,—we caught them, but before he got off he got out of the notion of taking them. Well, I'ts all for the best.

For the week ending Saturday [February] 25th.

4. These children are the daughters of Thomasson's sister Martha and her husband Alexander Johnson.

Taught school five days. Have a savier [severe] cold, and cough. Most of my scholars are in the "same row," and when we all get to coughing there is not much else to be heard. Wound up my school, Friday the 24, with a nice little examination. The scholars performed well. I am well pleased with the way they conducted themselves, and studied during the school. Believe it was the least trouble to me of any school I ever taught, One morning this week when I spoke of starting to school, Jodie said, "Make school here, make school here, Papa; Jane shan't tear it down." A month or so before the close of my school, I told the scholars that the best speller in each class should have a new book. Mary Mullies & Henry M. Campbell got them, but it was a close race, I tell you.

*Saturday [February] 25*—Cold and windy. I & Jodie went to Pa's, and Mollie & Jane went to Wm Coffin's to a quilting.

*Sunday [February] 26*—Not quite so cold as yesterday. Wind from the S. to-day.

*Feb. 27 M.* Brought old Med home. I am to give $20 for her. Mollie sick.

*Feb. 28 T.* Milus Dobbins came to see me early this morning about money.[5]

*Feb. 29 W.* A rainy day. Made a stable door shutter, etc.

*March 1 Th.* Fixed up part of our garden fence, plowed about half the garden, and Mollie planted her peas, sowed mustard seeds, etc.

*March 2 F.* Clear and warm. Went to Olin to be examined. Mr. Merritt, one of the Ex. Comm., asked me three questions,—one in Geography & two on Arithmetic—and renewed my certificate. Col. Campbell was in Town, so I presented my "Draft," got pay $45 and went over and paid it all over to Mr. Dobbins. I saw my note credited, got dinner[,] had the "gray mare" fed and then put off for home. Got there in good time.

*March 3 Sa.* Clear and windy. We went to meeting. Heard the "new preacher," Isaac F Keenens, preach for the first time in our lives. His text was, "But one thing is needful; and Mary hath chosen that good part, which shall not be taken from her," Luke X, 2. After sermon I was appointed, or added to the list of Class Leaders. Shall I accept of the

5. Thomasson refers to money he owes Mr. Dobbins for the land he purchased.

office? There *is* a work for each member of the church to perform, but is it *my* work to lead? I feel more like I needed leading myself, tha[n] I do like leading. I'm too blind, and if I know not the way, as I should, how shall I lead others therein? Help me, O Lord, to deny myself, to take up my cross, and to follow Thee. If the Leader follow the Great Shepard then may he be trusted to lead the sheep. But will the sheep follow? Some will, others will not, or at least very reluctantly, frequently straying from the fold, and generaly at a great distance behind. It sometimes happens, on a publick day at least, that we see members of our church not among the sober qui[e]t men of the crowd, but with the drunken rabble, at the tale end of some liquor cart, just where the devil wants them to be. Now, what should be done with such members? They say by their actions—which is the loudest kind of saying—that they are not followers of Christ, that they have gone over to the ranks of the enemy. If they *will* suffer the devil to lead them "captive at his own will," why not dismiss them from the church?

*March 4 S.* Had a singing at the school house in No. 4, called Mullie's school house. The house was crowded with young and middle aged people who behaved themselves better rather, than I looked for. A Mr. Bennitt, a music teacher, was there and helped me out greatly. Thanks to you, Mr. B.

For the week ending March 10.

This week I commenced sowing oats, sowed some more clover seeds, etc. I sowed a teacup full on a lot about three rods square, adjoining our south door yard. Guess if they all "come up" the clover will be thick. We staid at Pa's one night this week. Saturday was a very cold day. It snowed a little in the afternoon. Wind blew hard from the north west. The broken clouds flew by in rapid succession, droping as they passed a few flakes of snow. Such seems to me to be the coldest kind of weather. O how I like on such days to be shut in from the storm. "A few more bating winds and rains," and we shall all be in the grave—shut in f[r]om the storms and troubles of earth. But how shall we fare in the next world? (Dr. George F. Willson, just a short time before his death, in speaking of the next world, said he rec[k]oned he should have to take it as he took this,—*from the stump.* Said he had no one, when a boy, to help him along in the world. And the old man wept like a child.) Shall we be "cast adrift on the billows," or safely housed in heaven?

"Oh God forbid th[at] *we* should fall,
And lose our everlasting all,
But may *we* rise on wings of love,
And soar to the blest worlds above."

*March 11 S.* A clear, pleasant day. Staid at home.

*March 12 M.* Went to Pa's and got a small load of cow feed—straw & shucks—and by the time I got home with it the wind blew so hard I had to let it stay on the wagon till next morning. It turned very cold again this afternoon. This is certainly the "blowing month."

*March 13 T.* A very cold morning. The wind blows yet, but not so violently. Cut the timber, split the rails, hauled them, and finished our garden fence.

*March 14 W.* Sowed oats, planted part of our potatoes, etc.

*March 15 Th.* Finished planting the *taters,* after which I went to sowing oats.

*March 16 F.* I've not plowed a whole day this week, I thought to plow all day to-day, but lo! the "but sheep," as Vic, Ellin & Jodie call he[r] because she butted them, got out, and I had to go and hunt her. And then the cow failed to come home last night, and I must go and hunt her. So I was hindered about three hours from my plow, and did not [find] either the sheep or the cow. Mollie found the sheep and let her in again, but as for the old cow we wot not what has become of her.

*March 17 S.* Went over to Pa's, and finding the old cow, Lill, there[,] I led her home with a rope tied round her under jaw. That is the best way to lead cows I ever tried. Years ago I bought one of Uncle Flem. Thomasson, and undertook to drive her home, a distance of about twelve miles, with a line tied round the posture joint of one, I forget which, of her forelegs. She tried hard to get loose, and finally succeeded after skinning my hands terably, and draging me through the bushes no little. I lost two, 3 or 4 days hunting of her, whereas if I had had her fastened by the under jaw at first instead of the fore foot I might have conquered her so easily. I came off and left her.

*March 18 Sun.* A rainy day.

*March 19 M.* Rained some this morning too.

For the week ending March 24.

Weather cool, windy and dry. Finished sowing oats, and commenced breaking up my corn land. I also cut timber (old field pines) and split 200 rails. I cut & split in the forenoon, and plowed in the afternoon. On Saturday we went to Mr. B's. They have there now two babies. One is nearly as old as Jodie and can neither walk nor talk, tho' its a very fat baby. There [their] other one was just born a few days since, has a full head of black hair. I staid all night, or slept, at Shade's. Mollie & Jodie staid at her pa's.

*March 25 Sun.* Cool and windy. Came home in the afternoon. Have not missed anything yet but my curry comb, which has either "strayed or been stolen" undoubtedly. I had it Sat. eve. just before we left, and I think I hung it on the peg where I usually hung it. Where its gone, and how it fares I know not.

*March 26 M.* Windy & cool. Cut and split rails in the forenoon and plowed in the p.m.

*March 27 T.* A cold, frosty morning. This is cold weather for March. Fear it will injure my oats. Work, the same as yesterday, and a little samer for I built Mollie an ash hopper to-day.

*March 28 W.* Another big frost. Think most of the peaches are killed. Cut and split my usual number (50) of rails this forenoon and plowed the p.m.

*March 29 Th.* Another stout frost. Clear and a good deal warmer. Cut rail timber till breakfast, then plowed till dinner, and in the afternoon I went to the smith shop (Wm Louis') with my plows, and then on to Eagle Mills, where I bought

| | |
|---|---|
| thirteen pounds of iron at 5 1/2 cts,— | 71 1/2 |
| three do of coffee at 16 2/3 | 50 |
| one pound of soda at 10 cts | 10 |
| Half gallon mollasses at 50 cts | 25 |
| Total amount | 1.56 |

*March 30 F.* Went over to Pa's, got a bushel and a half of the three bushels of corn I bought of Caleb, and took it to mill. Split a few rails in the afternoon.

Pa this day received a letter from Rev. Morgan Bryant containing a receipt for Dr. Henderson's remedy for diseased lungs. It is a simple thing, but very good no doubt. I am trying it. It is this,—Take a teaspoon-full of the expressed juice of hoarhound, in a jill of new milk every morning, warm. This is *not* a "bitter pill."

*March 31 Sat.* This has been a very windy month, and the wind has at times (a good deal of the time too) been very cold. Mollie went to circuit preaching to-day at Mt. Moriah. I & Jodie stayed at home. I laid down 30 or 40 worm poles of the fence between my gates.

*April 1 Sunday.* Cloudy and rained a little.

We went over to Pa's. Rev. T. A. Nicholson was to have preached Aunt's funeral there (at Pa's) to-day but he was not here. He is to preach it the first Sunday in May. I[t] rained two small showers to-day. The ground is getting very dry and hard.

*April 2 M.* Cool and windy. Laid fence worm in the forenoon, and fenced in the afternoon.

*April 3 T.* Warm and windy. Pa came over and got some red lead, lamp black, etc. to paint his new wagon. Split rails, cut timber, etc.

*April 4* Warm. Rained a little this forenoon, but not enough to soften the ground so that I can plow. Cleaned out my stable while it was raining. Cut timber and split rails this afternoon. Mollie has four hens setting now, on about sixty eggs.

*April 5 Th.* Warm and windy. Finished getting the rails, 530, for the cross fence between my gates, cleaned up part of the hedge row, etc.

*April 6 F.* Went over to Pa's and brought home the meal, one an[d] a half bushels, which they had ground for us. Helped Mollie plant the garden beans, five rows of them, in the afternoon. Dry weather.

*April 7 Sat.* Warm, and a little cloudy. Went to the smith shop, W. L's., and to Eagle Mills. Bought three boxes of matches with four rabbit skins, and one quart of Hungarian grass seeds for which I paid ten cents. Got back home about, or a little before 12 o'clock. Sowed part of my grass seeds in the afternoon. Sowed them on land that was in corn last year, plowed them in with a scooter plow and then drug a pine top over them to pulverize the soil and level its surface somewhat.

*April 8* Easter Sunday, but no dyed eggs in our camp. "I, myself" used to think a great deal of a dyed egg at Easter. "When I was a child, I thought as a child;" but now, being a man, I had just as soon have a white egg as a black one, and sooner too.

It rained a light shower this morning about day dawn.

Our kinfolks, Mary & Williams, and Charles & Jane,[6] came to see us to-day. It rained lots this afternoon, so its wet and warm.

*April 9 M.* A very warm day. Our company all left this morning. Went over to Pa's and got a peck of seed corn.

*April 10 T., 11 W.* Warm, Plowed.

*April 12 Th.* Commenced planting corn.

*April 13 F.* Planted corn.

*April 14 Sat.* Plowed—listing for corn.

*April 15 S.* A rainy day, *I think.*

*April 16 M., 17 T, 18 W, 19 Th, 20 F, 21 Sat.* Planted corn, cool & rainy part of the week. I got one bushel of sweet potatoes of Shadrack Myers last week for which I paid him fifty cents. We planted part of them and bed[d]ed out the rest. On Th. 19 of this week—a cool rainy day—Caleb came after Mollie to go and help make his coat, or rather he brought the coat here for her to make, but she concluded to go to Pa's and make, or help make, the coat there, so we gathered up and off we went. Staid all night with Zeke & Mary for the first time in our lives. I had staid with them before but Mollie & Jody had not, i. e. at their house. On F. 20 I came home and hauled rails to make the cross fence between my gates. A warm day. Had a fine rain in the afternoon. Staid by myself at night. On Sat. 21 I took the wagon home after I finished hauling rails.

6. Thomasson probably refers to his sister Mary and her husband Ezekiel Williams; the other couple probably are Charles and Jane Benbow.

*April 22 S.* Went to the Quarterly meeting at Mt. Mariah. Heard the Elder, Rev. Bobbitt preach a choice sermon from "Be ye therefore perfect, even as your father which is in heaven is perfect." Matt. V, 48. Rev. Van Eaton, of Jonesville[,] preached the afternoon sermon from Matt. XIX, 27.

*April 23 M, 24 T, 25 W.* Planted corn. Finished the field towards Coffins.

*April 26 T., 27 F.* Plowed up my tobacco patch.[7]

*April 28 Sat.* A cool rainy day. Went to the shop & to Eagle Mills. Bought half a set of knives & forks, Mollie a pair of shoes, two pounds of shugar, bottle sweet oil.

*April 29 S.* Mollie sick.

*April 30 M.* Went to N[icholson]'s Mill and bought forty pounds of flour at 2 1/2 cts. per lb. Commenced plowing my corn, or breaking out middles rather where the land had not been broken up.

*May 1 T.* Rained a good deal last night, and some this morning. Put up most of the cross fence between my gates. Mollie got some cabbage plants at Coffins to-day & we set them out after supper.

*May 2 W.* Cool, cloudy & windy. Rained a few light sprinkles. Finished my cross fence, plowed some, etc. My clover looks fine. Its about a foot high, and is beginning to bloom. Raise clover, every body.

*May 3 Th.* Clear and windy somewhat.

7. Tobacco, like cotton, would be another cash crop from which Thomasson could earn some money. Given his long-standing objections to tobacco, the need for cash must have been pressing. The entry for May 16 shows, however, that he was unable to obtain any tobacco plants.

The manuscript Census of Agriculture for 1860 shows that he was growing neither tobacco nor cotton. The census taker reported that Thomasson had sixty improved acres and eleven unimproved acres on his farm, along with one horse, one milch cow, one "other cattle," six sheep, and six swine. The value of the farm was listed at $200, with $5 of farm implements and $165 of livestock. According to the census Thomasson was growing 15 bushels of Indian corn, 150 bushels of oats, 5 bushels of Irish potatoes, 5 bushels of sweet potatoes, and 1 ton of hay and would produce 25 pounds of butter, 20 pounds of beeswax, 15 pounds of honey, and 10 pounds of wool. The value of animals slaughtered was put at $5. Thomasson's entry appears in the Eagle Mills Post Office section of the district south of Hunting Creek.

*May 4 F.* Mostly clear and tolerbly warm. Plowed yesterday and to-day too till dinner. Heard a locust to-day—the first one for this year. His long and doleful cry of "Pharoah, Pharoah," reminds us of one whose relations are all lost, and he alone is left to tell of their sad fate. Poor Locust, how lonesome. We hear nothing much these nights but "Whipperwill, Whipperwill." Went over to Pa's this afternoon. Bought, at Madison's, half gallon molasses at .50 per gal . . . . . . . 25.

*May 5 Sat.* Warm weather. Replanted corn till dinner.

*May 6 S.* Clear (it was a little cloudy this morning) and quite warm. Staid at home and read our books. I read of Brown University in the state of R. I. It was, so says the book, founded in 1764. Has about 150 students, and 25,000 vol's in its library. Rev. Silas Livermore, Teacher of the School now at East Bend, it is said, is a graduate of this old institution. I also read a chapter in the *Chain of Sacred Wonders*—the meeting of Moses and Aaron. They met on Mount Horeb, and proceeded at once to Egypt where their brethren were in bondage. They led them out, and started thro' the wilderness for the land of Canaan. I guess they might have had a pleasant trip of it if they had only done right. When they were in want of food the Lord gave them quails for supper and manna for breakfast; and when they were thirsty the water gushed forth from the dry rock; How could a people thus blessed, sin against Him from whom they received those blessings? Yet they did sin, and hence they were made to wander in the wilderness for forty years, and to suffer many savere afflictions. "God is angry with the wicked every day," and He will by and by destroy them; but He will bless those who do his will.

"Thee, then, my God and King,
In all things may I see;
And what I do, in anything,
May it be done for thee!"

*May 7 M.* Clear and quite warm. Replanted corn in the forenoon. In the afternoon I sheared two of our sheep—the 2 we have at home. We have but 4 old ones and 2 lambs. Two of the old ones and one lamb are at Pa's yet. Tomorrow we are to go over there and fleece them. A swarm of bees passed over the field in which I was at work to-day. The first swarm I've seen or heard of this year.

We have at this time sixty young chickens.

*May 8 T.* We went over to Pa's and sheared the sheep. I sheared mine, and one of Pa's. Alexander sheared his, and Mr. Williams his, and helped Caleb shear Pa's. Had a fine rain in the evening. I'm told that in the neighborhood of Mt. Mariah there was a severe hail storm. The stones, it is said, were as large as hen's eggs.

*May 9 W.* Re-planted corn.

*May 10 Th.* A rainy day. Made Jody a little bench, etc.

*May 11 F.* Went to the shop in the forenoon and got old Med shod before with a bran new pair of shoes. Plowed in the afternoon.

*May 12 Sat.* Plowed. Rained a little in the afternoon.

*May 13 S.* Cloudy. We went to Pa's to meeting. Rev. T. A. Nicholson preached Aunt's funeral there to-day. His text was, "I am the resurrection and the life:" etc. John XI, 25 & 26.

*May 14 M.* Rained this morning. I went to Pa's and ground my new ax for the first time, got one bushel of corn and came home and went to replanting corn.

*May 15 T.* Plowed.

*May 16 W.* Do and finished planting corn in the afternoon. I had left a patch for tobacco, but plants being hard to come at I concluded to plant it in corn. Had a heavy rain this evening and a light sprinkle of hail.

*May 17 Th.* Ground to[o] wet to plow. Sprouted oats and wheat part of the day.[8] Old Med sick again. She had an attack Or the "*belly aken*", as Louisa would say, in the forenoon. I rubed her, and slaped her in the flank with my open hand a few times, rode her about in a piert trot, etc. till she was as I thought well again.

But just before sun set her disease returned. I tried the rubing, slaping and riding again but there seemed to be no virtue in it this time, so I sent Mollie & Jody over to Wm Coffin's, and I put out on the old gray to Pas

8. The meaning of this phrase, "sprouted oats," has remained somewhat obscure to the editor. However, since Thomasson refers in other parts of the diary to cutting the sprouts sent up from tree stumps in a recently cleared field, perhaps he means that he weeded his oat crop.

to get Mr. Williams to help me take her thro' a "co[u]rse of medicine." I had to lay whip and keep her in a smart trot to prevent her getting down to roll on the way. When I got to Williams' the whole family was in bed, but I rousted them in short notice, and I and Mr. W. commenced the "co[u]rse" by first drenching with sage tea & Mollasses; next with Jerusalem Oak root tea & milk; and lastly with dissolved Asafetida, rubing of her with a pole, etc. etc. between doses. She got better about midnight and between one and two o'clock we stabled her and went to bed.

*May 18 F.* Med seems well as ever. Had another fine rain this afternoon.

*May 19 Sat.* Clear and cool. Went to mill in the forenoon and replanted corn in the p.m. Mollie & Jody went to Williams' after guiney eggs. They got seven. Mollie came home sick, so I had to feed the pigs & chickens, milk the old cow, *etc, etc, etc.*

*May 20 S.* Clear and pleasant. Staid at home and read our books.

*May 21 M.* Plowed. Went to mill in the evening after our corn meal.

*May 22 T.* Plowed. Had a nice little shower of rain this afternoon. Our Beets, Beans & Potatoes look well; Onions fine; Peas tall; Cabbages only midling.

*May 23 W.* Plowed. Corn quite small.

*May 24 Th.* Plowed. We went over to Mr. Williams' this evening to go to Clark's cherry orchard to-morrow.

*May 25 F.* Rained lots last night. Still very cloudy and rained some this forenoon, so our cherry trip is all over for to-day. Came home after dinner and finished sprouting the May wheat, which is, I guess, at least one third if not half, smut. I would like well to know what causes the smut in wheat.

*May 26 Sat.* Cut some dry house logs,[9] sprouted oats, etc. etc. Had another light fall of rain this evening, with a brisk wind. Cloud came over from the North.

9. The "dry house" was a small structure in which people dried fruit. Thomasson refers either to logs from which to build a dry house or to logs for the fire built therein.

From May 26 to July 15 I wrote none in my Diary.

About the 30 of May, I think it was, we had a very heavy rain. June was a dry month. Wheat in this country is very sorry—not half a crop. Mine just about doubled itself, i.e. I raised about seven bushels from 3 1/2 of seed. This is poor incouragement for the farmer. What! Wheat only double itself when at least 8 or 10 to one was expected! Well that is indeed, a poor turnout, but the fault is ours. We have been blessed for a number of years with heavy wheat crops, so I guess some were ready to conclude that wheat would grow almost anyway. Hence but little attention was paid either to the seed or the soil, so a failure this year is no more than we might have looked for. "He that *tilleth* his land shall have plenty of bread," but he that only scratches and half cultivates need not expect much. So, Mr. Farmer, cheer up. Cultivate your soil thoroughly; Select your seed with more care; sow in "seed time," and then "in due season *you* shall reap if *you* faint not." Corn looks tolerbly well yet, but unless it rains soon the corn crop will be cut short too. Then we shall have another "hard year."

On Sat. 7 of July we had rain, but not enough to wet the ground deep. That was tax paying day. I went over to Williamsburg and made my return, paid my tax—$2.20, heard the candidates speak, etc.

*July 8 Sunday.* Cleared off to-day. We went to the Quaker church to monthly meeting. Heard Ann Benbow preach. She gave us, as I think, very good advice.

*July 9 M.* Went over to Pa's, and I & Mr. Williams made hay racks 14 or 15 feet long for Pa's new wagon.

*July 10 T.* Mr. W. came home with me last evening, and to-day we hauled at one load my whole crop of wheat to Wm Dennies thrashing machine.

*July 11 W.* The warmest day we've had. Plowed a little late in the p.m.

*July 12* Cool & dry. Plowed & hoed corn.

*July 13* Quite cool. Went to mill. Mr. Williams came over and borrowed two bushels of wheat of me, and we went on to mill together.

*July 14 Sat.* Very unwell.

*July 15 S.* Staid at home all day.

*July 16 M.* Commenced making hay.

*July 17 T.* Made hay.

*July 18* Planted 3 stack poles. Borrowed Wm Coffins wagon, hauled up and stacked our hay.

*July 19 Th.* A little cloudy. This has been a great week so far for hay making. Hauled and stacked our oats this forenoon, took C[offin]'s wagon home, etc.

*July 20 F.* Plowed and hoed corn, etc.

*July 21 Sat.* Commenced cleaning off a spot of branch bottom, above the spring, for a turnip patch.

*July 22 Sunday.* A little cloudy. I think it thundered some early this forenoon, so I hope we'll have rain to-day for it is very warm and very dry. The corn and garden stuff in genearal is drying up. Many of the corn tassels are dead and dry. The wheat crop failed, and if the corn fails too, what will the people do for bread? Had a fine rain, with some hail this afternoon.

*July 23 M.* Had more rain this forenoon which revived our drooping corn. There is a kind of ant, or woodlouse, which is doing bad work to many of the corn stalks. They commence at the root and cut the stalk out as hollow as a reed. I suppose the dry weather is the cause of their letting in on the corn.

*July 24 T.* Clear and pleasant. Plowed in my corn.

*July 25 W,* Clear and cool. Finished plowing my corn. It looks now like being dry weather again. I hear tho' a rumbling noise now and then which seems to come from the North East, but it is perfectly clear in that direction. It may be thunder, or it may be something else,—perhaps Mr. S. Myars' thrashing machine.

*July 26* Hoed corn in the forenoon. From the amount of rain that fell this afternoon I guess the noise I heard yesterday was thunder. The rain to-day was almost up to the May shower. Guess the corn, etc. etc. will grow now. Hope the corn crop will be good, as that is our dependance this year.

*July 27 F.* Rained a light shower to-day.

*July 28 Sat.* Cut some rail timber, made a stake & pole fence from the stable to the garden, etc. Mollie went to the Quaker Church and heard one Silas Carter, from Ohio, preach. She says he preached loud and long.

*July 29 S.* Rained a light shower last night, and is cloudy this morning.

*July 30 M., 31 T.* Dry weather. Worked on my cross fence, etc. etc.

*August 1 W.* Split rails, etc.

*Aug. 2 Th.* Election day. Cloudy in the forenoon. Went over to Williamsburg and voted for ______.

*Aug. 3 F.* Went over to Pa's, got Pone and the wagon and hauled my rails, poles, stakes etc to finish my cross fence. Mr. Williams helped me, and after we were done we went to D. Johnson's saw mill after a load of plank for him. Rained a light shower in the afternoon.

*Aug. 4 Sat.* Warm & dry. We went to see Charles & Jane this evening for the first time. They live at the saw mill.

*Aug. 5 S.* We went to Holly Springs to the "big meeting." Heard the Rev. Dr. Parks hold forth to a large crowd. His text was "Is there no balm in Gilead; is there no physician there? why then is not the health of the daughter of my people recovered?" Jeremiah VIII, 22. Dr. P. is by far the ablest Baptist preacher I've heard in this country, but there are a great many that seem to have but very little if any confidence in him. He "speaketh peaceably to his neighbor with his mounth," and tries to be every bodies man. Rained a light shower.

*Aug. 6 M.* Finished my cross fence, etc.

*Aug. 7 T.* Mollie went to her pa's. Noel came home with her.

*Aug. 8 W.* Commenced our dry house.

*Aug. 9 Th.* We, I & Noel worked on our dry house. I & Mollie were both taken sick last night.

*Sept. 6 Th., 7 F., 8 Sat.* Warm weather, very warm for Sept. Made hay, the weather being quite favorable for that work.

*Sept. 9 S.* Clouded up this forenoon, rained some and turned a few degrees cooler. We went to Mt. Moriah to meeting. The house was crowded. Heard two sermons—one at 11 by the preacher in charge, and the other at about 3 o'clock by Rev. W. L. Van Eaton. The first was a very good sermon from "Behold, I stand at the door, and knock," Revelation III, 20. Mr Van's text was "When a few years are come, then I shall go the way whence I shall not return." Job. XVI, 22.

*Sept. 10. M.* Cloudy and cool. I went to mowing, and Mollie went to meeting. Pa preached. At night we all went. There were nine or ten who came forward, at the close of the sermon, to be prayed for. One of them, Nettie Clark, was converted. The text to-night was, "He that, being often reproved, hardeneth his neck, shall suddenly be destroyed, and that without remedy," Prov. XXIX, 1. Keenins preached.

*Sept. 11 T.* We went to meeting to-day, and at night also. Pa preached in the day, and Tommy Nicholson at night. Had a good time. Many felt, I guess, that it was good for them to be there.

*Sept. 12 W., 13 Th.* Made hay, gathered fodder, etc. Meeting again tonight (Thursday)[.] We went out. Keerins preached from, "And I looked, and behold, a pale horse; and his name that sat on him was Death, and hell followed with him," Rev. VI, 8.

*Sept. 14 F.* Gathered fodder, etc.

*Sept. 15 Sat.* Do do in the forenoon, and in the afternoon I went to Eagle Mills & to Laurel Falls. We have had cool weather this week. Some say there was frost in places, and I guess there was, tho' I did not see any.

*Sept. 16 S.* Staid at home.

*Sept. 23 S.* We started over to Pa's this morning, but we did [not] get far on our way before we met Mollie's father & mother coming here, so we came back and did not go to Pa's till late in the afternoon. Then as Mr. & Mrs. B. went on home we accompanied them that far. Found Mr. Williams, who has had the fever for some days a little better than he had been. Pa and Ma are well nigh run down. They have a great deal to do, and but little help. I went over there the other day and found mother washing for Mary! Now Ma has not, since I can remember been used to washing for her own family. Old Aunt Mary done the washing as long as she was able, and longer too. Then sister Martha who is now in that good world where

"Sickness, sorrow, pain & death
Are felt and reared no more,"

took her place at the tub. I have assisted her many a time. And now for Ma to have to wash for Mary's family in her old days, I think it is hard, to[o] hard, to be borne. But Ma will do as she will do. She slaves herself more than is realy necessary I think.

*Sept. 24 M.* Came home after late breakfast, and went to work at my fodder. Warm with signs for rain.

*Sept. 25 T.* We went over to Pa's, Mollie to dye her cotton yarn; myself to help Caleb stack Williams' hay.

"Do not delay, till tomorrow,
Things that should be done to-day."

And again, "If you have a thing to do, do it at once," "for thou knowest not what a day may bring forth."

That hay was ready, and I think it might have been stacked one Saturday.

*Sept. 26 W., 27 Th.* Worked at my fodder.

*Sept. 28 F.* Mollie went to meeting.

*Sept. 29 Sat.* Cloudy and cool. Tied fodder.

*Sept. 30 S.* Still cloudy and cool. Mollie went over to Pa's this morning to see how they all are. I told her if Mr. Williams was bad to send Caleb after me. Caleb came and said they thought he (Mr. W.) was dieing when he left. I went over. Found him very low, tho' he still knew me. The Drs. Cash, Cranfill & Pa, have given him out. They say he cant get well. Poor man. If he is only ready to die, and sister Mary tells me that he says he is. Well I trust that he is and that he will soon be with Jesus.

O that we may all meet him in heaven.

*Oct. 1 M.* Mr. Williams is yet alive, but not any better that we can discover.

*Oct. 2 T.* Mr. W. is alive yet, and seems a little better, tho' we all thought last night that he would not live till day, and sent out for the neighbors to come in and lay him out.

*Oct. 3 W.* Mr. Williams commenced getting worse last evening, and continued to get worse & worse till to-day, Oct. 3, 1860, between the hours of 12 & one o'clock he died.

Poor man. No tongue in this world can tell what he suffered. Hope he is now at rest. "There is a rest remains for the people of God," and we hope our departed brother has entered into that rest. He had been for some years, I know not how many, a member of the M[ethodist] E[piscopal] Church. He told his wife frequently during his illness that he should not get well; that he was going to live with the Good Man. O that you, and I, reader may meet him in that bright world where parting is unknown, Amen.

*Oct. 4 Th.* Rained last evening, last night and this forenoon, a great deal.

*Oct. 5 F.* Brother Wiley, who came up Tuesday evening, left for home this morning early. We also came home. Mollie had not been here since Sunday morning, and I only a few minutes—long enough to feed the pigs, see to a few other things, etc—at a time.

*Oct. 6 Sat.* Gathered chestnuts, etc.

*Oct. 7 S.* Cool, cloudy and very damp tho' it did not rain much. We went over to Pa's and staid all day and all night.

*Oct. 8 M.* Cleared off warm. We came home this forenoon, and went to work.

*Oct. 9 T.* Cut some firewood, etc.

*Oct. 10 W.* Went over to Pa's and helped Caleb cut and haul a load of wood, haul over the poles and forks of two tobacco scaffolds, etc. In the afternoon I came home and hauled some rails & wood.

*Oct. 11 Th.* Cloudy & cool. Rained some last night. Hauled up and stacked my top fodder, stacked my clover hay, etc. Took the wagon home in the afternoon.

*Oct. 12 F.* Took our chestnuts, about 7 1/2 gallons to market. I only got 75 cts for them, which I laid out as follows:—

| | |
|---|---|
| one gal. molasses | .50 |
| one half gal jug | .10 |
| five pounds rye | .10 |
| A little black pepper— | .05 |
| | ·75 |

*Oct. 13 Sat.* Cloudy, cool and windy. Went to the shop and had old Med shod before.

*Oct. 14 S.* Cool, and rainy in the forenoon. The campmeeting at Union Grove is now in session. We talked of going over, but it is to[o] cold and wet to be out, so we staid at home all day.

*Oct. 15 M.* We had a killing frost this morning. I went over to Pa's this morning and helped Caleb cut tobacco, which, being green and frost bitten too, is hardly worth saving. We cut the most of it down, and piled it in and about the fence corners and left it. Guess the tobacco crop, as a general thing, is rather light this year, as it was planted late, and the early part of the season was dry.

A. C. Johnson[10] departed this life last evening at, I'm told, about seven o'clock. His disease was fever. Dr. Cranfill says it is the same ailment that once bore the name of "cold plague."

From what I've heard of Alexander's conversation, during has last illness, I think he is now in a better world than this. I hope so. Life is very uncertain. Ezekiel and Alexander were both stout, hearty men a few weeks ago. Now they are in their graves. The "pale horse" and his rider has passed over two strong men in the prime of life. The young and the old, the strong and the weak, the rich and the poor, all die. None are exempt from death. Then, as we know not the day nor the hour in which we shall be called hence, it is *very* important that we be always ready. To this end, O Lord,

"Help *us* to watch and pray,
And on thyself rely.

10. Thomasson's brother-in-law, Alexander Johnson, the husband of Martha Thomasson, who had died in 1859.

*Oct. 16. T.* Another very heavy frost this morning. Clear and warm after "early day." We went to the Buck Shoal grave yard and saw Alexander's body deposited there. Our relations are rapidly passing away, and we, too, will soon be gone.

*Oct. 17 W.* Warm, and somewhat cloudy in the afternoon. Mollie went to see her sister Jane Benbow, who has an heir—a fine son.

*Oct. 18 Th.* A rainy day.

*Oct. 19 F.* A stormy day. The wind blew very hard from the north east, and it rained nearly the whole day. The wind blew over one of our bee stands and made a smash of the honey comb. We saved the honey—a gallon or so of it. The poor bees are left in very destitute circumstances. Their house tilted over, honey all gone, (they had no bread) and winter, cold winter, close at hand. I would advise them to pay some one of their friends—a rich sister have—a visit, a Christmas visit, and stay all winter if it would be no interruption. Then in the spring they could swarm out in search of a new home, which being found they could go to work and make a fresh start in the world. But bees, like some people in this world, love themselves more than they do their neighbors, so, when the honey & bread seasons are over, they are "done taking in." These bees are much like a poor friendless family burned out of house and home. To perish with cold and hunger seems to be their doom.

*Oct. 20 Sat.* Another rainy day. I've made [no] preparations yet for sowing wheat more than sprout the ground. I walked out in the woods to-day and saw many a large, green tree stretched full length upon the earth. Giants of the forest, which have stood the storms of many winters, gave way to the blast on the morning of the 19th inst.

I have, perhaps, heard the wind blow as hard for a few minutes—yes, and perhaps, all day like it blew this time. It blows as hard I guess every March, but the trees then being striped of their leaves, it does not damage the timber so much. There is scarcely a tree in the woods which did not lose a limb on last Friday, and many of them lost their lives.

*Oct. 21 S.* It rained a great deal last night. No rain to-day, but it is not exactly clear yet. Mollie has been sick for three or four days, and I've had to do the cooking, etc. which however has not been a very heavy task as we have but little to cook. We have been living for several days on little roots, many of which are so small it takes two or three of them to make a mouthfull, which we call "sweet potatoes." We also have apples, beets, honey, molasses, now and then a chicken, a few eggs, etc. If we only had plenty of good milk and butter to add to our list we could live 'fat and full.'

*Oct. 22 M.* Went to mill, and to Pa's. Pa and sister Eliza are both sick.

*Oct. 23 T.* Cut fire wood, etc. Warm.

*Oct. 24 W.* We went over to Pa's. I helped Caleb haul one load of wood, and then came home, leaving Mollie & Jody at Pa's, and hauled two loads for us.

*Oct. 25 Th.* We came home this morning tolerbly early. I hauled ten two[-]horse loads of manure on a piece of ground I expect to sow in Holt wheat. I did not get done in the day, so I hauled 2 or 3 loads after night. Caleb had come after the wagon & team, about sun set, but as I was not ready to give them up he had to "Wait for the wagon."

*Oct. 26 F.* Hunted up our hogs; brought three of them home, and pen[n]ed one to fatten. We are meat hungry.

*Oct. 27 Sat.* Spread manure, went to Pa's after old Med, some see wheat, apples, etc. This has been quite a pleasant week. "Indian Summer."

"Of all the months that crown the year,"

it seems to me that October is one of the most pleasant. "When woods begin to wear the crimson leaf" the landscape is so beautiful, and the weather so delightful.

*Oct. 28 S.* A little cloudy, and some what cooler. Wind from the north east. I, this morning, finished reading a little book—the *Life and Essays of Dr. B. Franklin,* which alas! I ought to have read at least fifteen years ago. I think every young person in the world should read that book. Reader, If you have never read *Franklin's Life and Essays* procure a copy without delay and read it carefully. Then be industrious, honest and frugal all the days of thy life.

*Oct. 29 M.* A rainy day. We killed a shoat this afternoon. It was a few days over a year old, and only weighed about 50 pounds. Hogs get fat in the woods this fall as there is a great mast.

*Oct. 30 T.* Cleared off to-day, and Mollie went to Pa's to see how they are. She came back about night and reported, "all better." I cut crib poles in the afternoon.

*Oct. 31 W.* Warm, and a little cloudy. Plowed up a small portion of my old field on which I want to sow wheat next fall. This is the last day of October and I've not sowed any wheat yet.

*November 1 Th.* Wet and warm. Commenced sowing wheat this morning, but it set in to raining again about ten o'clock, so I had to "suspend."

*Nov. 2 F.* A very rainy day.

*Nov. 3 Sat.* Clear and warm. Pa came over and I went to Thomas Holcomb's after his leather,—a side of upper. Holcomb can tan, but he dont curry & block well.

We went home with Pa, and then on to Mr. Bell's,

*Nov. 4 S.* Clear and considerably cooler. We staid at Shades last night. Came home in the afternoon late.

*Nov. 5 M.* A stout frost, and the ground a little crusty. Sowed wheat, tho' the ground was full wet to plow.

*Nov. 6 T.* Clear & pleasant. Went to the Election and voted for J. C. Breckenridge & Joseph Lane.[11]

Called at W. Mullies' on my way home and got our new tub.

*Nov. 7 W.* Sowed wheat.

*Nov. 8 Th.* Do Cloudy cool.

11. Thomasson's vote may seem curious, since he clearly loved the Union and was not a fire-eating secessionist. But Breckinridge, worried about charges that he favored disunion, had reversed his decision to remain silent during the campaign and made a single speech flatly denying any connection with disunionism. Thereafter, his supporters often ridiculed the idea of secession and particularly stressed his loyalty to the Union in the upper South states of North Carolina, Virginia, and Kentucky. See Avery O. Craven, *The Growth of Southern Nationalism, 1848–1861* (Baton Rouge: Louisiana State University Press, 1953), vol. 6 of 10 vols., in *A History of the South,* ed. Wendell Holmes Stephenson and E. Merton Coulter, 341; and Donald E. Reynolds, *Editors Make War: Southern Newspapers in the Secession Crisis* (Nashville: Vanderbilt University Press, 1970), 88.

*Nov. 9 F.* A very rainy day. Went to mill.

*Nov. 10 Sat.* Clear and windy, but not very cold. Shade & Jane came to see us this evening, and brought us about a peck of fine sweet potatoes, of which we were very glad.

*Nov. 11 S.* Clear and warm. We all went to Zion to the "big meeting." Heard the Rev. Calwell Carter, a young chap, try to preach. His text was, "Sir, if thou have borne him hence, tell me where thou hast laid him, and I will take him away." John XX, 15.

*Nov. 12 M.* Clear & pleasant. Finished sowing wheat, if I do not get more to sow. I've sowed two bushels—all I had.

*Nov. 13* Got up our potatoes. We have about four bushels.

*Nov. 14 W.* Gathered corn.

*Nov. 15 Th.* We staid at Pa's last night. This morning I helped haul a load of wood, two loads of corn, etc. We came home in the afternoon.

*Nov. 16 F.* Hauled in the most of our corn to-day. We are cribing it in the shuck. I know not how it will keep.

*Nov. 17 Sat.* A rainy day. Rained lots last night, but not very much to-day. Took the wagon home, hauled a part of Mary's things down to Pa's,[12] hauled them a load of fire wood, etc.

*Nov. 18 S.* We went to Zion to the "big meeting" which did not brake till to-day, The Rev. G. W. Brown preached. His text was, "Behold the Lamb of God, which taketh away the sin of the world."

*Nov. 19 M.* Went over, armed and equiped, to commence school, but found no one there. The chair, desk & black board were gone, the house in miserable condition and the door locked. I sat on the door step a few minutes, waiting for some one to come. I soon concluded it was no use to stay there, so I left for home.

*Nov. 20 T.* Cut fire wood in the meadow.

12. Evidently Mary is moving to her parents' house, now that her husband Ezekiel Williams has died.

*Nov. 21 W.* Built a shuck pen, cut wood, etc. Cloudy, and looks like it might snow.

*Nov. 22 Th.* Got out some pines to haul, cut wood etc.

*Nov. 23 F.* A cold damp day. Got some split wood and bottomed Jodi's chair.

*Nov. 24 Sat.* A very cold, windy day. Got wood & pine for Sunday, warmed my fingers & toes, etc. This will do very well for winter weather.

*Nov. 25 S.* Clear and cold but not so windy as yesterday. We staid at home and no one came to see us.

*Nov. 26 M.* Got up this morning early before four o'clock even—and wrote several lines in my Diary. Commenced school in District No.—same place where I taught last winter. Had only 4 scholars.

*Nov. 27 T.* A rainy day. Had 7 scholars to-day.

*Nov. 28 W.* Still rainy. Six chaps out to-day.

*Nov. 29 Th.* Cleared off last night quite warm.

*Nov. 30 F.* Cool and windy. Taught school.

*December 1 Saturday.*

Cold and windy. Snowed a little several times during the day. I went over to Pa's early in the morning to go to mill but they said the mill was out of fix some way so there was no chance for grinding. I then went out to look for our sheep—the "old but[t] sheep" and lamb. I found them at the widow Bryant's near Shilo. I brought them home. In the afternoon Richard Mullies came to look at our land. We walked round it. I told him he could have it for three dollars an acre. He left for home saying that he would see me again.

*Dec. 2 S.* A tolerbly nice day. We staid at home, and read our books.

*Dec. 3 M.* Taught school. Had six scholars. Cold and cloudy. Commence snowing in the afternoon.

*Dec. 4 T.* No School to-day. Killed two of our hogs—the two we've had up for more than a year. They were tolerbly fat but not large. One of them weighed 108 and the other 114 pounds. Clark came after us to go and see Pa. We went and found him very sick. Old Dr. Parks staid with him to-night.

*Dec. 5. W.* Pa a little better this morning. I came home & fed, salted the meat, etc. and then went back to Pa's.

*Dec. 6 Th.* We came home.

*Dec. 7 F.* We went back to Pa's, and I went to the sale of Alexander's[13] property. Clark is Administrator. The hay & fodder went low. The corn brought 63 cents per bushel. The mare—a little black—which Alexander paid $100 for a few weeks before his death only brought $74.25. E. L. Roughton bid her off.

*Dec. 8 Sat.* And to-day was the sale at Mary's.[14] The hay went cheap, one stack of which she got for .25 cts. The sheep sold well. Frank Madison bid them off at one dollar and thirty five cents a head. Mary got the cow too for one dollar. Pa seemed worse again this afternoon, so I went down after Dr. Parks. He did not come but sent some medicine, and said he would be on in the morning. Pa has a very sore tongue, for which he is now trying an old woman's remedy, and he says it seems to have a good affect.

Here is the receipt.

Take equal quantities of sweet apple tree bark, persimmon tree bark, blackberry brier bark, and a branch of hyssop & sage. Boil down and add Alum & saltpetre—a lump of each the size of a sparrow egg. Thicken with honey, and wash the mouth as often as desired.

*Dec. 9 S.* I taught school but one day last week. Parks came on this morning, but did not leave Pa any more medicine.

*Dec. 10 M.* I came home early and went on to my school, leaving Mollie & Jody to come when they get ready. Cold, cloudy and damp.

*Dec. 11 T.* Clear and warmer than it has been for some time. Jody went

13. Alexander Johnson, Strong's late brother-in-law.

14. This sale follows the death of her husband, Ezekiel Williams, and her decision to move to her parents' house.

with me to school this morning, but I forgot the keys, so we had to come back home. I went over to Pa's. Think he is better.

*Dec. 12 W., 13 Th., 14 F.* Taught school.

*Dec. 15 Sat.* Snow fell last night to the depth of five or six inches. Got wood for Sunday, etc.

*Dec. 16 S.* Clear and warm over head. We staid at home *all* day.

*Dec. 17 M., 18 T.* Nice days over head. Snow melted some.

*Dec. 19* A dark rainy day. Snow melted rapidly. Creek got up.

*Dec. 20 Th.* Cleared off warm.

*Dec. 21 F.* A clear, warm day. Jody went with me to school to-day, but he would not spell any. He had to look about and see what was going on. He went, not to spell, but to see what was done at school. Sister Mary came over to-day and brought old Med home. Taught school all this week!

*Dec. 22 Sat.* A tolerbly nice day. We got Wm Coffin's wagon and went to market—Eagle Mills—with our glens cloth, a few beets, etc. The cloth we sold at fifty cents a yard. 17 × 50 = $8.50.
The beets we brought back. We got an oven, paid $2.15 for my blanket, etc. etc.

*Dec. 23 S.* A clear pleasant day. Mary went home, and Caleb came to bring old Med back home. We staid at home all day.

*Dec. 24 M.* A little cloudy like it was fixing for another snow. Mollie & Jody went to David Johnson's to get Jodie's first coat cut out. Eliza Johnson cut it, and charged ten cents.

*Dec. 25* Snowed a little in the latter part of last night. Cleared off today quite pleasant.

Got Joshua Sprinkle to help me kill our other two hogs to-day. One of them weighed 133, and the other 88 pounds. So we have this fall 444 pounds of pork, which is a good deal more than we ever had before at any one time. Caleb came over to-day to hunt rabbits. Only caught one. We weighed ourselves after dinner. Joshua Sprinkle weighed 162; myself 135; Mollie 117; Caleb 110; and Jody 40 pounds.

*Dec. 26 W.* Clear and cool. Taught school. Had twelve scholars. Mollie & Jody went over to Pa's.

*Dec. 27 Th.* Cool and a little cloudy. Had thirteen scholars to-day.

*Dec. 28 F.* About as yesterday only I think a little cooler. Twelve scholars to-day.

*Dec. 29 Sat.* A snowy day. Got wood, pine etc for Sunday. At night we cut up our sausage meat, a[]part of which Mollie made into 48 cakes; the rest she packed away in a large bowl.

*Dec. 30 S.* The ground is covered with snow and ice, and the trees are heavy loaded with sleet. It left off snowing some time in the night and sleeted. This was as rainy a day as we have often. Shadrack[15] came over to let us know that Ellin was very sick, and they thought she could not live. It was to[o] rainy and cold for us to go out that distance.

*Dec. 31 Monday.* Snowed last night. Started to school, but could not get over the creek. It has snowed and rained and snowed again till the waters are up. I came back and we went over to Pa's, and staid there the rest of the year. We stoped at old grand daddy Bell's as we went on. The old man is very low. Dont think he can ever get about any more. He was taken sick last week. Poor old man! He has been stout and has lived long in this world. He was drafted I think in the War of 1812 to go and fight for his country, but he hired a man to go in his place, so he staid at home. He has done a great deal of hard labor, but he will swing the axe and the sythe no more. He has plowed his last furrow, hoed his last hill of corn, and I trust he is now ready to be offered up. I guess from the many passages marked in his Bible that he read that book a great deal.

15. Shadrack is Mollie's brother, and he refers to the orphaned daughter of Strong's late sister, Martha T. Johnson.

# Diary 1861

*January 1* Clear and very cold early in the morning, but warmer about noon. Snow melted some. We staid at Pa's last night. Mollie & Eliza[1] set up till after midnight. We got home about noon.

*Jan. 2 W.* A cold rainy day. Taught school. Mollie and Jody went over to see old grand pap Bell. They had just got back when I came home from school at night.

*Jan. 3 Th.* Cleared off to-day quite warm. Creek over the foot log again this morning, so I had to cross on the big pine away up the creek. Had eleven scholars.

*Jan. 4 F.* Clear and pleasant. This was the day appointed by the president for prayer & fasting through out the U. S.[2] Old Dr. Parks came over to the school house to preach for us, but the people did not come out, so he left and went on to Pa's, but meeting was over before he got there.

Mollie went to see her grand pap, and to Pa's to meeting to-day. Caleb came home with her after a pig or 2 that I am to let him have.

*Jan. 5 Sat.* Clear and not very cool. I went to Boon Windsors this forenoon after one of our sheep that went off last summer. Boon was gone to Troy so I got Enos to go with me to the field in which the sheep were. We found them but could not get hold of mine. I came home and late in the afternoon I & Jody went over and Boon being at home we got our sheep. Jody rode old Med back and I drove the sheep.

*Jan. 6 S.* A very cold frosty morning. We went to old grand pap Bell's

1. Thomasson's sister, Nancy Eliza.

2. President Buchanan issued no official proclamation calling for a day of prayer and fasting. Perhaps Thomasson was referring to rumors that Buchanan might do so.

and to Pa's. We found the old man still alive, but I dont think he can last long. We found them all up at Pa's but Mary.

*Jan. 7 M.* Warm, cloudy and rained a little. Had nine scholars. We staid last night, and Mollie and Jody staid most of to-day, at grand daddy Bells. I left there this morning early; came on home and fed and went on to school. Paid Elison Johnson $1.75, the balance due him for the corn I bought last spring.

*Jan. 8 T.* Mostly clear, and warm.

*Jan. 9 W.* Still warm. Cloudy part of the day, and to night it is raining. Mollie went to see her grand pa.

*Jan. 10 Th.* Rained, snowed and hailed a small sprinkle now and then. The wind blew hard and cold. Had only 9 scholars. How strange it is that people will not send their children to school when its free!

*Jan. 11 F.* Cloudy and cold. Mollie went to see her grand pap. Had my same old number (9) of scholars. Worked on my slide at night.

*Jan. 12 Sat.* Mostly clear and quite warm for winter. Finished my slide, and hauled six loads of wood, and two loads of pine.

*Jan. 13 S.* Cloudy and real cold. We went to Pa's and to see old grand pap Bell.

*Jan. 14* Hailed and rained last night, and rained lots to-day. There is another heavy sleet now. Went to the school house, but as I had but one scholar I came back home and hung up our meat, made Jody a shoe at night, etc.

*Jan. 15 T.* Still cloudy and very damp, but not so cold as yesterday & day before. The sleet melted off. No school again to-day. I went to the creek but finding it over the foot log I came back, got old Med a[nd] rode over but the scholars were gone when I got there.

*Jan. 16 W.* Cleared off warm. Rained a great deal last night. Taught school to-day. Had eight scholars. Mollie went to see her grand pap, and to Pa's also. Sister Eliza came home with her.

*Jan. 17 Th.* A little cloudy, like it was fixing for more falling weather, and quite warm. Eliza went home this afternoon. I had nine scholars to-day. The creek was up yesterday, but foardable again to-day.

*Jan. 18 F.* A cool, rainy day. Had only eight scholars today.

*Jan. 19 Sat.* Mostly clear and quite warm. Went over to Pa's in the afternoon and brought our sheep, (the old but[t] and her lamb) home again. Received the Jan. No. of the *N. C. Planter.* Sold Newton Barnes two shoats for seventy five cents each, which settled my account with him. So we are now even.

*Jan. 20 S.* Mostly clear, and a good deal colder than yesterday was. Pa came over to see us.

*Jan. 21 M.* Cloudy and cold. Had thirteen scholars to-day. Bought a basket of P. Campbell's negro boy, which he says holds a bushel, for 35 cents. Hung up the balance of our meat to-night. Mollie went to Wm Louis' shop to-day and got old Med shod before.

*Jan. 22 T.* A little cloudy and quite cold. Had only ten scholars to-day. Mollie went to her grand paps, but was at home again before I got home from school. The old man seems to be somewhat on the mend. Wants to eat steady.

*Jan. 23 W.* A very snowy day. At night the snow was shoe top deep. I took about two bushels out of our loft. We have lots of snow this winter. Hail and rain also plenty. Had six scholars to-day only.

*Jan. 24 Th.* Cloudy, and warmer than yesterday. Rained and sleeted last night. Think this is rather the heaviest sleet we have had this winter. Went to the school house, but had no scholars, I came home and hauled wood on the slide.

*Jan. 25 F.* Cloudy and not very cold over head. Snow melted some. Only had four scholars today. Took a sleigh ride this evening over to Pa's.

*Jan. 26 Sat.* Colder than yesterday. Rained, hailed and snowed. Its well for us that we have but few such days as this has been. We came home in the afternoon.

*Jan. 27 S.* A clear day, and tolerably warm overhead. We staid at home all day.

*Jan. 28 M.* Clear and warm after the morning. Had 13 scholars, Mollie & Jody went to Romes.

*Jan. 29 T.* Clear most of the day, and quite warm in the afternoon. Had fourteen scholars.

*Jan. 30 W.* Clear and warm, Wind from the southwest. Had 19 scholars to-day.

*Jan. 31 Th. and last day of January.* Clear most of the day, and tolerbly cold. Had 17 scholars to-day. Jody can spell in words of three and four letters.

*February 1 F.* A rainy day, but not very cold. Had 17 scholars.

*Feb. 2 Sat.* Rained a great deal last night and to-day. I think this may very well be rembered as the wet winter. It is not *very* cold, but very wet. Mended my shoes, etc.

*Feb. 3 S.* The sun shone out to-day about 12 for a few minutes, but it was soon cloudy again and commenced raining just before night. Wind from the North East. Mrs. Nelly Johnson came to see us to day for the first time. We loaned her *Daw's Works,* to read, with which she seemed to be much pleased. The Miss Coffins, Ruth and Liddia, paid us a visit this afternoon too. We loaned them *Moore's Works,* some Magazines and the *Young Bride's Book.*

*Feb. 4 M.* Cleared off warm. Had thirteen scholars. Mollie & Jody went to Pa's, as they came home they got a lot of little cedars to plant out.

*Feb. 5 T.* A. C. Johnson came here to night to get me to write a Deed. He has sold his land to E. A. Windsor. I wrote the deed.

*Feb. 6 W.* Taught school.

*Feb. 7 Th.* A very windy day, but not very cold.

*Feb. 8 F.* Stanbery Hicks staid with us last night. We went to the sale of Alex. & Ezekiel's property at Pa's, but did not buy much.

*Feb. 9 Sat.* Went out in the forenoon to hunt our hogs but did not find them. Mostly clear and warm. Yesterday was a little cloudy and cold. We went this afternoon to Mollie's Pa's. We had not been there since Nov.

*Feb. 10 S.* A rainy day. Came home in the afternoon. Mealie[3] came home with us to stay a week or 2.

*Feb. 11 M.* Still rainy. Had but 6 scholars.

*Feb. 12 T.* Clear and warm. Had 13 scholars.

*Feb. 13 W.* A little cloudy and warm. Had 8 scholars.

*Feb. 14 Th.* Rained very hard this afternoon. I got wet going home from school.

*Feb. 15 F.* My school would have closed to-day if I had lost no time. The Com. was out and signed my papers.

*Feb. 16 Sat.* Rained again to-day. Went to Troy & Eagle Mills. Got at the latter place half gallon molasses and 1 1/2 oz. Rubarb at .15 cts per oz. Handed my school papers over to Wm Colvert and took a receipt for them. W. C. is going to court next week and he is to bring my money. When I got home I found Caleb there with his dogs and gun.

*Feb. 17 S.* Cloudy and cold. One of our lambs was sick to-day. We gave it eggs, fat, salt, sweet and castor oil.

*Feb. 18 M.* Our lamb got well. Cold weather.

*Feb. 19 T.* Snowed very fast this afternoon for an hour or two.

*Feb. 20 W.* Mollie got her cloth (apron piece) out today.

*Feb. 21 Th.* Windy and not so cold.

*Feb. 22 F.* School closed. R. I. York got the prize book.

*Feb. 23 Sat.* A rainy day. Went to Eagle Mills and got my money ($45.) forty five dollars, Paid off my book acct. and ten ($10) dollars on my note. Bought a paper of composition for 12 1/2 cts.; 25 cts. worth of sugar and 45 cts worth of cotton yarn. 12 1/2 + 25 + 45 = .82 1/2 cts

*Feb. 24 S.* Cold and windy. We went over to Pa's and staid all night.

*Feb. 25 M.* Clear and warm after the morning. Came home.

3. This name might be a nickname for Mollie's infant sister, Emma.

*Feb. 26 T.* Hauled out manure on our garden in the forenoon & went to Eagle Mills and Troy in the afternoon. Ma came over and got me to go for Mary. I got for her one gal. molasses, 50 cts (4 1/2 lb) worth of sugar and 40 cts (2 lbs) worth of coffee. For the Articles I got for us see pass book.

*Feb. 27 W.* Clear and quite warm. Plowed up our garden in the forenoon, and commenced breaking up my corn land in the afternoon.

*Feb. 28 Th.* Still clear and warmer than yesterday. Plowed some in the forenoon and went to the Elect. in the p. m. and voted "no convention."[4] I paid Enoch Coffin (.50 cts) for the brick (200) I got of him last August.

*March 1* Clear and warm. Planted our onions, went over to Pa's at night and had a corn shelling. Shade and Bob Bell came over from Buck Shoal and helped us. We shelled out 7 1/2 bushels.

*March 2 Sat.* Another very nice day. We have had a week of as nice weather as I ever saw I think at this season of the year.

*March 3 S.* Not well. Pleasant weather. Mollie's "mom," as her children all call her, came over to-day.

*March 4 M.* Went to Campbells Mill, for the first time, this morning but the old mill was "out of order" so I got no meal. Commenced sowing oats. Turned cooler this evening.

*March 5 T.* A very cold, windy day. Sowed oats.

*March 6 W.* Went after our meal, sowed oats, etc. Not so cold to-day.

*March 7 Th.* Warm. Sowed oats.

*March 8 F.* Pleasant. Fixed up the lot fence, finished the garden fence, etc.

*March 9 Sat.* Rained last night. Got up a lot of good pine in the forenoon, and planted our potatoes in the afternoon.

4. North Carolinians voted on the question of whether or not to convene a convention that could consider secession. At the same election they chose delegates who would sit in the convention, if it took place. The voters rejected the convention, but had it been held, Unionist delegates would have been in a large majority.

*March 10 S.* Clear, windy and cold. Real March weather. We went over to Wm Coffins in the afternoon.

*March 11 M.* Clear and warmer than yesterday was. Sowed oats, planted apple trees, etc.

*March 12 T.* A little cloudy in the forenoon and quite cool. Clear in the p. m. and warmer. Sowed oats. Lill had her third calf this morning. It's a fine big bull. Jody is now begining to read. He is in his fifth year.

*March 13 W.* Warm and a little cloudy. Sowed oats. Three pecks a day is about all I can sow and plow in. I sow them thin, plow close and old Med walks slow.

*March 14 Th.* Plowed in oats till about 10 o'clock when it commenced raining.

*March 15 F.* A cold, windy day. Snowed considerably in the forenoon. Commenced grubbing our turnip patch. Eliza and Vic.[5] came over to see us in the afternoon and staid "till next day."

*March 16 Sat.* Somewhat cloudy and cool. Went to the shop, and got my plow lode; to Eagle Mills, and got 5 cts worth of ginger, and the P. O. R. on Agriculture for 1855; to Troy, and got a No. one pocket knife; Price $1.00 but I only paid .85 cents for it. I also got a bottle of "Ayer's Cherry Pectoral" for which I paid $1.00. I went to Charles Benbows and took dinner with them. Charles gave me a small bottle of his bitters.

*March 17 S.* Cool, cloudy and a little rainy. Caleb rode over on "Blaze," Clark's filly. Jo. Bell also came after Mealie[6] and brought Mollie's dog—a little black puppy. Quite rainy in the afternoon.

*March 18 M.* A very snowy day.

*March 19 T.* Clear and tolerbly warm. If the snow had not melted nearly as fast as it fell yesterday I think it would have been six or seven inches deep this mornng. It snowed all day and very fast most of the time, but the ground was barely covered this morning for a few hours. It cleared

5. Thomasson probably means his sister, Nancy Eliza, and Victory Johnson, the orphaned daughter of his sister Martha and Alexander Johnson.

6. The 1860 census showed Joseph M. Bell as the fourteen-year-old son of Mollie's parents, and "Mealie" (if this was the nickname for Emma) was three.

off last night and frosted, but I see the peaches are not all killed. But few are in full bloom. Those that are are killed, the others are not. We went to Pa's in the afternoon and I went to mill.

*March 20 W.* Cloudy and cool or cold I might say. We came home in the forenoon, In the afternoon Mollie went to Coffins and I and Jody cut down pines to dry for fire wood. Andy Haith and Martha Johnson were married this afternoon, I guess.

*March 21 Th.* A cold, windy day, with now and then a light shower of snow. Cold weather for March.

*March 22 F.* Clear and pleasant. Sowed oats and plowed them in. Elison Johnson had a choping to-day. I was invited but not being able to use the ax all day I did no[t] go.

*March 23 Sat.* Cloudy and rained a little. Finished sowing oats this afternoon. I and Jody went squirrel hunting. We were out about an hour, and caught a little gray squirrel, and a great big old "possum." What luck we had! All in hearing of home too!! This the first "possum" that ever Simon, our little yellow fice dog, treed that I know of.

*March 24 S.* Clear and windy, but not cold. I think most of the peaches are killed. They are not yet in full bloom.

*March 25 M.* Clear and warm. Plowed. Our old cow, Lill, sick, or lost her cud, or something else. We made her two cuds and and Mollie poked them down her throat with a stick while I held open her mouth. The first was made of Elder bark and salt, the second one of Elder bark, Tansy & salt. They had the desired affect. About mid night I went up to see how she was and found her chewing away on them.

*March 26 T.* Cloudy, warm and windy. Plowed. Planted the balance of our potatoes. We did not measure them very nicely, but I guess we've planted at least two bushels. The old cow seems better today.

*March 27 W., 28 Th.* Plowed.

*March 29 "Good Friday."* This has been a great week for plowing. The ground tho' is getting a little hard now. Hope we'll have rain about Easter, but not such a flood as I dreamed of the other night, and also one night a few weeks ago. I dreamed that the whole earth, as far as I could see, was covered with very muddy water, about a foot deep and running very swiftly. I went today in the forenoon to Nicholson's mill. Took a bushel of corn, had it ground for feed, and bought forty six pounds of flour at 3 cts. 46 × 3 = 1.38,—one dollar and thirty 8 cts. Paid one dollar. The 38 cts I am to pay the next time I go there to mill. Plowed in the afternoon.

*March 30 Sat.* Had a fine, warm rain last night. Clear, warm & windy. Plowed some in the forenoon. In the afternoon we (I, Caleb & Jody) went to B. Pitt's and got half a bushel of sweet potatoes for seed, for which I paid 38 cents.

*March 31 Easter Sunday.* Clear and cooler than yesterday. We had, for dinner today, a Butter Pudding, made of half pound of flour, 4 eggs, one pint of milk, salt and soda. It was good.

These are gloomy times, and seem to be growing darker and darker every day. Every mail brings us some bad news,—news of disunion and war. Six of the slave states have now gone out of the union, and events seem to be ripening very fast to take out the remaining nine. And is this union to be utterly and forever destroyed? This glorious union—the price of the blood of our fathers—is it to be abandoned as a thing of no worth? No. "Never give up the ship." *Never! Never!*

For the week ending April 6.
Monday, the first, was cool & cloudy. Bed[d]ed out our sweet potatoes, plowed most of this week. One Friday evening the 5 we went over to Pa's, and Saturday morning I went to mill. Cool & rainy Sat. 6.

For the week ending April 13.
Sunday was cool and rainy. Caleb came after us to go and see sister Mary. Her second child, a little boy, was born last night. We went over and found her very bad, but she got better awhile after night. Old Dr. Parks was with her. Monday was a very cold, rainy day. We staid at Mary's. Tuesday we came home. I swaped my lot at the Zion Institute to W. B. Madison for one of Colt's Repeaters. Cut out and closed my shoes this evening. On Wedn. I made my shoes. On Th. we built a hen house in the A. M. In the p. m. I cut briers & bushes in the meadow,

fence corners etc, & Mollie went over to see Mary & her boy. On Friday I mended my old shoes, and commenced a pair for Jody. On Sat. I went to Eagle Mills and to Charles Benbow's. This was a windy day. Charles & Jane came home with me.

*April 14 S.* Clear, windy & cool. I loaned Charles the *Scrap Book.*

*April 15 M.* Cloudy and a little warmer. Commenced laying off corn rows. Rob[b]ed our bees a little for the first time.

*April 16 T.* A rainy day. It commenced raining yesterday evening. I half soled Mollie's shoes, etc. etc. etc.

*April 17 W.* Cleared of off windy & cold. Hauled wood and rails, etc. I fear the fruit will all be killed to night.

*April 18 Th.* A few degrees warmer. Commenced planting corn about four o'clock this afternoon. Planted eight rows and there came a shower of rain.

*April 19 F.* Cold and windy. We planted corn half the day—the P.M.—tho' it was so cold.

*April 20 Sat.* Not so cold as yesterday, but cool still. We planted corn. Pa came over after the old wagon. He told us the war had commenced in Charleston some two weeks ago. *Where* and *when* will it end? And what will be the result? Freedom to the slaves? I think so. The South has commenced the war, and now I do not believe the North will ever listen to any terms of peace which do not include "Freedom to the slaves."

*April 21 S.* A very nice, warm day. Yesterday morning we had a big frost. Think most of the peaches are killed. This is Mollies birthday she being 24 years old today. None came to see us, and we went to see no one, but staid at home all day.

*April 22* Clear and warm, but windy. Planted corn.

*April 23 T.* Quite warm, Finished planting the field south of our cabin this morning. I turned old Med in the meadow yesterday evening to pick grass while I was at supper and she undertook to cross the branch on an old rotten bridge and fell thro'. I and Mollie took hold of her feet and turned her over so that she could get up and get away. It was a dangerous "fall in" for the old mare in her present condition, but I dont think she is seriously injured.

*April 24* Warm and windy, and in the afternoon we had a hastey, heavy shower of rain, after which it cleared off somewhat cooler. Ridged up ground, planted corn.

*April 25* Clear and warm. Ridged up corn land.

*April 26 Friday.* Finished planting corn this forenoon. Hauled manure on the garden this afternoon.

*April 27* Rained a little today. I plowed the garden, cut a little ditch, cleaned out the spring and knocked about generaly. We set out 80 cabbage plants this evening.

*April 28* We went to Zion to meeting and heard one Rev. Dodson preach. The day was cool & windy. All the talk these days, or a great deal of it at least, is about war. The preacher today said "We are involved in war." War is a dreadful thing, and all men ought to love and do all they could to keep peace. O that nations would learn to war no more.

*April 29* Clear and warm. We went over to Pa's this afternoon. It seems that our nation is going to war with itself! The South against the North! What folly!

*April 30* A very windy day. I went from Pa's to mill early this morning, and we came home after breakfast.

*May 1 W.* Clear and cool.

*May 2 Th.* Cool. We went over to Pa's again this evening. Clark was down and he & Caleb & Clark's hireling had all be[e]n fishing in the branch and caught 252 little fish!

*May 3 F.* I went over to the mill this morning and bought 100 pounds of flour for which I paid $3 1/4. We came home after dinner. I've been quite unwell this week. Not able to do a days work.

*May 4 Sat.* Dry and windy. We have had a great deal of windy weather this spring. Cut sprouts. This was muster day at Williamsburg, but I did not go.

*May 5 S.* Cool and rainy. We staid at home all day. Caleb came over in the afternoon.

*May 6 M.* Still cool and rainy. David Dobbins made a call on me this evening for money, but he did not get any. He said he was a volunteer and must have $100! What for? Does he have to bear his own expenses?

*May 7 T.* Cut a few sprouts in the corn field, and commenced replanting corn.

*May 8 W.* Windy, windy. We have of late windy weather, and direful times. Pa came over this morning & *trimed* my shotes.

*May 9 Th.* Replanted corn.

*May 10 F.* A little rainy in the forenoon.

*May 11 Sat.* Still and warm. Plowed our potatoes. This is general muster day at Williamsburg & Yadkinville. I did not go. We heard the drums at Wmsburg which is 4 miles off I think.

*May 12 S.* Clear and warm. This a beautiful morning.

For the week ending Sat. May 18.
On Monday (the 13) of this week we sheared our sheep at home, and on Tuesday, (the 14) [sheared] them over at Pa's. Sister Mary came home with us and staid till Sunday.
W. Th, & F. I plowed. Sat. took old Med to the shop and got her shod.

*May 19 S.* Cloudy and rained a little. We took Mary home.

*May 20 M., & 21 T.* Cloudy & cool. Plowed.

*May 22 W., & 23 Th.* Clear & cool. Replanted corn, finishing Thursday evening.

*May 24 F.* Warm after the morning, which was quite cool. We went to Eagle Mills & to Charles Benbows. We took 21 1/4 pounds of bacon to E. M. for which we go[t] 12 1/2 cts per lb. 2.65. We took in exchange for our bacon

| | |
|---|---|
| one gallon molasses | .50 |
| eight pounds of sugar 12 1/4 | 1.00 |
| one lb. soda | .10 |

| | |
|---|---|
| Mollie a pair of Sunday shoes | 1.35 |
| and a box caps | 5 |
| | $3.00 |

We went over to Charles'[,] got dinner, staid awhile & then came home.

*May 25 Sat.* Went to muster, but was not well so Cap. Clegg let me off. They called for volunteers.

*May 26 S.* We went to Zion to meeting and heard the Rev. S. Swain preach.

*May 27 M., 28 T., 29 W., 30 Th.* Warm & dry. Cultivated my corn over. This is my first trial with the cultivator. I like it well.

*May 31 F.* Sprouted oats, etc. etc.

*June 1 Sat.* Cloudy.

*June 2 S.* Went to the X roads and heard the Rev. Dr. Parks preach. Text, "Aquaint now thyself with him and be at peace," etc. to the end of the verse. Job. XXII, 21.

*June 3 M.* Went to Eagle Mills and sold my repeater to Wm Colvert for $25. I took up my note $12.60 and the balance in goods.

*June 4 T.* Finished sprouting oats. We had rain Sunday evening & Monday evening. Set out some—about 150 potatoe slips, also a great many beet plants.

*June 5 W.* Planted a portion of our potatoe patch in corn. Went to Madison's store, settled with them.

*June 6 Th., 7 F., 8 Sat.* We went to Mr. Beall's Sat. 8. We had not been there since the 9 Feb.

*June 9 S.* Rev T. A. Nicholson preached Alexander Johnson's funeral at Ashley Johnsons to-day. We were there. The sermon was good. Text, "God is love."

*June 15 Saturday.* This has been a warm, dry week. Worked in my corn. Pa & Ma staid with us Thursday night the 13 for the first time in a great while. Sunday 16. We went to Zion and heard the Rev. Jackey Night try to preach. He took a big text, viz. Genesis 10, XLIX.

*June 22 Saturday.* Another week of hot, dry weather. Hoed corn. Corn has not suffered much yet, but oat & garden stuff have. Oats will be "short" this year.

*June 23 [Sunday]*[7] We went to Zion & heard the Rev. S. Swaim preach. In the afternoon we have a severe little storm of wind, hail & rain.

*June 24 T.*[8] Hauled rails and made fence in the forenoon. In the afternoon it rained a nice shower.

*June 25 W.* Finished cutting wheat. Rained again this afternoon.

*June 26 W.* Warm weather. Finished the cross fence between the oats & clover etc. Mollie, Jody & Ruth Coffin went to John Jennings to get June apples.

*June 27 Th., 28 F., 29 Sat., 30 S.* This week was warm up to Friday evening when it turned cool. It rained some this week too. We staid at Pa's Friday night. Came home Sat. morning, and in the afternoon we went squirrel hunting but did not find any. Sunday 30.[9] We staid at home.

*July 1 Monday.* Commenced cutting oats. Rained in the afternoon.

*July 2 Tu.* Clear cool. Cultivated corn.

*July 3 W.* Do.

*July 4 Th.* Cut & cultivated corn.

*July 5 F. & Sat. 6.* Cloudy & rainy.

*July 7 S.* We went with Coffins to their "monthly meeting." They had no preacher. Alex. Benbow read a letter from Daniel Martin of Iowa, formaly of N. C.

*July 8 M.* Cut oats. Rained about 2 this afternoon and caught some of my oats down. We went to Pa's after the rain and I went to mill. James S. Grant & Charney Cole both died on Sunday the 7 inst.

7. Thomasson wrote "Saturday" by mistake.

8. This day would, of course, have been Monday, and June 25 and 26 are both listed as Wednesday. Evidently Strong Thomasson was filling in his diary, from memory, some days after the events.

9. Here he gets back to the correct date.

*July 9 T.* Rained again this afternoon. Cut a few oats & had pretty tight work to get them up before the shower. I am building Mollie a da[i]ry at odd times this week.

*July 10 W.* Cut a few more oats. Had more rain. The way the corn cabbages etc grow now is nice. Saw the first corn silk for this year to-day. Its in our corn on the branch near the old bridge.

*July 11 Th.* A little cooler. Finished cutting oats. We can see a comet these nights when the clouds are out of the way. I saw it the first time Monday night the first inst. at about two o'clock. It then appeared to be about an hour high, and directly in the North. It now appears to move from south east to North west, falling back a little every 24 hours.

*July 12 F.* Cool. Went to Eagle Mills and to Troy. Covered the dary in the afternoon. The comet is still visable.

*July 13 Sat.* Clear & cool. Hilled up our sweet potatoes. Mollie moved into her dary. A fine roomy one it is too.

*July 14. S.* Clear and quite cool for July. Caleb brought old Med & Jack home. We staid at home.

*July 15 M.* Warmer. I & Jody took a load of wheat to S. Myars and got it thrashed. Pa and brother Wiley came over in the afternoon and staid an hour or two. Wiley came up from Davie yesterday.

*July 16 F.* A rainy day.

*July 17 W.* Went to the tax paying and paid my tax $2.20 for 1860. Got home about 2 p. m. and went to mill.

*July 18 Th.* Six years ago to-day I & Mollie were married. Got up my oats this forenoon, and in the p m. it rained & rained.

*July 19 F., 20 Sat.* Hoed corn.

*July 21 S.* We staid at home till late in the evening, when Caleb came over and got us off home with him.

*July 22 M.* At Pa's. I went to John Johnsons & to Mr. Bells money hunting but did not get a dime. It rained a great deal this p. m.

*July 23 T.* Brought Mollie & Jody home and Eliza to stay with them while I help Caleb get up the hay. We hauled two big loads this p. m.

*July 24 W.* Today we hauled up 3 big 2[-horse] loads. It was all in the rain, and about a load not so much as cooked.

*July 25 Th.* Great time to make hay. Caleb came over and helped mow my lower meadow.

*July 26 F.* Mowed some, put up one stack of hay, etc. Caleb went home or started home in the p. m. I guess he got there.

*July 27 Sat.* Took a load of wheat to Wm Denny's and got it thrashed. I think it is at least half smut. Got up hay in the p. m.

*July 28 S.* Rained a light shower last p. m. Jane Benbow came to see us last evening and staid till this p. m. R. Albea & wife came this evening.

*July 29 M.* Warm and cloudy. Finished mowing.

*July 30 T.* Cured my hay. Rained a fine shower late in the p. m.

*July 31 W.* Got up the last of my hay.

*August 1 Th.* Had old Med shod before with new shoes. It rained lots this afternoon. Rome [Albea] came after Cathran. He left her here Monday morning and she staid with us till this afternoon. A four days visit.

*August 2 F.* Commenced clearing off our turnip patch.

*August 3 Sat.* Split a few rails and burned off the turnip patch. Very warm weather for such work.

*August 4 S.* We staid at home all day by ourselves.

*August 5 M.* Fenced and plowed up the turnip patch.

*August 6 T.* Sowed the turnip seed this A. M. Plowed them in & smoothed off the ground by draging a brush over it.

*August 7 W.* Cloudy in the forenoon. In the afternoon I & Jody went to Nicholsons Mill with wheat and corn. Mollie went to Joshua Haithcocks and staid till we came back.

*August 8 Th.* Somewhat cloudy. At night we went to Eagle Mills and heard one of my old school mates, the Rev. A. Weaver, preach a very fine sermon. His theme was prayer, from Luke XI, 9, 10, 11, 12, 13. We staid all night at C. Benbows.

*August 9 F., 10 Sat.* Caleb came over last night to let us know that Clark and wife are both at Pa's sick. They have got the tyfoid fever.

*August 11 S.* Staid at Pa's last night & till late this evening.

*August 12 M.* Not well.

*August 13 F.* Had a fine rain. Went over to Pa's in the p. m.

*August 14 W.* Rained some this A. M. I and Caleb made cider. We came home after dinner.

*August 15 Th.* Mollie sick with a cold. I & Jody not very well.

*August 16 F.* Cook & Felts thrashed Coffins wheat this forenoon, and I stacked the straw.

*August 17 Sat.* A rainy day. Made Mollie one shoe.

*August 18 S.* A showery day. We staid at home.

*August 19 M.* Rained a great deal last night. Made Mollie's other shoe this A. M. Think without an accident the corn crops are safe. I've heard some talk of the army worm in the neighborhood, but I hope they will do no great damage. We went over to Pa's the p. m. to see the sick.

*August 20 F.* I & Caleb made a run of cider in the forenoon. We came home in the p. m. Hot weather.

*August 21 W.* Cloudy. Commenced shrubbing off a piece of ground near my stable which I aim to sow with wheat.

*August 22 Th.* Shrubed in the A. M. and helped Mollie wash in the p. m.

*August 23 F.* Sick.

*August 24 Sat.* Not much better. We went to Pa's this evening. Clark is very weak but some better he thinks. Nancy is on the mend.

*August 25 S.* Clear & pleasant. Came home in the afternoon. Clark rested tolerbly well last night. Hope he'll soon be well.

*August 26 M.* Dug potatoes.

*August 27 T.* Cloudy and rainy.

*August 28 W.* Still rainy. Read this morning a few pages in *Valley of the Amazon,* Part 2d. This is a big world—much larger than the old Dutchman thought it to be. South America itself is a large place, and in some respects quite a pleasant country, but in others not so much so. The same, I suppose, is true of every country on the globe.

*August 29 Th.* Rained a fine shower this afternoon. We staid at Pa's last night, and this morning I went over to Mr. Bells. Noel[10] has the fever. Brother Clark is not much, if any, better. We came home in the evening after the rain. We have had a great deal of rain of late.

*August 30 F.* Not well.

*August 31 Sat.* Mollie broke herself down washing today, so we did not go to Pa's.

*September the first. Sunday.* Clear and warm. We went to Pa's this forenoon. Found Clark very weak and with some fever yet. Nancy is on the mend. Staid at Pa's all day.

*Sept. 2 Monday.* Cloudy this morning. Clark seems worse. We came home.

*Sept. 3 T.* Clear and warm. Finished getting in our crop of potatoes. We have five bushels only, which is but a small yeald for two bushels of seed in good ground. We consumed some thro' the summer but not a great many. It was too dry in May & June for potatoes. I & Jody had a good supper this evening of apple pie, molasses, butter & milk. We ate harty, but Mollie, poor gall, not being well these days,[11] refused to partake with us and was off to the field after a mess of green corn, for which she and old Med have an extraordinary appetite. Well, old Med is a faithful old nag. She has served us long and well and now that she is old, poor, & teeth bad it would be hard not to let her have a bite of soft corn now and then. As to Mollie she *will* have her share. Well, the best way to do,

10. The 1860 census listed this sixteen-year-old as "Noah."

11. Mollie is pregnant with her second child.

is to take the world fair and easy, and let the females have their own way in some things at least. I might get a kick or a bite from old Med, tho' I dont think she was ever known to do such tricks, and a "curtain lecture" from Mollie if I were to refuse them a little corn. Its half past 9 p.m. so I must be off to bed. Good night.

*Sept. 4 W.* Tried to plow a little but my plows were to dull.

*Sept. 5 Th.* Warm weather. Made some rails.

*Sept. 6 F.* Got my plows from the shop and plowed some.

*Sept. 7 Sat.* This afternoon we went to Pa's. Found brother Clark very bad. He had been getting worse all the week.

*Sept. 8 S., 9 M., 10 T., 11 W., 12 Th.* We staid at Pa's these days. I only coming home twice, & Caleb twice, to milke our cow and feed the hogs. On Thursday morning the 12th of Sept. brother Clark departed this life. He was sick eight weeks, five of which he lay at Pa's. We believe that Clark has gone to heaven where we hope to meet him. We came home Thursday p. m. I lost so much sleep this week I was sick 3 or 4 days, and did not get to see Brother buried. I kept my bed Friday & Saturday most of the time. A couple of cold baths, rest and sleep had the desired affect, so I was able to commence pulling fodder Monday morning 16th.

*Sept. 21 Saturday.* This has been a great week for saving fodder,—warm & dry. I finished pulling Friday 20th yesterday. A good deal of my fodder dried up. I pull to the ear and cut tops. It rains moderately this p. m. We dressed to go to Mr. Bells, but the rain prevented us.

*Sept. 22 Sunday.* We went to Mr. Bells. Hariet[12] has the fever.

*Sept. 23 M.* We came home. Cut tips Tuesday & Wednesday. Thurday we were at Pa's. I helped them kill a beef. We got 24 pounds—half a 1/4 Thursday evening and Friday it rained a great deal. Jody has something the matter with his right leg. On Sat. morning the 28 went after Pa, but did not find him at home. Jody's leg is very bad. He has high fever, and his leg is very sore and swelled tight. Pa came about 12.

*Sept. 29 S.* Pa & I went to Mt. Mariah to hear Tommy N. preach, but

12. Hariet is probably the Bells' infant daughter.

he was not there; so Pa preached. Pa & Ma came to see us last evening and staid till this p. m.

*October 5 Sat.* This week I made another stack of good hay, got up my top fodder, took our black wool to the machine—Eagle Mills—hauled some rails, etc. etc. On Wednesday of this week it rained. To-day, Sat. Oct. 5 we had our first mess of turnips for the season. I sowed the seed August 6. See page 52.[13] We have a patch of nice turnips.

*October 6 Sunday.* Cloudy forenoon and quite warm for Oct. We stay at home all day and read our books. I read the concluding chapters of "Acts of the Apostles" and commenced reading the "Epistle to the Romans." I also read Josephus' account of Solomon's end, and of the reigns of Rehoboam & Jeroboam. Solomon was a great and good man, superior to all other kings in happiness, riches and wisdom. He had 1000 wives not one of which it seems was good woman, (see Eccl. VII, 28) hence he, in his late age, was deluded by them, and led off from the worship of God, as taught in the Law of Moses, to the worship of idols. Women have great influence over men. Adam said "The woman whom thou gavest to be with me, she gave me of the tree, and I did eat." See Gen III, 12. O that they would always exert that influence for good, and exercise themselves, "to have always a conscience void of offence toward God, and toward men."

This is Sunday, Nov. 10, and I have not written any in my Diary since Oct. 6. I sold my land to Alvarian Johnson for $180, but he backed out, so we are here yet, and Mollie says she is glad of it. Well, its all for the best.

On Sat. & Sun. [October] the 19 & 20 was the quarterly meeting at Mt. Mariah. We went 2 days and one night. We heard the P[residing] E[lder] Rev. W. H. Bobbitt preach two choice sermons. His text on Sat. was, the 6, 7, 8, & 9 verses of the XIII of Luke. On Sunday the latter part of the 9 verse of the XII of Hebrews, "Shall we not much rather" etc.

*[October] 21 Monday.* This week I got up my corn. On Sunday 27 Wesly Roby was to have preached brother Clarks funeral but, being ingaged with the Campellites—Hugate and others—in a scriptural discussion,

13. Thomasson refers to the pages of his handwritten diary.

he disapointed us. The Rev. N. Gwin, a fine young preacher, attended in Roby's place. He preached a good sermon from "What shall I render unto the Lord for all his benefits towards me" Psalms, 116, 12.

*[October] 28 Monday.* This week I went to plowing again. Mr. Bell—Mollies pa—came over and made us two door shutters. So we can "shut the door" now.

*Nov 4 Monday.* This week I hauled out manure, commenced sowing wheat on the 7, finished shucking out our corn etc. etc. This has been a very warm fall so far. We have had but little frost yet, but it rains and rains. Our turnips are fine—by far the best we have ever raised. Our cabbages and beets too are good, sweet potatoes only moderate. They were planted late. On Thurday 7th I commenced sowing wheat. We have fine warm weather, with a good rain or two every week.

*Nov. 10 Sunday.* A nice day. We staid at home most of the day only walking over to Wm Coffins a few minutes in the afternoon. I borrowed the *History of the Martyrs.* Its an old Book, but a good one. It was published the year I was born.

*Nov. 11 M.* Cloudy in the A. M. and warm. Hauled rails, made a lot for my hogs and put them up.

*Nov. 12* Mostly clear and quite warm. Hauled rails, made fence etc. Who ever saw as late a fall as this? We have not had a thoroughly killing frost yet.

*Nov. 13 W.* Clear and warm. Hauled rails till 12. While we were at dinner Mr. Bell steped in. He came to help me fix up our old houses, but as I was not ready to work on them, he helped me fence, move move my old stable, etc.

*[November] the 14 & 15*—Th. & F. were warm days.
Thursday night it rained.

Saturday [November] 16 was a cold windy day, being the first cold weather we've had this fall it pinches close.

*November 17 Sunday.* Calm and pleasant. We had this morning a deadening frost. Looked and felt a little like winter.

*Monday [November] 18* Sowed wheat.

*Tuesday [November] 19* Do.

*Wednesday [November] 20* Rained in the forenoon. In the afternoon I hauled out manure.

*Thursday [November] 21* I went over to Pa's and swaped horses with him for a few days.

*Friday [November] 22* Took Bet to the shop and got her shod. It rained in the p. m.

*Saturday [November] 23* I and Alexander Benbow went to Statesville and back. We left home between 4 & 5 this morning and got back about 11 in the night. We had a cold, hard drive. We both obtained cirtificates to teach "Common School."

*Sunday [November] 24* Cloudy and cold. Rev. Wesley Roby preached brother Clark's funeral at Pa's today. His text was the 26 of the 73 Psalm. "My flesh and my heart faileth;" etc. The sermon was a choice one.

*Monday [November] 25* A cold windy day. Cut and hauled a load of wood, pulled up our turnips and hauled them to the house, etc. Shade & Jane & Jane Benbow came to see us today.

*T [November] 26* Not quite so cold. Put away the turnips, the potatoes, etc. in the A. M. and in the p. m. I hauled out manure on the wheat.

*W. [November] 27* Cloudy & cool with a little rain this afternoon. I & Jody staid at Pa's last night & Mollie staid at Charles'. I brought home a load of plank, laths, boards etc from my old shop.

*Th. [November] 28* Warm and rained some in the afternoon. I finished sowing wheat today. I sowed 2 1/4 bushels only.

*F. [November] 29* Warm and cloudy. Made a stable door shutter, hauled a load of leaves into the cow lot, etc.

*Sat. [November] 30* Cold and windy. Last night there was a real storm of wind and rain. It rained a great deal and the wind blew very hard.

*December 1 Sunday.* A cold day. I went to Mt. Mariah to meeting. Heard the Rev. T. A. Nicholson preach the widow Hethcox funeral. His text was "But man dieth, and wasteth away; yea, man giveth up the ghost, and where is he?" Job XIV, 10.

*Dec. 2 M.* Cloudy & rainy in the p. m. Commenced School in Dis. No. 4. Had 7 scholars—little boys.

*Dec. 3 T.* Clear and cold. Jody went with me to school today.

*Dec. 4 W.* Clear and very cold. School small. Pa came over to ask us to the choping to-morrow, but I have to go to school.

*Dec. 5 Th., 6 F.* Weather moderated. School small.

*Dec. 7 Sat.* Warm, and somewhat cloudy. We went to Pa's. I went to mill etc. As we came home in the afternoon with a load of plank, meal, feathers, etc. etc. old Med fell down broad side, and I had to unhitch her before she could get up again. Mollie and Jo. were scared bad enough. So was I. Charles Benbow came up in the time of the scuffle, and asked if she—old gray—had lay down to rest. After I got things loose she got up with out any serious injury to herself, the harness, or the wagon.

*Dec. 8 S.* Warm and cloudy. We staid at home all day. Caleb came over in the eve[n]ing and staid all night.

*Dec. 9 M.* Clear and very warm for December. Had 12 scholars today.

*Dec. 13 Friday.* This has been a week of fine weather. Taught school.

*Dec. 14 Sat.* Mostly clear, and cold. Caleb staid with us last night, and helped me kill a black guinea pig this morning. It weighs 72 pounds even, was over a year old, and tolerbly fat. I cut and hauled wood and pine this p. m.

*Dec. 15 S.* We staid at home all day. Two of John Jennings' girls—Rebeca & Elizabeth came over in the afternoon.

*Dec. 16 M.* Taught school.

*Dec. 17 T.* Killed our hogs. Alfred Campbell helped me. They weighed as follows 72 + 104 + 110 + 113 = 399. I sold sister Mary 47 1/2 pounds of pork, then 399 - 47 1/2 = 351 1/2 pounds of pork for us.

*Dec. 18* Taught school.

*Dec. 19 Th.* Covered the north side of our old house. Charles Benbow and Enos Windsor helped me. We got it done 1/4 before 2 p.m.

*Dec. 20 F.* Taught school.

*Dec. 21 & 22 Saturday & Sunday.* Sunday it rained and sleeted.

*Dec. 23 M.* The trees, fences, bushes, etc etc. are heavy laden with ice this morning.

*Dec. 24 T.* Dismissed School at 12 o'clock to-day for Christmas.

Christmas passed off without much noise. A good many of the rowdies, whose delight it was to disturb the peace and quietude of our neighborhood on former occasions like this, are now in the army. Their room is good company. The scarcity of powder will account for the little shooting, and no accidents, that I heard of, which are generaly the result of a careless use of firearms.

# Diary 1862

*Diary for January 1862.*

I taught School this month, only losing one day, and that the last day of the month. Cause,—I was sick.

Weather—It has generally been warm. We had one or two heavy sleets, a little snow, rain, and a few cold days.

*Diary for February 1862.*

I taught School about half of this month, and quit. In this month we had a good deal of rain and hail, a heavy sleet or two, a little snow, and one day of hard wind on which a great many "old trees" were blown down in the fields, some of *my* fence, a few green "giants of the forest" were uprooted. Our baby—we have not named him yet—was born the 12 of this month. He's a fine boy. Weighed nine pounds. Mary (sister) staid with us a week or so and left. Mary Hampton then came and staid with us three weeks.

Sunday 23d was a nice day. Pa came over and staid with us all day. Jane & Catharan, Mollie's sisters, paid us a visit to-day too.

*Diary for March 1862.*

Saturday the first day of this month it hailed, rained, snowed. It was a bad time to be out, but there was a general parade at Williams Burg. I was there. A call for 164 volunteers was made. Three companies, viz.—Youngs, Cashes and Grayhams turned out their quota, but the other three, viz.—Williamsburg, Fraylies and Taylors Springs failed. A bounty of $100 is now offered.[1]

Sunday [March] 2d was cool and cloudy.

Mon. [March] 3d was a very rainy day. I went to the shop but did not find the smith.

*Tuesday [March] 4* I choped some in the new ground.

*Wed. [March] 5* I hauled wood and rails. To-day there was a draft at Williamsburg![2] Ten men in the Wmsburg Co viz.—Martin Jacks, F. M.

1. At this time the Confederacy faced a crisis of military manpower. With the winter coming to an end, heavy fighting would soon resume, yet the Confederate army was melting away. Many of the Confederacy's soldiers had enlisted in the spring of 1861 for one year; now they intended to go home when their terms expired. The central government offered bounties, furloughs, and other inducements in an attempt to spur reenlistments. The central government also called on the states for volunteers.

North Carolina's convention, which had continued to meet after enacting the state's ordinance of secession, responded to this appeal. On February 19, 1862, the convention passed an ordinance to raise North Carolina's quota of Confederate troops. The ordinance directed the governor to call for volunteers to meet the needs of the central government. He also was to determine the quota of volunteers needed from each county, making allowance for the number of troops already in service from each county.

Evidently the war was not very popular in Iredell County. The bounty of one hundred dollars represents an unusual effort to induce volunteering, for the Convention provided for a bounty of only fifty dollars. Local authorities apparently sweetened the inducement in the hope of raising the needed troops voluntarily. For the convention's action, see *Ordinances of the State Convention Published in Pursuance of a Resolution of the General Assembly [Ratified 11th February, 1863]* (Raleigh: W. W. Holden, Printer to the State, 1863). For background information on the Confederacy's crisis, see Paul D. Escott, *After Secession: Jefferson Davis and the Failure of Confederate Nationalism* (Baton Rouge: Louisiana State University Press, 1978), 63–64.

2. Because sufficient volunteers were not forthcoming, militia officers filled out their quota by a draft. The fact that this and similar drafts took place in North Carolina is not generally known. Studies of the Confederate States have noted that Georgia, Texas, and South Carolina either threatened or instituted a draft at this time, but the standard histories of North Carolina (and anything written previously by this editor) make no mention of any drafting in North Carolina.

The records of the Adjutant General's Department, however, indicate that Iredell County was not the only county in which a draft took place. In a letter dated February 22, 1862, Adjutant General J. G. Martin informed Sergeant R. M. Sherrill of Lenoir, in Caldwell County, "The law requires that each County of the State shall contribute its proportion of men for the war, and those counties deficient will be drafted until its [*sic*] quota has been made up. The draft will be made in Caldwell Co. at an early date."

North Carolina's militia law specified that when the central government requisitioned a detachment of militia from the state, every captain of infantry was to list on his roll all eligible men and they would be "subject to draft." See *Revised Code of North Carolina, Enacted by the General Assembly at the Session of 1854* (Boston: Little, Brown and Company, 1855), chap. 70, sec. 75,

Barnard, Jos. W. Haith, Littleton Hutson, W. L. Shoemaker, Wm Trivett, Alexander Hames, J. E. Forcum, A. M. Slate, W. C. Lindly, were drafted.[3] There are but few men in this country that ever witnessed a draft before, as this is the first since the War of 1812, and there were but few drafted then. *Then* our fathers had England to contend with. *Now* we have "all the world, and the rest of mankind."

*Thursday [March] 6* It snowed most of the afternoon very fast. I went to Eagle Mills.

*Friday [March] 7* A very cold, windy day. March.

*Sat. [March] 8* Clear and not so cold. I went to mill, came back by Pa's. Ellen came home with me to see Jody & the baby.

*Sun. [March] 9* Clear and warm.

*Monday [March] 10,* and no oats sowed yet. Warm, cloudy and rained fast for an hour or so about the turn of the day. I have not plowed a furrow yet I intended plowing the garden to-day, but the rain prevented me. I fixed up the fence, set out some peach and apple trees, set the cabbage stalks and "Hanovers" for salad, etc.

*Tues. [March] 11* Plowed up the garden in the A. M. and went to Jesse Tetroes choping in the p. m.

---

p. 422. The same provision was present in *Revised Statutes of the State of North Carolina, Passed by the General Assembly at the Session of 1836–1837,* 2 vols. (Raleigh: Turner and Hughes, 1837), 1:417, and the practice probably goes back to revolutionary and colonial days.

For the letter of the adjutant general, see Adjutant General's Department, Letter Book, December 10, 1861–April 23, 1862, AG 19, North Carolina Division of Archives and History. For actions by other states, see Escott, *After Secession,* 63–64.

3. What was Thomasson's status under the state draft? According to the *Revised Code* of 1854 cited above, there were three categories of men between the ages of eighteen and forty-five that were exempt from the draft: those "exempted by the second section of the act of congress of one thousand seven hundred and ninety two"; "judges of the superior courts of law and equity"; and "ministers of the gospel, regularly ordained." Thus he was exempt as a minister. (The *Revised Code,* in section 2 on pages 398–99, also exempted certain people, including "teachers and pupils of common schools, while engaged as such" from ordinary military duty—but not from service "in case of invasion or insurrection in the State.")

The General Assembly, in its 1860–61 session, had revised the militia code, but its revision strengthened the exemption for "ministers of the gospel" by decreeing that they were not to be enrolled in the militia. See *Public Laws of the State of North Carolina Passed by the General Assembly at its Session of 1860–1861* (Raleigh: John Spellman, Printer to the State, 1861), 40.

*Wed. [March] 12* Commenced sowing oats, planted our peas, a few onions, etc. Tolerbly warm. Alex. Benbow came by on his way to Statesville after his School money, and as I have no way to go, I sent in my Draft too.

*Thursday [March] 13* A cool rainy day. We finished planting onions this morning, and I planted 7 short rows of potatoes.

*Friday [March] 14* A rainy day. Went to the shop, to Eagle Mills and to Troy. Paid T. Holcomb two dollars ($2) for tanning sole leather. Pa and Vic came over this evening.

*Sat. [March] 15* It rained a great deal this A. M. I went with Pa over to P. Campbell's to buy wheat, but he did not get any.

*Sun. [March] 16* Windy and cool. We staid at home till 2 or 3 p. m. and then went to Jesse Teatroes and staid an hour or so. This was the first Sunday visit we ever paid them.

*Mon. [March] 17* Clear and warm. Plowed a little in the p. m.

*Tues. [March] 18* Had a heavy frost this morning, but the day was warm. Sowed a few oats.

*Wed. [March] 19* Cloudy and cool. Sowed a few oats and planted some potatoes in the A. M. and in the p. m. it rained. Who ever saw so much rain in March? Two or three days every week!

*Th. [March] 20* Still raining.

*Fri. [March] 21* Cool and a little windy. Enoch Coffin finished his job of clearing and spliting rails to-day, and I paid him off. He cleared 1 3/4 acres of ground, and split (so he said, I did not count them) 1200 rails. I gave him $3 per acre for the clearing, and 25 cents per hundred for the rails. He boarded himself. Dont think he more than cleared his fifty cents a day, if that.

*Sat. [March] 22* Cold and windy. Hauled 12 loads wood off my new ground, went to Benj. Coffins and got half a gallon molasses. I gave half bushel of potatoes for them. Potatoes sell this spring at $1 per bushel.

*Sunday [March] 23* Cold weather. We staid at home. Brother Caleb came over in the afternoon and staid an hour or so.

*Mon. [March] 24* Still cold and windy. Planted potatoes.

*Tues. [March] 25* Finished my potatoe patch, about 1/20 of an acre. I opened trenches with the plow, droped in the "tater" and then filled said trenches with rich earth from a gully that was filled 3 years ago with pine brush, leaves, etc, and from pine woods. I hauled 10 one horse loads. I plowed in oats in the p. m.

*Wednesday [March] 26* A stormy day. Thunder and lightening, rain, hail and snow. It did not rain nor hail very much, but I do not think I ever saw snow fall faster nor in larger flakes. I do not remember ever seeing such a day as this before. Pa says it used to thunder and lighten in snow clouds when he was a boy. But few snow clouds visit us these days, and when one does come we don't look for thunder and lightening, but for something colder.

*Thurs. [March] 27* Went to mill.

*Friday [March] 28* Finished sowing my winter oats.

*Sat. [March] 29* Cloudy and cool.

*Sun. [March] 30* Do do Shadrack Bell and wife paid us a visit today.

*Mon. [March] 31* Sowed oats, sowed oats.

*April the first.* A nice, warm day. Joshua Sprinkle choped for me to-day.

*April 2 Wed.* Cloudy. Finished sowing oats.

*April 3 Th.* Clear and warm. Went over to Campbell's mill and engaged five bushels of corn, at $1 per bushel, for Pa. We went to Pa's this evening.

*April 4 F.* I & Jody came home with a load of hay, and Mollie went on to her pa's.

*April 5* Rained a little in the forenoon. Ellison Johnson and I run us a gallon or so of very nice tar to-day.

*April 6 Sun.* I & Jody cut out this morning after Mother. I left him at

Pa's to play with Vic, Ellen & Lenny and went on to Mr. Bell's where I found Mollie & baby all safe. Shadrack and his wife were there also, so we had a real chat.

> We talked of *war,* and talked of *peace*
> We talked of *men,* and talked of *geece.*[4]

The 7, 8, & 9 were cool, rainy days.

The 10, 11 & 12 variable, 13 cloudy & cool.

*Saturday April 19* Rain, rain, rain. Made 2 pairs of shoes—mine & Jodies—this week, fenced one side of my new ground, etc.

*Sat April 26* Enoch Coffin worked three days for me this week. We finished the new ground & pasture fences, split a few rails, some firewood, etc. etc. Enoch left Friday morning. I went to plowing, & after dinner we commenced planting corn. Jody droped the corn. It rained Friday evening and Saturday.

Cleared off Sunday morning 27.

*Monday April 28* Cloudy & cleared off. I got 2 1/2 bushels of corn of E. A. Johnson and went to mill.

*Thursday April 29* we expected to have planted a little more corn, but when I went out to give old Med her breakfast I found her unable to get up. I got 3 hands and we tried to help her up, but it was all in vain. She never got up any more. She died on the night of the first day of May.

Wed. the 30 of April & Thursday the first of May it rained. Friday 2 A. M. I invited some hands to help me roll logs, and p. m. of the same day I went over to Pa's. Sat. 3 4 of my nieghbors came in and we had a "log rolling." Got thro' about eleven o'clock.

4. By this time Thomasson probably was wondering whether the Confederate government would require him to fight in the war. In April the Confederate Congress passed a conscription law, the first approved by a national government in American history. Fortunately for Thomasson, however, an exemption act accompanied the law. Two provisions of this act apparently enabled him to stay out of the army. "Ministers of religion in the regular discharge of ministerial duties" were exempted, as were "all teachers having as many as twenty scholars." The text of the exemption law appears in an enclosure to a letter from Jefferson Davis to Governor Joseph E. Brown of Georgia, April 28, 1862, in Allen D. Candler, ed., *The Confederate Records of the State of Georgia,* 6 vols. (Atlanta: Chas. P. Byrd, 1909–11), 3:201–2.

*Sunday [May] 4* Cloudy and rains moderately. Who ever saw so much rain? I think it has rained, on an average, two or three days in every week since the first of March. It is now the 4 of May and there is but very little corn in the ground yet. Think this spring is at least a month later than springs used to be.

*Monday [May] 5* A rainy day. Mended a pair of shoes for Mrs. S. Johnson.

*T [May] 6* Clear and cool. Went up to Col. M. Willburns and paid the Interest ($9) for three years past on my note. I also paid him $2.60 interest on Pa's note.

*W. [May] 7, T. [May] 8, F. [May] 9, Sat. [May] 10.* Clear and warm. Hired Ellison's mule and planted corn. Planted the sugar cane seeds yesterday 10th.

*Sun. [May] 11* Warm and dry.

*M [May] 12* Planted our sweetpotatoes.

*T. [May] 13* Caleb helped me haul wood and rails off my new ground.

*W. [May] 14* Cloudy, and rained some in the forenoon. Fired the log piles in the new ground, sheared the sheep, etc.

*Th. & F. [May] 15 & 16*—Cloudy days with now and then a light shower,—just enough to soften the ground on top for the corn to get thro'. Its coming up nicely too.

*Sat. [May] 17* Rained lightly to-day too. Four days this week of damp weather. Well "Its all for the best," Whatever pleases the Lord should please us, be it "rain or shine." "Whatever is, is right," in refferance to the dealings of God with man.

*Sun. [May] 18* Pleasant, growing weather.

We planted our new ground corn the 26 & 27 days of this month.

## *Diary for June & July 1862.*

The first week in June was very rainy. The waters got up and cleaned out some of the creek bottoms to the clay, or as deep as they had been broke. Such heavy rains are providential, and for some good, but we cant see exactly how men are benefited by having their farms so completely washed away. One thing, however, we do know, and that is that *all things* work together for the spiritual welfare of all those that love God. O for that spirit that prompted Job when he said, "The Lord gave, and the Lord hath taken away. Blessed be the name of the Lord."

The wheat crop has failed in this and adjoining counties. The straw cured up like hay and some of it fell down. Some say a worm in the stalk causes it to fall. The red rust killed the blades about the time it was heading out. I sowed 2 1/4 bushels and raised 12. I & brother Caleb hauled it to Shade Myars' machine the last week in July and had it thrashed. The oats are a failure too, all that were sowed last spring, those sowed last fall and winter were very good. I sowed, last spring, 2 bushels of winter oats, and saved 30 bundles and 2 of spring or black oats and saved 14 doz. Dry weather in June and the rust cut off the oat crop in these parts. Corn is small, but looks well. Hope we shall have a heavy crop. $2.50 are offered for wheat, 1.50 for corn and 30 cts for bacon, & one dollar for oats. I bought 2 bushels of winter oats of Pa last week for which I gave 10 lbs bacon. I bought 3 bunches of cotton yarn at $3 per bunch! Money is plenty, and food and raiment scarce.

*Diary for August 1862.*

Two or three weeks of dry weather in this month cut off the corn crop very much. Dont think I ever knew a failure of every thing before. Short wheat, (many farmers raised less than they sowed, and that of an inferior quality) short oats, short corn, and short every thing else. A famine is feared among all classes, but more among those who have all to buy & nothing to sell. This is indeed a hard time with them. The hireling now, with a large family to support, has to work two days for four pounds of bacon, three days for one bushel of corn, and six days for one bushel of wheat! And I fear the time is not far off when meat and bread will not be had at any price. Well if the Lord will let famine, pestilence or any thing else come that will bring the dreadful war now raging in our land to a close. There is perhaps nothing better calculated to humble a nation than famine and nothing better calculated to produce famine than war and short crops. These we have, in the South, to a fearful extent. Famine, famine, and it may be seven years famine, seems to stare us in the face. Thy will, O Lord, be done on earth as it is done in heaven. As our days may demand so let our strength be. Amen.

*Diary for September 1862.*

The first week in this month Col. Thomas Holcomb & I went to Salem, Forsyth, N. C. to get our wool carded. We carried wool for six of our neighbors too. We drove into town about night, and early next morning our rools were ready.

# Other Documents

Occasionally, when Basil Armstrong Thomasson wanted to write on a blank sheet of paper, he used pages from his homemade diary. Leafing forward from the entries he had just recorded, he would draft a story, a letter, a Sunday school lesson, or something else that was on his mind. Days or weeks later, when his diary caught up to these pages, he usually did not remove the document he had created but instead wrote around it. Thus, Thomasson's diary contains a variety of documents that are informative and sometimes revealing.

The text of these documents is given below, along with a heading indicating where in the diary each appeared. In each case, the document was created weeks or months before the diary entry closest to it.

The final four documents were found not in the diary but in a box of miscellaneous documents (mostly family deeds and receipts) owned by Mrs. Jean Harris Thomasson. The first is a draft of a letter written by Basil Armstrong Thomasson; it is not known whether a version of this letter was ever sent. The second reproduces most of a short draft of a Sunday school lesson or sermon. The third was written out by Thomasson on a separate slip of paper, and the fourth was a leaf from a sermon or Sunday school lesson.

## *Following Sept. 6, 1855*

"Oh! Mother, dear Mother," said little Annie as she lay upon her dying pillow and gazed wishfully at her mother who was standing by her bedside and tenderly smoothing with her soft hand the glossy curl of her afflicted little daughter, "What makes Pappy so cold, and unfeeling towards his little *pet bird* as he used to call me when I'd climb upon his knee soon as he would take a seat in the *big rocking* chair and sing to him my childish songs? And when he enters or leaves the room why, oh! why does he slam the door so hard and make so much noise? Does he not know, Mother, that his little *bird* is sick and that she will soon take her flight from earth to that heaven of which I have heard you so often talk and sing, Mother?" "Oh! my little daughter it pains my inmost soul to see the coldness with which your pappy has treated us all for a few months. He knows that you are ill but a cold indifference has taken the place once occupied by love in his heart and he seems not to care for the affliction even of his little daughter; but go to sleep, my child, and let not your pappy's conduct disturb you just now." The above conversation took place in the house of Mr. B. who was once an industrious and wealth gaining farmer, one of the kindest of husbands and an affectionate father. But falling in company with a drinking crowd one evening as he entered a certain would be pleasant little village were it not for two or three grog shops and the noisey peace disturbing gang that usually attend such places, he was by them induced to join in and take a social drink. He drank and drank again and was soon changed from the tidy gentleman to a raving mad man. In this condition the shop keeper forced him out into the street to face the cold blast of a stormy night. He started on his way homeward but the street being to[o] narrow for him to travil in his present condition, he fell over the pavement close by the door of a kind man who hearing his bitter oaths and deep distressing moans, came to his aid and kindly lifted him from the frozen pavement and laid him before a warm fire. Here he lay and sobered, then rose and returned to his home a ruined man. The demon of the still had triumphed over his manly spirit and he was now undone. The appetite for strong drink was formed within and notwithstanding the many solemn vows made to his loving wife, her entreaties, and all the force of moral suasion that could be thrown around him by the best of friends, he never went from the village where he received his first glass, the muster ground, court green or from any public gathering where liquor was bought and drunk, sober again.

## *Following Sept. 8, 1855*

The Plagues of Egypt Exodus

*First,* the water turned to blood; "and he lifted up the rod and smote the waters that were in the river, in the sight of Pharoah, and in the sight of his servants, and all the waters that were in the river were turned to blood." Exodus VII.20

*Second,* the plague of frogs. "and the frogs came up and covered the land." VIII.6

*Third,* the lice. "all the dust of the land became lice throughout all the land of Egypt." 17

*Fourth,* the swarm of flies. "And there came a grevious swarm of flies into the house of Pharoah and into his servants' houses and into all the land of Egypt." 24.

*Fifth,* "there shall be a very grevious murrain." "And the Lord did that thing on the murrain, and all the cattle of Egypt died; but of the cattle of the Children of Israel died not one." IX. 6

*Sixth,* The plague of boils. "And they (Moses and Aaron) took ashes of the furnace, and stood before Pharoah; and Moses sprinkled it up toward heaven, and it became a boil breaking forth with blains upon man & upon beast."

*Seventh,* The hail storm. "And the Lord rained hail upon the land of Egypt." 25. "And the hail smote throughout all the land of Egypt, all that was in the field, both man and beast; and the hail smote every herb of the field, and broke every tree of the field." 25. "Only in the land of Goshen, where the Children of Israel were, was there no hail." 26

*Eighth,* The locusts. "And when it was morning, the east wind brought the locusts." X. 15. "And they did eat every herb of the land, and all the fruit of the trees which the hail had left." 15.

*Ninth,* The darkness. "And there was a thick darkness in all the land of Egypt three days." 22.

*Tenth,* The death of the first born. "The Lord smote all the first born in the land of Egypt." (XII. 29) of both man & beast.

## *Following Sept. 19, 1855*

*Utility of the Sunday School System*

The occasion requires us to speak of the Sundy School system. This is, without controversy, one of the most efficient agents employed by the great Head of the church in accomplishing the prophecy of the saying, "and all thy children shall be taught of the Lord." The Sunday-School system! There is something hallowed and delightful in the very title! If I were to treat of this system in all its bearings,—of the history of its origin and progress,—the wisdom and benevolence of its principles,—the self-denial, prayer, holiness, and energy required for its successful prosecution, and all the hallowing influences it exerts upon the condition and destiny of man in both worlds, days might be profitably employed in the discussion. But I shall confine myself to one point—the utility of the S. school system. Some may consider this like an attempt to illuminate a sunbeam, or demonstrate a self-evident proposition which no one questions. They will say, "It is a trite subject and stands in need of no further argument and illustration."

The business of education in general is a trite subject; yet who will deem it useless further to explain its principals, display its advantages, and urge its claims. The fundamental doctrines of the gospel are [a] hackneyed theme, yet who will say they require no further enforcement of illustration?

No one will dare it, while so many thousands favoured with its ministrations close their eyes against its light, and 3/4 of our fallen race have never heard its offers of mercy.

Temperence is another subjct upon which there has been a great deal said; yet is this any reason why we should not say more? Because the battle ground is becoming so familiar, and consiquently a little contemptable, should the soldiers of the "cold water army" lay aside their arms, and look out for a new field of labor? By no means, while our *land* is even cursed in consequence of intemperance, and while 30,000 little mounds are annually raised above the level of our green earth, which hide from our view the bodies of those that have fallen victims to this dreadful *enemy* to the peace, and happiness of society in general.

Neither dare we to say that the utility of the S. School system is sufficiently known and appreciated, while the Sabbath is the chosen day, among our little folks, to take the shining fish from our clear and rippling brooks, and to rob the birds of their beautiful eggs and nicely built nests; and while it is the day set apart, by those that have the care of youth, to spend in idle conversation and to "eat, drink, and be merry," and while hundreds of millions in all lands are sighing for a participation in its blessings.

The limited extention of the benefit of this system in comparison with the moral wants of the world is a sufficient proof that few have made any proper estimate of their magnitude.

The doctrines of the gospel, the temperance cause, and utility of Sabbath-schools are, indeed, old themes, but age does not render them unworthy of our attention. They are themes of too great importance to wear out, or to become even "thread bare." They will last while time lasts, and be remembered in eternity, by these who enjoyed, or who might have enjoyed their blessings. Then let us cherish our Sabbath-school as a glorious and lasting institution, which has been brought about, not only for the good of those that have not time to attend school on any other day, but for the especial benefit of *all* classes,—the old as well as the young, and the rich as well as the poor. Then let us come out to our S. school every Sunday if possible, bringing with us our books, and not show contempt to so great an institution by staying at home, or by coming out without books, and expecting only to sit here as idle spectators. Let us all engage in the work of education. Let us all become teachers. If we cant teach others, let us try & teach ourselves, and thus all become both teachers and pupils, and cooperate in this "work of faith, and labor of love.["] In this humble attempt to illustrate the utility of Sunday-schools, we propose,

I. To notice their beneficial influence upon *the scholars;* which we will do next Sunday.

Lecture 1st delivered in Aylesbury Sabbath school, May 28th, 1854.

B. A. Thomasson

## *Following May 31, 1856*

Log Cabin, May 1, 1856

My brother dear
Your favor's here
The *Advocate* I mean
For letters none
I've had this noon
The cause I can not wien.

Perhaps that *bur**
Has all *my shur*
Of time to writing given?
And if so be
I'de like to see
Or hear of better *doin*.

*You spoke of in a previous letter.

Its not often I get my machine screwed up tight enough to grind poetry, or even fine prose, but when I do get her set she's some, as you'll see by the following, if the above has not assured you of the fact.

Ten months to day,
Have passed away,
Since Mollie and I were married.
Time on swift wing,
Has brought us spring,
And many things both rare and varied.

The woods are green,
Flowers are seen,
All covered and sparkling with dew.
The singing birds,
And lowing herds,
Comingle their music anew.

And thus we see,
All nature agree,

And join in praising her Maker,
While man, poor man!

You know about as much how he lives as I do, so here I stop.

. . . . "nothing lovelier can be found
In woman, than to study household good,
And good works in her husband promote."

*Paradise Lost,* Book IX, page 241.

". . . . smiles From reason flow,
. . . . and are of love the food."

". . . . solitude sometimes is best society,
And short retirement urges sweet return."

What half of this page was left blank for I do not now (Sunday night, July 3d 1859) remember. If women in general would "study household good" more, and spend less of their time and less of their husband's money trying to keep "in fashion," I think we would have better times. Try it, ladies, and let us see the result.

## *Following June 23–27, 1856*

Log Cabin, June 1, 1856

Dear bro. As to adopting the plan of E. B.[1] there is, just now a difficulty in the way. And, there is more than one, at least two that are not easy to get over. The first is, I was not taught, from the cradle till the time of my leaving home, to do things in a systimatic way. The business of the day was tangled, dispatched in a disorderly manner, and there was scarcely ever *head* or *tail* to any thing. And thus I've been taught, by *example,* which is the *loudest* teaching *in the world,* to go blindly and orderless to work.

The second is Pa & Ma are interested in the work. You know that it always has been the case that if the farm and house work did not

1. Evidently Strong's brother wants to leave school and study at home, with Strong as his teacher. This letter sets forth reasons why this would not be a good idea.

progress as rapidly as they thought it should there wer[e] signs of displeasure in their faces. Time spent in study seems to[o] much like a waste, hence we must work, work, work. For illustration, I now, and have all the spring worked early and late, yet you say that Pa told you I didn't work much. Now suppose I get up in the morning at 5, work till 9, then sit in the shade with my books till 3, then work till 7, and what will he say next time he sees you? It is true, I did but little out door work last winter, but I was not idle by any means. If I was not at work on the farm, as Pa perhaps thought I ought to have been, I was at work on my mind. Now I think more of that if it is a little, weak one, than I do of this, or any other farm, and you know that where one's treasure is ther[e] will his heart be also.

You say that unless I work I will always be poor. That I know. But there are two kinds of poor people. Some I know that toil hard, yet they are poor. Others that seem not to work at all, yet they are rich. Why is this great difference? Some say that Fortune favors some, and frown[s] upon others, in other words God makes some rich and others poor. This is true. But why is it so? The Bible says that God is no respecter of persons. Then what is the difficulty? Here it is, *We don't work right.* We are to[o] forgetful of D. C's mottoe, "First see that you are right" etc. Hence the poverty of most of [word omitted]. We toil daily for the body, scarcely ever bestowing even a passing thought upon that little object within which if properly cultivated, would be of more real servise to us than the U. S. Mint, It's like a man having a mine of gold and quarry of lime-stone on his land, passing over the gold which would make him rich, and working his lime which only yealds him a scant subsistance. Now the *lime* and the *gold* must both be brought into service, but let that which is of the greatest value receive the most attention. God is impartial. Hence if we are poor, in every sense of the word, it is our own fault, for "What soever a man soweth of that shall he also reap."

I have written more on this subject than I expected to when I commenced, but the difficulties mentioned are not intended for excuses. I have only mentioned them as obsticles in the way, over which it is hard to crawl. These, with others I might mention, make up the grand reason why you can not study at home.

I intend to study no matter whose face wrinkles, and next year some one may tend the farm that will work. If I labor just enough to procure for myself and family food and raiment and am therewith content, and devote the ballance of my time to study whose business is it?

I wish, in this letter, to give you another sermon on the use of tobacco and coffee. You may be tired of the subject? If so abandon their use and I'll say no more, unless, it is by way of exertation. While men sin it is the duty of others to preach, after they "cease to do evil" they should be exorted to steadfastness. We have an account in the New Testament of a certain rich man which had a steward; and the same was accused unto him that he had ["]wasted his goods." Let this be our text. Now we are not informed as to this steward's character more than that he was a wasteful man. He might have been a perfect moddle in uprightness in every other respect, so far as the dealings between him and his master were concerned. We grant that he was as he is not charged with any thing else. But notwithstanding his many virtues there comes in a report to the rich man that he (his steward) has "wasted his goods." The rich man calls to his steward and after going over, perhaps, a long catalogue of good traits attached to his character, informs him that he has heard an evil report concerning him, and inquires "How is it that I hear this of thee?" The steward makes no reply, and the rich man in order to test the matter demands an account of his stewardship. The account is rendered, the steward found guilty, and we hear the sentence pronounced —"Thou mayest be no longer steward." We are not informed how this steward wasted his master's goods or to what extent. Whether he threw them away, suffered the accounts to remain uncollected till they were out of date, or gathered in his master's money and with it bought some unnecessary article we know not, but we are informed that he wasted the goods for which the stewardship was taken away from him, for he says "within hiself," "I am resolved what to do" and he goes and does it.

We have seen the effect of the sin of wastefulnes in this steward's case[:] the stewardship taken away. Now let us examine ourselves and see what we are doing. Well we are all stewards having one Master, a Lord—"a rich man," the Creator of all things, and to whom all things belong. Then we are all officers, and our office an honorable one. But do we honor it by faithfully discharging our duty? We are in possession of our Master's goods. Do we use them with an eye single to his glory? or do we waste them? Let us see. Our Lord has given us money. Do we apply it to a good use or to a bad one? Now you admit that the use of coffee and tobacco is only a habit, and a bad one at that. Then every cent of money spent for these articles *is wasted.* Now what do we do with our money? Do we clothe the naked, feed the poor and minister to the real (not the supposed) wants of mankind as we have opportunity? or do we send *our* Lords money for articles which are worse than useless to us, and thus waste his goods? If we do[,] the stewardship will be taken from us. To illustrate this point, suppose a friend of mine was to give me $5 and tell me that if I would apply that $5 to certain purposes, at the end of one year he would give me $25. Now suppose that I was to waste this money insted of using it as I was requested, could I then at the end of the year expect to receive the $25? Now all that we have has been given to us by a friend, he has told us what to do with it, and has promised more if we apply what we have rightly, but if we do not we can't expect, nor will we receive the promised reward.

I know of men who spend from 5 to $10 a year for tobacco and coffee, while they give but very little if any thing for support of the gospel, or for any charitable purpose. How will the case stand with them when an account of their stewardship is demanded? Our poor widows instead of casting their two mites into the treasury of the Lord, as did the widow of old, spend them for coffee. What will become of their stewardship? And you, my brother, if you do not spend your money for these articles you use them, and if a man steals a loaf of bread and I eat it[,] knowing that it was stolen[,] I am guilty with the thief. Now, to secure your stewardship, you had better be careful how you waste your master's goods. Better quit at once the things you know to be wrong, and hoping you will I close my ______ what is it?

B. A. Thomasson

## *Following Dec. 13, 1856*

No. and names of all the white children, entitled to the benefits of the Common School law, in District No 30, Oct. 2d 1856.

| *Males* | *Females* |
|---|---|
| 1. H. Jacks | 1. Jane Lindsey |
| 2. Isaac Jacks | 2. Mar. Lindsey |
| 3. C. Jacks | 3. P. Johnson |

4. J. Myars
5. A. D. Lindsey
D. R. Jacks
J. L. Jacks
Jessey Denney
John Denney
10. Samuel Barron
Sicaro Pitts
Alvis Myars
Alvin Myars
Josiah Myars
15. C. Johnson
A. Johnson
T. E. Johnson
C. L. Johnson
W. Johnson
20. T. Johnson
J. Johnson
D. Johnson
M. Denney
S. Denney
25. J. J. Winecauf
Wm. Grant
James Grant
C. E. Thomasson
Hiram Minish
30. Wiley Minish
Wm. Grant
Henry Green
Josiah Green
Thomas Pardue
35. May Morgan
Miles Chamberlain
Jo. Commer
Jarvis Reed
Clinsis Denney
40. D. W. Windsor

4. El. Jacks
5. Ma. Jacks
Lo. Jacks
Mar. Barron
Mar. Harben
Sus. Denney
10. Ny Jefferson
C. J. Pitts
Viana Wood
M. Wood
Mar. Myars
15. M. A. Myars
El. Myars
J. Myars
Ca. Myars
El. Johnson
20. Mar. Johnson
A. M. Johnson
Em. Burgis
Ca. Johnson
La. Johnson
25. Sa. Denney
A. Denney
Cath. Denney
H. A. Benbow
Mar. Hudspeth
30. Han. Money
M. E. Grant
M. C. Thomasson
N. A. Thomasson
Tilda Minish
35. Ch. Green
Ma. Green
El. Green
Luv. Autrey
E. A. Autrey
40. Sus. Autrey

| | |
|---|---|
| D. L. Beall | Sor. Autrey |
| W. W. Beall | Sy. Baggarly |
| 43. A. W. Beall | Nan. Evans |
| 44. Alexander Soss | El. Evans |
| | 45. Mar. Reed |
| | N. A. Beall |
| | E. R. Beall |
| | 48. C. A. Beall |

Males 44
Females 48

92

B. A. Thomasson, Teacher
E. A. Windsor
R. W. Beall } Com.
Branson Pitts

Oct. 2d 1856

*Following Dec. 20, 1856*

Hamptonville, N. C.
Oct. 10th 1856

Mr. C. H. Wiley.[2]

Dear Sir: My object, in asking your attention, or a few minutes, on the receipt of this is, first, to propose the publication of another "Reader," for beginners; and secondly, the plan of the work, publication, etc. We especially in the western part of North Carolina, need a Reader for young learners, that can be sold as cheap as *Webster's Spelling Book.* Then teachers, by buying by the wholesale, could supply the Districts in which they teach, and sustain no real loss; for they would thus be casting their bread (a small portion of it) upon the waters, and they would be sure to find it not many days hence. Now I propose to prepare, and have published on nice white paper, and in tolerbly large, clear type, "My Little Reader," the wholesale price of which will be about three cents per copy. And that I may be able to succeed in getting out this much

2. The superintendent of North Carolina's common-school system.

needed work soon, I earnestly solicit your aid in procuring advertisements of our Colleges, Academies, and Books, as I wish to have enough aded to the work to pay, at least, half the cost of publishing.

The Book will then be sold so *low,* that teachers can well affoard to supply their Districts, at their own cost, which they will doubtless do; and thus "My Little Reader" will be one of the best of advertising mediums.

B. A. Thomasson

To the Rev. C. H. Wiley, S[tate] S[uperintendent of] C[ommon] S[chools]

### THE FIRST READER,

### LESSON I

Tom has fed his dog.

A dog can run.

The sun is up.

See the old man.

Do not run in his way.

Put on the hat, and do not let the dog, nor the pig get it.

Do not go in the way of a bad boy, but do as you are bid.

Do no ill to any one.

### LESSON II

I can read in my new book.

My book is new, and I must not soil it.

A good boy will not tear his book.

I must be a good boy, and let Tom read in my book when I see him.

I must be kind to all the boys.

If I see a boy do a bad trick I must not do so too.

A good boy is loved by all who know him.

*Following Dec. 26, 1856*

### MY LITTLE READER

#### LESSON I

| | | |
|---|---|---|
| MAN | HOUSE | MOUSE |
| BOY | BARN | TREE |
| CAT | PIG | COW |
| DOG | RAT | BOOK |
| GUN | HORSE | LOT |

I saw a man and a boy.

The man had a gun, and the boy had a new book.

The dog can run and bark at the pig, and the cat can eat a rat.

The mouse lives in the house.

The barn is near the tree.

The cow is at the barn.

The horse is in the lot.

#### LESSON II

| | | |
|---|---|---|
| JOHN | SUN | YOU |
| ATE | ROAD | CORN |
| PEAS | BEES | HIVE |
| WAX | OATS | FIELD |

John has fed his pig.

The sun is up now.

I will try to spell and read well.

Shut the gate, and do not let the pigs come in from the road, and root up the corn and peas.

Bees live in a hive, and make wax.

Corn and oats grow in the field.

MY LITTLE READER

LESSON III

| TOAD | WORM | FLY |
|---|---|---|
| ANT | KITE | SLATE |
| TOM | GIRLS | PIPE |
| LAND | LIFE | GOD |

A toad will eat up a worm or an ant or a fly; but it will not bite you.

I have got a new kite, and a new slate for Tom.

Boys and girls should not use a pipe.

All that move on the land get life from God.

LESSON IV

| HATS | COATS | FUR |
|---|---|---|
| WOOL | SHEEP | FRIENDS |
| MILK | THEY | THEIR |
| BREAD | THEM | SHE |

Boys wear hats and coats.

Some hats are made of fur, and some of wool.

Wool grows on the sheep.

Boys should be kind to the sheep, for they are great friends to them.

Boys should be kind to the cow too, for she gives the milk in which they soak their bread.

Hamptonville, Yadkin Co. N. C.
Oct. 10th/56

Messers. A/ S. Baines & Co.

Will you be so good as to send the subscriber a catalogue of your publications, (specially those recommended by Mr. Wiley to be used in our C. S) together with their lowest wholesale prices to Agents.

Also inform us for what sum you would publish 500 copies of "My Little Reader," designed for young learners, and containing Twenty Five

Lessons in Spelling and Reading, about Fifty Pages on nice white paper, and in tolerbly large, clear type, and bound in the style of *Webster's Spelling Book.*

Respectfully yours,
B. A. Thomasson,
Hamptonville,
Yadkin Co.
N. C.

P. S. I wish to have the above work published for gratuitous distribution.

## *Following Dec. 30, 1856*

Let us, brethren & sisters, review the chapter which has been read in your hearing, and for a short time prayerfully, & carefully examine our own hearts.

The gift of speaking various languages, "the gift of prophecy," and that of understanding all mysteries, and all knowledge, and the faith of miracles was considered, by the Apostle, so far above the gift of property that he takes no notice of that at all, but comes at once to things of more importance; and yet those gifts are worthless, in a spiritual point of view, without having combined with them the one thing needful—charity.

As charity is the principal thing spoken of in the chapter we will take it for our subject, and first consider what it is. Of charity there seems to be two kinds, or some differance at least, between the charity here spoken of, and what the world calls charity. Aiding the poor, relieving the sufferer and administering to the wants of mankind is charity, and good as far as it goes, but it is not the charity the Apostle speaks of in this chapter, for he says, "And tho' I bestow all my goods etc," nor is it able of its-self to secure us "A home in heaven." This charity seems rather to be the fruit of the genuine, which is simply "love to God, and to-our fellow men."

Let us, in the second place, examine the workings of this charity, or this love as we may now call it.

We are informed that it suffereth long and is kind; are we, brethren & sisters kind as we should be? Do we treat all, both friends and foes kindly? If we do not we are destitute of this love for this book clearly proves to us that those who possess it are "kind."

And again, charity envieth not; Are we uneasy at the prosperity of others, and do we envy them their position in life, no matter how exalted it may be? If so this love is not in our hearts, for "charity envieth not."

Again, "charity seeketh not her own." Here I fear many of us fall short, for we are to[o] careful in seeking our own worldly comfort. The Apostle says "Let no man seek his own, but every man another's wealth." Do we do this? I fear not for we mix to[o] much with the world. Here, in order to do this, see clearly the necessity of coming out from among the world, and being "kindly affectioned one to another with brotherly love, in honor preferring one another.["]

And again, "charity thinketh no evil." Do we give loose rein to our thoughts and are we ever ready to impute to others evil designs whether their actions will justify us or not? If so there is a lack, in our hearts, of that love which "thinketh no evil," but is ever ready to put the best construction upon every thing.

Charity "Rejoiceth not in iniquity, but rejoiceth in the truth." Do we rejoice at the vices, and cklamities even of our worst enemies? I fear that many of us are guilty here, for we seem so willing to aid in spreading evil reports concerning even our friends sometimes. This is wrong. We should be slow to believe flying reports to the injury of others; ya, we should not rejoice "in iniquity," but in the truth we should glory, and aid in the destruction of iniquity, and the establishmen[t] and promulgation of the truth.

Charity will also endure while eternity shall last for "Charity never faileth," but tongues shall cease, prophecies shall fail, and knowledge shall vanish away. Then let us seek after this love "as the dry land thirsteth for the rain" that it may be in each of our hearts a well of living water springing up into everlasting life; let us cultivate this love as a rare and tender plant—one which the least neglect will cause to droop & die—for if we have it not in our hearts we are become as sounding brass or a tinkling cymbal—destitute of holiness—without which "no man shall see the Lord."

Southern States stand committed in the providence of God to maintain it.["] Now are these things so? If they are, has the South taken the proper course? It seems to me that the South is gathering thorns for what she deems flowers, and that she is killing off her darling institution just as fast as Greeley, Beecher, or even the president of the abolition horde himself could wish her to.

May 4. The above was written some days ago since which we have heard the news of war. The preacher in his sermon last Sunday said "We are involved in war." Now if it were some other nation about to invade our land we might feel more secure, but to think of our own nation going to war with itself, the South against the North, it is awful to think of. How can people be so thoughtless and so wicked? May the God of heaven save our nation.

## *From Draft of a Sunday School Lesson Found in the Box of Family Documents*

"For the children of this world are in their generation wiser than the children of light." Luke XVI + part of the 8 v.

This is the language of our savior . . . in refference to a rich man and his steward. . . . and then [the rich man] demanded that his steward should give an account of his stewardship, "for thou, since thou hast wasted my goods, mayest be no longer steward." Now after he has followed an avocation for years, found it to be a good one, and feels that it has been and is still his only means of support, he is sure to form an attachment to it, and feel very loth to give it up, no matter by whom he may be commanded so to do. Notice, for instance, some of our office holders, who have been in office for years and to whom it has become as natural "to serve the people" as they call it, and to rake the dimes [?] for this service, which [they do?—page torn] many times very ill, and could [page torn] displeased with, into their own [page torn] thus create a great "vast of [page torn]. Notice them, I say, when the people demand a change. See how loth they are to give up the means of their support. See how hard they strive, and not infrequently use vile means in order to procure votes enough that they may not be driven from their office, after which the[y] have formed such a *hankering,* and be compelled to seek a livelihood in another direction.

## *Resolutions Copied by Thomasson*

Resolutions of the Rev. Joseph Benson, who was born at Melmerby, in the County of Cumberland, England, on the 25th of January, 1748.
1. "To rise at five o'clock in the morning, and to go to bed at nine at night.

Never to trifle away our time in vain conversation, use less visits, or in studying any thing which would not be to my advantage."
2. "To be careful to maintain private prayer, and not be content without communion with God in it. To spend from four to five every morning, and from five to six every evening, in devout meditation and prayer; and at nine in the morning, and at three in the afternoon, to devote a few minutes to prayer."

Basil A. Thomasson

December 29th 1850.

## *Portion of a Sunday School Lesson or Sermon*

The devil, ever since the creation of man, has made it his business to destroy him—both soul and body. And, notwithstanding the great friendship he so often professes for man, he is ever seeking his ruin. Look, for instance, at the temptation in the garden. "And he said unto woman, 'Yea, hath God said, ye shall not eat of every tree of the garden?[' "] "And the woman said," etc. Then the devil replied, "Ye shall not surely die," etc. Now what was the object of all this? We answer, the destruction of man. But see how artful the enemy was. He enters the arena of man's happiness and get[s] in conversation with the woman. Why didn't he approach Adam directly? He knew that Eve was the "weaker vessel," and that if he could win her over she could manage Adam and his hellish end would be accomplished.

# Index

adaptation, 213
agriculture, 4, 6, 9, 14, 25, 31, 36, 40, 41, 54, 55, 56, 64, 80, 84, 88, 92, 93, 96, 97, 99, 101, 104, 113, 123, 127, 129, 137, 171, 172, 195, 198, 199, 204, 226, 234, 235, 238, 239, 241, 246, 251, 255, 259, 261, 267, 275, 276, 278, 279, 284, 290, 306, 309, 325, 326, 328
amusements, 14, 38–39, 64–65, 73, 169, 205, 217, 247, 271

barter, 54, 56, 123, 227, 275, 305, 308–9
Beecher, Lyman, 351
borrowing, 78, 176, 217
Breckinridge, John C., 291

conscription, 321, 322 (n. 2)

debating society, 67, 73, 76
draft, 321, 322 (n. 2)

economy, local, 4, 5, 9, 13, 14, 16, 116, 123, 149, 168, 192, 202, 204, 217, 225, 236, 237, 243, 253, 257, 262, 267–68, 283, 293, 295, 327

Greeley, Horace, 351

health, 5, 12, 15–16, 20, 30, 50, 57, 62, 76, 86, 90–91, 99, 106, 144, 146, 153, 203, 220, 231, 233, 242, 268, 285–86, 288, 294, 307, 315, 349–50
Henry, Patrick, 3

Know Nothings, 76

labor, hired, 28, 29, 31, 51, 79, 82, 143, 146, 153, 234, 243, 324, 325. *See also* sharing work
lending, 157

musters, 29, 101, 128, 139, 182, 195, 308, 309, 321

neighbors, relations with, 97, 143, 174, 220, 228, 236, 247, 262, 284, 295, 300
notes, financial, 5, 7, 9, 16, 23, 26, 55, 56, 58–59, 66, 78, 79, 91, 145, 148, 149, 157, 167, 188, 200, 217, 225, 238–40, 268, 298, 301, 309, 327

Penn, William, 221
politics, 88, 91, 155, 178, 249, 282, 291
Puryear, R. C., 88, 91

road work, 30, 31, 114, 132, 174, 175, 197, 205, 208

Scales, Alfred, 88, 178
schools, common (public), 3, 4, 9, 10, 15, 18, 45–47, 49, 52, 61, 90–92, 148, 152–53, 158, 165, 179, 206–7, 270, 318
schools, private or subscription, 17, 57
school teaching, 3, 7–12, 16–19, 45–56, 142–47, 150–58, 160–61, 176, 178–79, 183–85, 187–88, 199, 206–9, 211, 215, 217–18, 221–23, 227, 234, 250, 256, 263–64, 266, 269, 270, 284, 292–98, 301, 319–21, 339–45, 347–49
sectional crisis, 71, 305, 306, 307, 350–51
sharing work, 4, 15, 19, 25, 35, 52, 53, 124, 130, 154, 205, 236, 237, 270, 278, 280, 326
slavery, 29, 141, 306

taxes, 90, 166, 194, 237, 280, 311
temperance, 24, 59, 68–69, 128, 182, 220, 228–29, 233, 248, 255–56, 330–31, 334
Thomasson, Basil Armstrong: attitudes of, on economy, 21, 56, 61, 79, 116, 132, 151, 160, 176, 230–31, 239; attitudes of, toward learning, 3, 42, 83, 124; attitudes of, toward women, 5, 12, 13, 18, 25, 26, 29, 36–37, 51, 53, 54, 64, 68–69, 71, 81, 83, 131, 135, 157, 203, 219, 225, 229, 236, 238, 241, 253, 255, 262, 264, 274, 279, 280, 285–86, 316; and clerks at store, 66, 67, 71; comments of, on social structure, 8, 11–12, 37–38, 70, 87, 249, 303; efforts of, to sell a wagon, 43, 55, 65, 82, 97, 147; efforts of, to start a blacksmith shop, 16, 23, 26, 27, 33, 38, 57, 103; efforts of, to start schools, 16, 19, 21–23, 43–44, 57; marriage and family life of, 89, 90, 93, 98, 101, 106, 107, 120, 124, 128, 133–34, 135, 140, 143, 163, 164, 173, 203, 215, 218, 219, 221, 227, 229, 232–33, 238, 241, 247, 248, 253, 254, 255, 262, 264, 270, 274, 279, 289–90, 299, 303, 311, 314–15, 321; relations of, with parents and relatives, 4, 5, 7, 11, 13, 17, 20, 21, 26, 40, 49, 54, 55, 60, 90, 97, 111–12, 155, 161, 163, 164–65, 167, 175, 178, 182, 183, 198, 202, 220, 221, 231, 232, 238, 242, 248, 249–50, 251, 255, 259, 269, 275, 285, 294, 297, 300, 313, 315, 317, 321, 325, 336–37, 349–51; religious experience and views of, 4, 5, 6, 10, 18, 24, 34, 40, 48, 49, 57, 74–75, 94, 96, 121–22, 132, 138, 139, 140, 144, 151–52, 167, 174, 175, 176, 182, 188, 189, 193, 194, 198, 203, 212, 213–16, 218, 228–29, 233–34, 237, 248, 259–60, 270–71, 276, 282, 285, 286–87, 309, 310, 316–17, 318, 322–24, 345–46, 351–52; self-education of, 3–4, 5, 7, 11, 15, 19, 26, 28, 29, 31, 35, 42, 50, 61, 72–74, 82, 87, 114, 117, 119, 122, 130, 141, 145, 149, 150, 156, 163, 187, 190, 192, 216, 219, 223, 234, 290, 314, 316, 317, 336–38
Thomasson, Mollie Bell, courtship of, 6, 9, 14, 17, 18, 19, 21, 23, 26, 27,